Quantum Mech

Quantum Mechanics: An Introduction for Device Physicists … *rth Edition* provides a complete course in quantum mechanics for students of semiconductor device physics and electrical engineering. It provides the necessary background to quantum theory for those starting work on micro- and nanoelectronic structures and is particularly useful for those beginning work with modern semiconductor devices, lasers, and qubits.

This book was developed from a course the author taught for many years with a style and order of presentation of material specifically designed for this audience. It introduces the main concepts of quantum mechanics which are important in everyday solid-state physics and electronics. Each topic includes examples which have been carefully chosen to draw upon relevant experimental research. It also includes problems with solutions to test understanding of theory.

Fully updated throughout, the fourth edition contains the latest developments, experiments, and device concepts. It contains completely new material on quantum computing and information processing, including qubit implementations and quantum sensing.

Key features:

- Fully updated throughout with the latest research from the field
- Presents a different view of quantum mechanics from an engineering perspective
- Takes a modern view of the understanding of quantum mechanics as it affects real-world technologies, including nanotechnology and quantum information processing.

David K. Ferry is Regents' Professor Emeritus in the School of Electrical, Computer, and Energy Engineering at Arizona State University. He was also a member of the graduate faculty in the Department of Physics and the Materials Science and Engineering program at ASU, as well as Visiting Professor at Chiba University in Japan. He came to ASU in 1983 following shorter stints at Texas Tech University, the Office of Naval Research, and Colorado State University. In the distant past, he received his doctorate from the University of Texas, Austin, and spent a postdoctoral period at the University of Vienna, Austria. He enjoyed teaching (which he refers to as "warping young minds") and continues active research. The latter is focused on semiconductors, particularly as they apply to nanotechnology and integrated circuits, as well as quantum effects in devices. In 1999, he received the Cledo Brunetti Award from the Institute of Electrical and Electronics Engineers and is Fellow of this group as well as the American Physical Society and the Institute of Physics (UK). He has been Tennessee Squire since 1971 and Admiral in the Texas Navy since 1973. He is the author, coauthor, or editor of some 40 books and about 900 refereed scientific contributions. More about him can be found on his home pages https://search.asu.edu/profile/35634 and http://dferry.net/.

Quantum Mechanics

An Introduction for Devices Physicists and Electrical Engineers

Fourth Edition

David K. Ferry

CRC Press
Taylor & Francis Group
Boca Raton London New York

CRC Press is an imprint of the
Taylor & Francis Group, an **informa** business

Designed cover image: Alexandre Gondran

Fourth edition published 2027
by CRC Press
2385 NW Executive Center Drive, Suite 320, Boca Raton FL 33431

and by CRC Press
4 Park Square, Milton Park, Abingdon, Oxon, OX14 4RN

CRC Press is an imprint of Taylor & Francis Group, LLC

First edition published by Institute of Physics

Third edition published by CRC Press 2021

Library of Congress Cataloging-in-Publication Data
Names: Ferry, David K. author
Title: Quantum mechanics : an introduction for devices physicists and electrical engineers / David K. Ferry.
Description: 4th edition. | Boca Raton, FL : CRC Press, 2027. | Third edition published by CRC Press 2021. |
Includes bibliographical references and index. |
Identifiers: LCCN 2026010773 (print) | LCCN 2026010774 (ebook) | ISBN 9781041206088 hardback |
ISBN 9781041194668 paperback | ISBN 9781003717195 ebook
Subjects: LCSH: Quantum theory
Classification: LCC QC174.12 .F47 2027 (print) | LCC QC174.12 (ebook)
LC record available at https://lccn.loc.gov/2026010773
LC ebook record available at https://lccn.loc.gov/2026010774

ISBN: 978-1-041-20608-8 (hbk)
ISBN: 978-1-041-19466-8 (pbk)
ISBN: 978-1-003-71719-5 (ebk)

DOI: 10.1201/9781003717195

Typeset in Times
by KnowledgeWorks Global Ltd.

Contents

Preface to the First Edition

Most treatments of quantum mechanics have begun from the historical basis of the application to nuclear and atomic physics. This generally leaves the important topics of quantum wells, tunneling, and periodic potentials until late in the course. This puts the person interested in solid-state electronics and solid-state physics at a disadvantage, relative to their counterparts in more traditional fields of physics and chemistry. While there are a few books that have departed from this approach, it was felt that there is a need for one that concentrates primarily upon examples taken from the new realm of artificially structured materials in solid-state electronics. Quite frankly, we have found that students are often just not prepared adequately with experience in those aspects of quantum mechanics necessary to begin to work in small structures (what is now called mesoscopic physics) and nanoelectronics, and that it requires several years to gain the material in these traditional approaches. Students need to receive the material in an order that concentrates on the important aspects of solid-state electronics, and the modern aspects of quantum mechanics that are becoming more and more used in everyday practice in this area. That has been the aim of this text. The topics and the examples used to illustrate the topics have been chosen from recent experimental studies using modem microelectronics, heteroepitaxial growth, and quantum well and superlattice structures, which are important in today's rush to nanoelectronics.

At the same time, the material has been structured around a senior-level course that we offer at Arizona State University. Certainly, some of the material is beyond this (particularly Chapters 10 and 11), but the book could as easily be suited to a first-year graduate course with this additional material. On the other hand, students taking a senior course will have already been introduced to the ideas of wave mechanics with the Schrödinger equation, quantum wells, and the Krönig–Penney model in a junior-level course in semiconductor materials. This earlier treatment is quite simplified but provides an introduction to the concepts that are developed further here. The general level of expectation on students using this material is this prior experience plus the linear vector spaces and electromagnetic field theory to which electrical engineers have been exposed.

I would like to express thanks to my students who have gone through the course, and to Professors Joe Spector and David Allee, who have read the manuscript completely and suggested a great many improvements and changes.

David K. Ferry

Preface to the Second Edition

Many of my friends have used the first edition of this book and have suggested a number of changes and additions, not to mention the many errata necessary. In the second edition, I have tried to incorporate as many additions and changes as possible without making the text overlong. As before, there remains far more material than can be covered in a single one-semester course, but the additions provide further discussion on many topics and important new additions, such as numerical solutions to the Schrödinger equation. We continue to use this book in such a one-semester course, which is designed for fourth-year electrical engineering students, although more than half of those enrolled are first-year graduate students taking their first quantum mechanics course.

I would like to express my thanks in particular to Dragica Vasileska, who has taught the course several times and has been particularly helpful in pointing out the need for additional material that has been included. Her insight into the interpretations has been particularly useful.

David K. Ferry

Preface to the Third Edition

At the turn of the new century in 1900, the world of (what we now call classical) mechanics was comfortable. Newton's laws worked very well, and scientists and engineers knew how to apply them. This was true for small objects such as screws and wires as well as astronomical objects subject to gravity. The latter, as well as Coulomb interactions, was apparently well understood in terms of the instantaneous force at a distance. Yet, within a scant five years, this comfortable understanding would be completely overturned in a manner that would take years to sort out. This would be true for very small objects, such as atoms, and for very large objects, such as the solar system, although the change would come from different theories and people.

At the small scale, the change would come before the end of the year and begin with Max Planck and the quantization of the photon required for his new theory of black-body radiation. Planck was not working in isolation, as he had colleagues who had carried out new careful measurements on black-body radiation, measurements which clearly showed that Wien's earlier theory just was not correct. Planck was able to fit the new data exceedingly accurately by recognizing two new fundamental constants—one he termed the Boltzmann constant and the other came to be known as the Planck constant. From Planck's new theory would eventually come a first approximate new theory for atomic structure and then a full new theory that worked for atomic structure but also even for quantum gravity—the new quantum mechanics would appear only slightly less than three decades later. But the new quantum mechanics would be argued for longer, and even in the last third of the new century, Richard Feynman would declare that "... nobody understands quantum mechanics." How can this be? Understanding of a subject requires that a philosophy has arisen which describes how the new theory fits with the reality of the existing world. Perhaps the problem with the new theory was best explained by Murray Gell-Mann, who said, "The fact that an adequate philosophical interpretation has been so long delayed is no doubt caused by the fact that Niels Bohr brain-washed a whole generation of theorists into thinking that the job was done 50 years ago." Bohr had developed the first approximate theory for atomic structure and had then propagated strongly the view that the new theory was magic, and we could only find out what was true by measuring it. We will examine this somewhat further in the first chapter when describing the various understandings that led to the well-known Bohr–Einstein debates.

At the large scale, or perhaps I should say the high-velocity scale, Newton's laws were forever changed by the introduction by Albert Einstein of his relativity theory. Now, there was a maximum velocity that matter could achieve, and this dramatically changed space and time forever. What was left was the understanding that Newton's laws worked fine, provided they were used within a well-defined set of parameters that were neither too fast nor too small. Nevertheless, it is not at all clear that the two theories work together. Relativity theory rolls mechanics, electromagnetics, and gravity all together. From this, we have learned to consider four forces in nature—the electromagnetic, strong, weak, and gravitational forces. A form of quantum field theory has developed to treat the first three, but inclusion of gravity has been quite difficult, and many of the assumptions in quantum field theory are not applicable in many everyday problems.

Leaving all that aside, one can find a great deal of satisfaction in working with a form of non-relativistic quantum mechanics that provides satisfactory applications in many real-world problems. In fact, most of the population employs tools, particularly electronics tools, that depend upon the quantum properties of matter, even if they do not realize it. The information revolution that began around the middle of the last century depends entirely upon the development of the transistor and integrated circuits, and the continued development and expansion of the latter. Yet, the transistor and integrated circuit themselves explicitly depend upon the quantum properties of the materials from which they are made. Thus, an understanding of quantum mechanics is essential to the engineers and physicists who design these devices. And the purpose of this book has been, and continues to be, to provide an adequate understanding of quantum mechanics. And, despite the expressed doubts of Feynman, it is my belief that an adequate understanding of quantum mechanics exists for this purpose. The success of the first two editions of this book, and of many other fine books on the topic, supports this view. The difference in this book is the focus on device physics and electrical engineering applications to which the theory is applied. That is, what is very important to these people may be quite esoteric to physicists in other fields, and vice versa.

An example of the last comment is quantum tunneling. A physics colleague, who has written a fine textbook on quantum mechanics, was told by his colleagues to leave tunneling out of his new edition, as it was too esoteric and unimportant. However, the topic is crucial to considering leakage in modern transistors, and it is of fundamental importance to Josephson junctions, which may well be the basis for future quantum computing. Hence, it is a topic of fundamental importance to device engineers and physicists. In fact, many of the changes in this third edition have been made to allow us to describe the new applications that are required for quantum computing, including a discussion of qubits (as well as Josephson junctions).

I would like to thank Prof. Dragica Vasileska for many comments arising from her reading of various sections of this manuscript.

David K. Ferry

Preface to the Fourth Edition

This book has always taken a somewhat different view of quantum mechanics, mainly because engineers (as well as device physicists) need different things from quantum mechanics. To explain this, let me revert to a discussion I had many years ago with the author of a well-known quantum mechanics book used by physicists. He remarked that his colleagues were urging him to leave tunneling out of his new edition because it was not all that useful to them in their fields of research. Yet, tunneling is at the heart of our everyday life, since the sun would not shine if tunneling did not exist. Moreover, tunneling is crucial to the device world, as it is at the heart of many processes, ranging from leakage current in MOSFETs to single-electron devices at the nanoscale. So, tunneling is not only needed by device engineers, but its treatment has to be expanded over that of most quantum mechanics books. This expansion has appeared through all the previous editions but has been expanded even more in this edition due to its appearance in the operation of some qubit implementations used in quantum information processing.

Other changes that appear here have come from suggestions by users of the previous editions. One such example is the treatment of optical absorption in semiconductors, which has been added as an example of time-dependent perturbation theory. Such optical absorption is clearly the driving technology in devices ranging from sensitive optical detectors to solar cells for energy conversion. More problems have also been added to the chapters, but I still believe that most instructors have their own favorite problems, so the problems here can be considered to be suggestive of others.

Then, Chapter 11, on quantum information, has been greatly expanded with discussion of many typical qubit implementations presented as examples of quantum devices. This material is certainly in response to the rapid expansion of quantum topics over the past two decades, and the presence of quantum computers as real systems available for use by the community. But this is not the only expansion of material. The importance of the property of duality (wave and particle description) appears at many places throughout the book. This property is likely one of the most important aspects of quantum mechanics, even though it is seldom recognized as such. Yet, almost every introductory discussion in most textbooks actually introduces duality without discussion when they immediately move the wave description to solve a problem, even the most simple one, such as reflection from a barrier. The undiscussed rationale for this is simply the fact that, once this is done, the mathematics becomes simple classical boundary value problems that have been solved an enormous number of times over the decades. Indeed, most solutions give rise to named functions, often presented to the student as being new to quantum mechanics.

As a result of the expansion of the material in many parts of the book, there is now a great deal more material than is needed for a single one-semester course, which was the original concept for the book. This means that the instructor can largely shape his own course through picking and choosing which material to include. Or, as an alternative, the instructor can stretch the course to two semesters and still have enough material available. This would particularly be true were the second semester to focus on devices and quantum information, although this might need to seek out deeper understanding and treatments of many of the devices introduced. Nevertheless, the principal aim is a one-semester course at the level of an upper-division undergraduate or a first-year graduate student who is being introduced to quantum mechanics for the first time. The fundamental principles of quantum mechanics should not be difficult for a student at this level, provided he/she is not blindsided by the mathematics that appears. Fortunately, most engineers have been subjected to a course in applied mathematics, which often is a course on boundary value problems, and electrical engineers see this mathematics again in electromagnetic field theory (where life outside of rectangular coordinates quite often appears). They are also familiar with complex mathematics from AC circuit theory, so most of the mathematics has some degree of familiarity.

Finally, I am indebted to Professor Dragica Vasileska, who has used all of the previous editions in her courses and has provided invaluable feedback on the material. Her impact on this new edition cannot be overestimated.

David K. Ferry

About the Author

David K. Ferry is Regents' Professor Emeritus in the School of Electrical, Computer, and Energy Engineering at Arizona State University. He was also a member of the graduate faculty in the Department of Physics and the Materials Science and Engineering program at ASU, as well as Visiting Professor at Chiba University in Japan. He came to ASU in 1983 following shorter stints at Texas Tech University, the Office of Naval Research, and Colorado State University. In the distant past, he received his doctorate from the University of Texas, Austin, and spent a postdoctoral period at the University of Vienna and the Boltzmann Institute for Solid State Physics, Vienna, Austria. He enjoyed teaching and continues active research, still collaborating with colleagues at home and abroad. His current work is focused on semiconductors, particularly as they apply to nanotechnology and integrated circuits, as well as quantum effects in devices. In 1999, he received the Cledo Brunetti Award from the Institute of Electrical and Electronics Engineers and is Fellow of this group as well as the American Physical Society and the Institute of Physics (UK). He has been Tennessee Squire since 1971 and Admiral in the Texas Navy since 1973. He is the author, coauthor, or editor of some 40 books and about 900 refereed scientific contributions. More about him can be found on his home pages https://search.asu.edu/profile/35634 and http://dferry.net/.

1 Waves and Particles

1.1 INTRODUCTION

Science has developed through a variety of investigations more or less over the time scale of human existence. On this scale, quantum mechanics is an extremely young field, existing essentially for only just over a century. Even our understanding of classical mechanics has existed for a comparatively long period—roughly having been formalized with Newton's equations published in his *Principia Mathematica*, in April 1686 (1687). Up until the very end of the nineteenth century, mechanics in the form of Newton's laws were comfortable, being understood and accepted by all who dealt with the motion of reasonable bodies. With the turn of the new century, however, mechanics would be shaken to its roots. The change began already on December 14, 1900, when Planck (1900) presented his new theory of black-body radiation to the German Physical Society, and this change would affect mechanics in physically small structures (and lead to quantum mechanics). Mechanics of large bodies would similarly be changed in 1905 with the arrival of Albert Einstein's (1905a) relativity theory. In this book, we will deal with only the first of these two great breakthroughs.

Black-body radiation had been studied for many years prior to Planck's discovery, but it lacked an adequate theory. One of the early theories was due to Wien (1896), but it was approximate, and it was discovered that it really didn't fit the observations, nor did newer observations fit the old. After a student in Tübingen pointed out this fact, new and very careful experiments were performed at one of Berlin's national laboratories (Lummer and Pringsheim, 1899), and fortunately, Planck was able to interact with the latter group. This provided the data that allowed Planck (1900) to finalize his theory in what is now known as the Planck black-body radiation law:

$$I(f)df \sim \frac{f^3 df}{\exp\left(\frac{hf}{k_B T}\right) - 1}, \tag{1.1}$$

where f is the frequency of the radiation, T is the temperature, and I is the intensity of the radiation. In addition to these parameters, Planck defined two new constants which came from comparing the data to the theory. One of these, he named the Boltzmann constant, $k_B = 1.380649 \times 10^{-23}$ J/K. The second, he did not name, but it has since come to be known as Planck's constant $h = 2\pi\hbar = 6.62607015 \times 10^{-34}$ J-s. (While these constants historically were more uncertain, the redefinitions that occurred in the international system of units after 2019—the SI base units—have given us these values.) In coming to his radiation law, Planck had to make one more connection and that is a relation between the frequency and the energy, as

$$E = hf = \hbar\omega. \tag{1.2}$$

(The last two forms differ only by moving the factor of 2π from Planck's constant to the radian frequency, which leaves the reduced value of the constant.) In this last equation, E is the energy, and f is the frequency of the radiation. While Planck had given us the idea of quanta of energy, which he did while reaching his result, he was not comfortable with this idea. However, it took only a decade for Einstein's theory of the photoelectric effect (discussed later) to confirm that radiation indeed was composed of quantum particles of energy given by (1.2).

The idea of (1.2) leads us to conclude that Planck had quantized the radiation with which he was dealing, although this is not evident from the equation alone. One also has to consider the amount of power that is provided by the radiation. For example, let us assume that the light from the sun (which

DOI: 10.1201/9781003717195-1

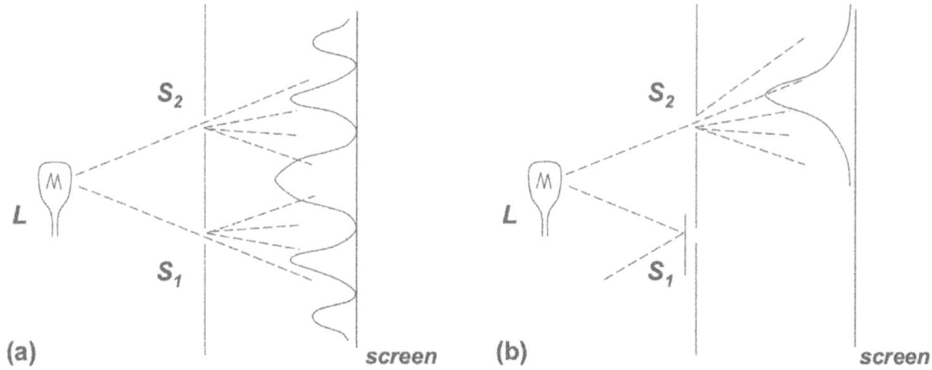

FIGURE 1.1 In panel (a), we illustrate how light coming from the source L and passing through the two slits S_1 and S_2 interferes to cause the pattern indicated on the "screen" on the right. If we block one of the slits, say S_1, then we obtain only the light intensity passing through S_2 on the "screen" as shown in panel (b).

is sometimes viewed as a black-body radiator) arriving at the surface of the Earth is, on average, about 0.1 W/cm^2. If the average photon has an energy of 2.5 eV, then the number of photons arriving at the surface of the Earth is about 2.5×10^{17} per second per square centimeter. It is this combination of quantities, power and energy, that together requires the light to be quantized. That is, about 2.5×10^{17} photons/cm^2 arrive every second from the sun, and this requires that the power falling on the Earth's surface must jump with the arrival of each photon. Here, Planck has required that the photons are *particles* and not waves.

To be truthful, this requirement of light being particles is not such a revolution, as the ancient Greeks always believed that light was composed of small corpuscular units. Even the great physicist Newton tended to believe that light was corpuscular in nature. It was only the work of Young (1809) in the late eighteenth century that showed that light would give rise to an interference effect, a property of waves. Young showed that if one creates a source of coherent light (a single frequency) and passes this light through two separated slits, the wavelike property of the light will create an interference pattern, as shown in Figure 1.1.

This now presents a controversy. Is light a wave or is it a particle? Certainly, Young's experiments showed wavelike properties. If we block one of the two slits, so that light passes through just a single slit, the interference pattern disappears, and one sees just the normal passage of the light waves. The interference pattern formed by passing light through two different paths so as to create two different phases of the light wave is an essential property of the single (original incident) wave. When one can see such an interference pattern, it is said that we are seeing the wavelike properties of light. As engineers well know, the wavelike view was reinforced once again in the middle of the nineteenth century with the arrival of Maxwell's equations for electromagnetic waves. Planck's results challenged this view. In support, explanations of other experiments would soon lend support to the particle view.

1.2 LIGHT AS PARTICLES

Proof of the fact that light was a particle came quickly enough, and it came as an explanation of the photoelectric effect. It was known since the time of Becquerel (1839) that when light was shone upon the surface of a metal, or some other conducting medium, electrons could be emitted from the surface provided that the frequency of the incident light was sufficiently high. Subsequent studies termed the effect the photoelectric effect, as it could be used to convert light or radio waves to normal electricity. Much of this work was directed toward converting such waves to DC behavior in order to produce better detectors. However, the effect remained unexplained until 1905.

1.2.1 The Photoelectric Effect

There were several factors that describe the emission of particles from a solid via the photoelectric effect that were curious. First, if one approached from the low frequency side, no effect is observed until a critical frequency is reached. Then electrons began to be emitted for further increases in the frequency of the light. The velocity of the emitted electrons depends only upon the wavelength of the incident light, and *not upon the intensity of the radiation.* In fact, the energy of the emitted particles varies inversely with the wavelength of the light waves. On the other hand, the *number* of emitted electrons does depend upon the intensity of the radiation, and not upon its wavelength. Today, of course, this is not surprising at all, but this is after it was explained in the Nobel Prize–winning work of Einstein. What Einstein (1905b) concluded was that the explanation of this phenomenon required a treatment of light in terms of its "corpuscular" nature; that is, we need to treat the light wave as a beam of particles impinging upon the surface of the metal. In fact, it is important to describe the energy of the individual light particles, which we call *photons*, using the relation (1.2).

The photoelectric effect can be understood through the consideration of Figure 1.2. However, it is essential to understand that we are talking about the flow of "particles" as directly corresponding to the wave intensity of the light wave. Where the intensity is "high," there is a high density of photons. Conversely, where the wave amplitude is weak, there is a low density of photons. In the figure, most electrons sit at, or below the Fermi energy in a metal. In addition, there is a surface barrier that keeps the electrons from falling out of the metal, and this barrier is known as the *work function*, usually denoted as E_W. This level, at the top of the work function, is denoted the "vacuum level" in the figure, and typical values are in the range 4–5 eV. So, the work function is the difference between the vacuum energy level and the Fermi energy.

When a photon enters the solid, it can scatter with the electron just as in any other particle–particle interaction. Since both the photon and the electron are particles, this collision conserves energy and momentum. Typically, the photon is massless, a requirement to be traveling at the speed of light. Hence, since the heavier particle usually wins in any collision, the result is that all of the photon's energy is transferred to the electron. However, this electron still cannot escape from the solid unless the transferred energy is larger than the work function. So, we connect the critical energy mentioned above to the work function. The first requirement mentioned above reduces to

$$\hbar\omega > E_W. \tag{1.3}$$

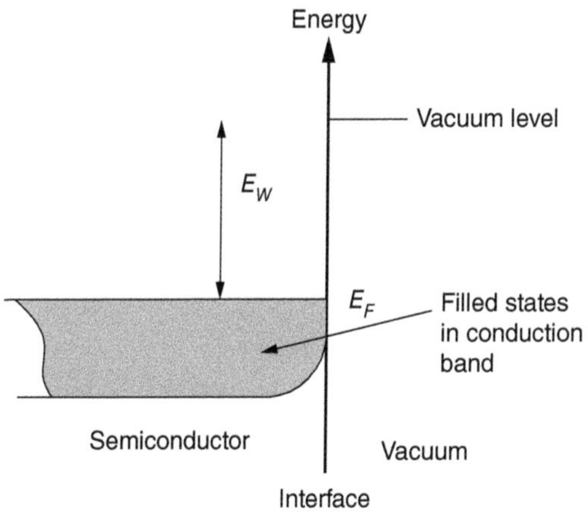

FIGURE 1.2 The energy bands at the surface of a metal. An incident photon with an energy greater than the work function, E_W, can cause an electron to be raised from the Fermi energy, E_F, to above the vacuum level, whereby it can be photo-emitted.

If the electron does gain enough energy to escape from the metal, it will have to be emitted over the vacuum level, so it will lose an energy equal to the work function in gaining its escape. Thus, the particle now is emitted with the energy:

$$E = \hbar\omega - E_W. \tag{1.4}$$

This explains the second problem, in that the electron is emitted with lower energy. However, also, there remains one electron for one photon, so if more electrons are required, then more photons must be used to illuminate the metal. Thus, the number of electrons emitted depends upon the number of photons absorbed, or upon the intensity of the light.

Now, at this point, we have a conundrum. Is light a wave, as suggested by the interference phenomena of Young, or is it a particle, as suggested by Planck and the photoelectric experiments? Needless to say, this would be investigated.

1.2.2 Waves and Particles: Duality

Thomson (1907) studied the ionization of gases under ultra-violet and X-ray illumination and noticed that only a small fraction of the gas molecules were being ionized. This led him to conjecture that the energy of the light was localized in Planck's packets, which also had to be *physically small in dimension*, so that these corpuscles of light missed the majority of the molecules. Acting upon a suggestion of his, Taylor (1909) studied the properties of very weak light.

Taylor thought that light striking a metal surface, or a photographic plate, would actually cause a "... series of bright specks on a dark background." He supposed that if the plate was moved further from the source, "... we shall diminish the number of these [specks] ... but not the energy in the individual units." This went beyond Planck's quantization by also suggesting that the photons were physically quite small. By attenuating the light in his measurements, Taylor could get a situation in which the photons came through one at a time and were separated from one another by a large distance. He used a flame passing through a single slit as his source and then diffracted the light around a needle. He attenuated the light with various smoked glass windows, which he calibrated first to determine the attenuation of each window. With the weakest light, the photographic plate had to be exposed for 2,000 hours. Taylor states that he saw no diminution in the diffraction pattern for any of his exposures, thus confirming Thomson's ideas. Even though light arrived at each photographic plate as a single photon at a time and exposed a single spot on the plate, the net pattern formed over time matched the expected wave interference patterns. Consequently, the light in this experiment was acting as both particle (upon striking the plate) and wave (when passing the needle). From this, and from later experiments on two slits, we know that what is observed is each photon creates a single point on the screen behind the two slits. That is, a single photon impacting the screen does not create the interference pattern. However, after thousands of photons impact the screen, each one going through the two slits as an individual photon, the interference pattern emerges. However, while a single photon does not create the interference pattern, it must at the same time be aware of the pattern that will be created by a larger number of photons. In a sense, the photon must sense both the wave and particle behavior in its flight. Thus, it was clear that light had both wave *and* particle aspects and could exhibit both in a single experiment!

A corollary of the acceptance of light as particles is that there is a momentum associated with each of the particles. It is well known in field theory that there is a momentum associated with the (massless) wave, which is given by $p = hf/c$, which leads immediately to the relationship:

$$p = \frac{hf}{c} = \frac{h}{\lambda} = \hbar k, \tag{1.5}$$

where in the last form we have used the well-known relation $k = 2\pi/\lambda$.

It is finally clear from the interpretation of light waves as particles that there exists a relationship between the "particle" energy and the frequency of the wave, and a connection between the momentum of the "particle" and the wavelength of the wave. Equations (1.2) and (1.5) give these relationships. The form of (1.5) has usually been associated with de Broglie, and the wavelength corresponding to the particle momentum is usually described as the *de Broglie wavelength*. However, it is worth noting that de Broglie (1937) referred to the set of Equations (1.2) and (1.5) as the Einstein relations! In fact, de Broglie's great contribution was the recognition that atoms localized in orbits about a nucleus must possess these same wavelike properties. Hence, the electron orbit must be able to incorporate an exact integer number of wavelengths, given by (1.5) in terms of the momentum. This then leads to quantization of the energy levels and particle momentum.

It is not just the mathematical relations above that connect the wave and the particle. What Taylor demonstrated is that the photon exhibited both wave and particle properties during its flight from the flame to the photographic plates. It was not either/or, it was both! The photon passed the sharp edge and behaved as a wave but arrived at the plate as a particle. The photon had to be considered as both a wave and a particle; the behavior it displayed depended upon the measurement being made. De Broglie (1923, 1924) would address this by proposing that both wave and particle were present—there would be a guiding wave upon which the particle rode. This was certainly a novel idea and one that makes a great deal of sense. Some people thought it too simple to be correct.

1.3 ELECTRONS AS WAVES

In the previous section, we discussed how light, or electromagnetic waves, can behave as either a wave or a collection of particles. It is fair to then ask whether (what we normally think of as) particles, such as electrons, behave as waves. This question remained unanswered for several years. In 1921, Clinton Davisson and Charles Kunsman (1921, 1923), working at the AT&T research laboratories in America, had noticed that the scattering of electrons from Ni and other metals showed unexpected peaks in the angular variation of the scattering. In sending a beam of electrons into a crystal, such as Ni, they found these unexpected results. If the energy of the entering electrons is sufficiently large, they can knock other electrons out of their atomic orbits, and these electrons can be emitted in complete analogy to the photoelectric effect, with the incident electron replacing the photon. However, Davisson and Kunsman also observed electrons with almost the same energy as the incident beam being reflected from the crystal over a large range of angles. Studies of the angular distribution of the emitted/scattered electrons showed the appearance of peaks much like diffraction peaks, apparently due to the periodic nature of the atomic arrangement. In these studies, the scattering cross-section had an oscillatory behavior with peaks and valleys as the energy was varied as in the case with Ni. Walter Elsasser (1925) proposed that these effects were the result of the diffraction of electron *waves*, and his paper was published only after the editor consulted with Einstein. Just after this, George Thomson (1927), the son of J. J. Thomson, obtained clear diffraction patterns of electrons striking thin metal films. Sending the electrons through the thin films, he observed the diffraction rings of the exciting electrons. Thus, it was finally clear that the electrons could show wavelike properties.

From these experiments, one can observe that there is a very similar behavior between light and electrons. Both can appear as either a wave or a particle. Consequently, there are times when it is clearly advantageous to describe particles, such as electrons, as waves. In the correspondence between these two viewpoints, it is important to note that the varying intensity of the wave reflects the presence of a varying number of particles; the particle density at a point x, at time t, reflects the varying intensity of the wave at this point and time. For this to be the case, it is important that quantum mechanics describe both the wave and the particle pictures through the principle of superposition. That is, the amplitude of the composite wave is related to the sum of the amplitudes of the individual waves corresponding to each of the particles present. Note that it is the amplitudes, and not the intensities, that are summed, so there arises the real possibility for *interference* between the waves of individual particles, just as interference appears for light waves. Thus, for the presence of two (noninteracting) particles at a point x, at time t, we may write the composite wave function as

$$\Psi(x,t)=\Psi_1(x,t)+\Psi_2(x,t). \tag{1.6}$$

This composite wave may be described as a *probability wave*, in that the square of the magnitude describes the probability of finding an electron at a point (we will go further into this point below).

It may be noted from (1.5) that the momentum of the particles goes immediately into the so-called *wave vector k* of the wave. A special form of (1.6) is

$$\Psi(x,t)= Ae^{i(k_1x-\omega t)}+Be^{i(k_2x-\omega t)}, \tag{1.7}$$

where it has been assumed that the two components may have different momenta (but we have taken the energies/frequencies to be equal). This form is, of course, familiar as it is the natural form found in electromagnetic theory. For the moment, only the time-independent steady state will be considered, so the time-varying parts of (1.7) will be suppressed. It is known, for example, that a time-varying magnetic field that is enclosed by a conducting loop will induce an electric field (and voltage) in the loop through Faraday's law. Can this happen for a time-independent magnetic field? We know that a DC magnetic field can affect the motion of electrons through the Lorentz force. The question is how this arises for a wavelike electron.

A particularly remarkable illustration of the importance of the phase is the magnetic Aharonov-Bohm (1959) (AB) effect. The AB effect was predicted in the mid-twentieth century and was verified by a number of experiments almost immediately. In this effect, we consider a ring structure with an input lead on one side and an output lead on the other. The AB effect is an interference between electrons passing on either side of the ring and interacting at the output. The vector potential for a magnetic field passing through the annulus of the ring is azimuthal, so that electrons passing through either side of the ring will travel either parallel or antiparallel to the vector potential, and this difference produces a phase modulation, which then gives an oscillating conductance.

The AB effect can be illustrated with an experiment conducted on a graphene ring (graphene is a monolayer semiconductor composed of pure carbon atoms) from Huefner *et al.* (2010). The structure and data are shown in Figure 1.3. Panel (a) is an image of the graphene structure—the light gray areas are the patterned graphene. The original layer was deposited upon an oxidized Si wafer as a

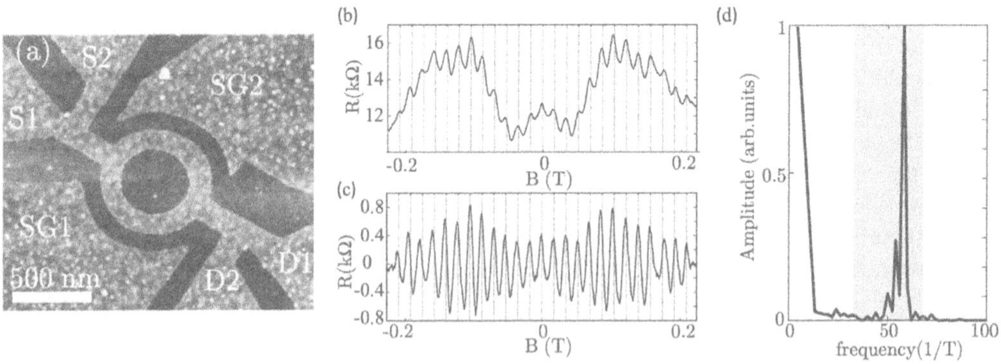

FIGURE 1.3 (a) Scanning force micrograph of the graphene ring. The ring has an inner radius of about 200 nm and an outer radius of about 350 nm. (b) The raw data for The four-terminal resistance across the ring as a function of magnetic field. (c) A background has been removed from the data of (b) to enhance the oscillations. (d) The Fourier transform of (b) to show the single-frequency nature of the oscillations. (Reprinted from Huefner *et al.* (2010) under the Creative Commons attribution license.)

substrate. The raw data is shown in panel (b), while having the background removed to highlight the oscillations is shown in panel (c). The latter is Fourier transformed in panel (d) to highlight the regularity of the oscillations.

The behavior can be explained simply. The vector potential will be considered to be directed counter-clockwise around the ring. (We adopt cylindrical coordinates, with the magnetic field directed in the z-direction and the vector potential in the θ-direction.) To understand the phase interference, we note that the wave momentum is affected by any electric field via (this follows classical physics, discussed in Section 1.6, where the time derivative of the momentum is given by the force)

$$\hbar \frac{d\mathbf{k}}{dt} = F = -e\mathbf{E} \rightarrow \hbar \mathbf{k} = \hbar \mathbf{k}_0 - e \int \mathbf{E}\, dt, \tag{1.8}$$

where $\mathbf{E}$ is the electric field (not to be confused with the energy). In electromagnetics, we know that in the proper gauge, the electric field is related to the vector potential as

$$\mathbf{E} = -\frac{\partial \mathbf{A}}{\partial t}. \tag{1.9}$$

Equations (1.8) and (1.9) can now be combined to show that the proper momentum arises from the Peierls' substitution in which the normal momentum vector is replaced by $p - eA / \hbar$. This means that the proper phase for the two parts of the wave function is determined from

$$\phi = \phi_0 + \frac{1}{\hbar}(\mathbf{p} - e\mathbf{A}) \cdot \mathbf{r}, \tag{1.10}$$

so that the exit phases for the upper and lower arms of the ring can be expressed as

$$\begin{aligned} \phi &= \phi_0 + \int_{\pi}^{0} \left(k + \frac{e}{\hbar} A \right) \cdot a_\vartheta r\, d\varphi \\ \phi &= \phi_0 - \int_{-\pi}^{0} \left(k - \frac{e}{\hbar} A \right) \cdot a_\vartheta r\, d\varphi \end{aligned} \tag{1.11}$$

The two waves from (1.11) will now have a phase difference:

$$\delta\phi = \frac{e}{\hbar} \int_0^{2\pi} A \cdot a_\vartheta r\, d\varphi = \frac{e}{\hbar} \int B \cdot a_z\, dS = 2\pi \frac{\Phi}{\Phi_0}, \tag{1.12}$$

where $\Phi_0 = h/e$ is the quantum unit of flux, and Φ is the flux enclosed in the ring. (Stoke's theorem has been used between the second and third terms of the equation.) The phase interference term in Equation (1.12) goes through a complete oscillation each time the magnetic field is increased by one flux quantum unit (h/e). This produces a modulation in the conductance (resistance) that is periodic in the magnetic field, with a period h/eS, where S is the area of the ring. This periodic oscillation is the AB effect, and in Figure 1.3 results are shown for such a semiconductor structure. The frequency in panel (d) is approximately 60/mT (or approximately a 0.01667 T period). Using the above formula, this gives an effective radius of the ring of about 280 nm, which is midway between the inner and outer radii of the actual ring.

The important point here is the interference that occurs in the ring structure. This interference is not quantum mechanics. Interference has been discussed since Young, more than two centuries ago. It occurs with waves, whether those waves are classical or quantum mechanical. The quantum mechanics that is in play here is the fact that we considered the particles (electrons) in graphene to be waves. The duality of these electrons is the basic quantum features that have been employed. Duality is the new feature that quantum mechanics has brought to the AB effect.

The preceding paragraphs describe how one can "measure" the phase interference between the electron *waves* passing through two separate arms of the system. In this regard, these two arms serve as the two *slits* for the optical waves of Figure 1.1. Observation of the interference phenomena shows us that the electrons in this experiment must be considered as waves, and not as particles. The analysis above has actually been entirely classical, using techniques that are standard in electromagnetic theory. While quantum mechanics is mentioned in the title of the AB paper, the only quantum mechanics has been treating the electron as a wave rather than as a particle. This is the heart of the matter. Once more, we have a confirmation of the *correspondence* between waves and particles as two views of a coherent whole.

1.4 REALITY AND CAUSALITY; DUALITY ONCE AGAIN

The anomaly of light and electrons having both wavelike and particle-like properties was confusing to the world of the early twentieth century and, not in the least, to the scientists of the day. It challenged the understanding of our views of mechanics and electromagnetics. Up to this time, one view held that particles were definitely only particles, while in a second view, optical and radio waves were definitely waves. It would only get more complicated as the spectra of light absorption and emission from atoms were considered.

By 1910, it was well known that the absorption of light by atoms, and the subsequent emission of light from these atoms, was definitely quantized into a spectrum of discrete frequencies or wavelengths. This spectrum had been measured for a great many atoms and characterized by many scientists. One of the early pioneers was the Swedish physicist Anders Jonas Ångström. In the middle of the nineteenth century, he developed tools with which he studied the emission spectrum of the sun and determined that hydrogen, as well as other atoms, was in this spectrum (Ångström, 1862). In 1868, he published what is known as the great map of the solar spectrum (Ångström, 1868). Working with the data from these publications, Johann Balmer (1885) devised a formula to empirically fit this data. Balmer was a Swiss mathematician, and this seems to be his greatest contribution, done when he was already in his 60s. He wrote the formula in terms of the wavelength of light λ as

$$\lambda = \frac{Bn^2}{n^2 - 2}, \tag{1.13}$$

where B is a constant, and $n = 3, 4, 5, \ldots$. Of course, he was not clear why the second number in the denominator was 2, but it fit all the data in the visible spectrum to which he had access.

Working almost independently, Johannes Rydberg (1889), a Swedish physicist, developed a somewhat more general formula, as he was trying to fit a wide range of materials. He recognized that Balmer had found a special case of a more general formula that he had created. Rydberg's formula reads

$$\frac{1}{\lambda} = RyZ^2\left(\frac{1}{m^2} - \frac{1}{n^2}\right), \tag{1.14}$$

which reproduced Balmer's formula when $m^2 = 2$. Here, Ry is known as the Rydberg constant

$$Ry = \frac{e^4}{8\epsilon_0^2 h^3 c}\frac{m_e m_p}{m_e + m_p} \sim 13.605eV \tag{1.15}$$

which he thought was the same for all materials (as it turned out to be), and Z is the atomic number (the number of protons in the atom). As Balmer's series was mainly in the visible, it is natural to understand that it was this part of the hydrogen spectrum that was the first one found and remained the only known set of emission lines for a great many years. Other series would be found as time passed onward. The Rydberg constant gives the ionization energy of the single electron in hydrogen (needed to excite it away from the nucleus).

To try to explain these observations, Nagaoka (1904) modeled a planar atom in which the electrons all resided on a circle around a central positive nucleus. The latter had shown that this was stable provided the attractive force between the electrons and the nucleus was large enough, and the electrons all had about the same velocity. Nagaoka also had considered already that the electrons may lie in many rings, all of which may or may not lie in the same plane, so it could be a planetary-like model or a more spherical model. He noted that the many rings would align better with studies of the optical absorption measurement in atoms, which showed a series of very sharp lines in the spectra, which he connected to the rings of his atomic model. Subsequently, Niels Bohr would work on the problem. He had gone to Manchester to work with Rutherford as a postdoctoral fellow. There he studied Rutherford's planetary model of the atom but was disturbed by the question of why the electrons didn't simply collapse into the nucleus as they radiated their energy. Certainly, he knew of Nagaoka's work with the electrons outside the nucleus. This led him to ask whether the electrons would be restricted into very well-defined energy levels using the same idea of quantization that Planck had used. He was still working on the problem when his fellowship expired, and he returned to Denmark. He continued to work on his atomic model, and the first paper was finally published (Bohr, 1913). Surprisingly, Bohr had not learned about the Balmer and Rydberg formulae until about this time but learned just in time to use the experimental results in his paper. This would be a fortunate addition, because it clarified his thoughts and provided the final justification for the paper. The result of using Planck's quantization of the photon energy also led to the result that the angular momentum of the electrons in their motion around the nucleus was also quantized, but for reasons which he did not understand.

The truth, however, is connected to the fact that this was not a solid theory. Rather, it was a hand-waving exercise which fortuitously gave an understanding of the source of the Rydberg formula. It explained the series of lines and fit the experiments. However, if you believed that reality existed without the need for observers, then Bohr's work was unsatisfying. So, now the dilemma arises. For Bohr adopted the view that it was the experiment *and the apparatus with which it was performed* that provided the reality of the result. It was the reality of the measured atomic spectra that defined the reality of his theory. Reality arrived with the measurement. Different measurements could have different realities. If the measured atomic spectra defined reality, then his theory was good to go. He no longer needed to explain why the momentum was quantized, or why the electrons did not collapse into the nucleus. There was no need to have any underlying theoretical derivation—his approach fit the experimental results and thus was as real as it gets. Not only did Bohr redefine what reality meant, he also sent the idea of causality to the dustbin. His version of the state of quantum effects was that one did not have to comprehend the nature of the atom; we only had to measure it. Quantum phenomena now required an observer to provide reality, and this observer provided the needed evidence to interpret the experiment. And it would be another decade before the proper quantum theories arrived; by then, Bohr had cemented his philosophical view into the quantum community, mostly by a process a later physicist would refer to as "brain washing" (Gell-Mann, 1979). As students and postdocs passed through Copenhagen, they were indoctrinated (or brainwashed) into Bohr's philosophical views. Bohr did not deem it necessary to examine whether there was a duality between waves and particles.

In the early 1920s, Louis de Broglie (1923) developed his alternative to the wave *or* particle picture. His inspiration was the experiments of Taylor mentioned above. This led him to propose that the corpuscles (modern-day photons) of light were coupled to a non-material light wave of the appropriate frequency. He viewed the wave as guiding the particle in the mode of a "pilot" wave (it is still true that ships take on a pilot to guide them into, and out of, port; in this sense de Broglie's pilot wave guides the particle to its destination). This approach was a wave *and* particle viewpoint and would be a major sticking point between the on-coming schism that would arise between the two principal groups working on the new quantum theory. One camp would be led by Bohr, while the other would involve Einstein and be based upon what would be called the wave theory.

Following his initial work, de Broglie (1924) realized that his wave and particle approach could also be applied to particles. And this meant that the particle wave had to have a momentum associated with the accompanying wave. Einstein had already shown that the photon had a momentum given in (1.5), which meant that the particle would have a corresponding wavelength given by its momentum as

$$\lambda = \frac{h}{p}, \tag{1.16}$$

which arrives directly from (1.5). This important result provided precisely the reason for Bohr's angular momentum to be quantized. If the electrons in the atom acted as waves, then it was important that the motion around the circumference of the orbit was exactly a multiple of the wavelength, in order to fit the wave function into the orbit. Now, there was a fundamental reason for Bohr's results.

In 1925, we would get the first version of the new quantum mechanics from the work of Heisenberg (1925). At the time, Heisenberg was a postdoc in Copenhagen working with Bohr, but also holding a position in Göttingen with Born. His theory was one of kinematics, the mechanics of moving bodies. In principle, he was dealing with measurable quantities, but his work became known as matrix mechanics. Importantly, he developed the quantization of the angular momentum through

$$\oint p\,dx = J = nh, \tag{1.17}$$

where p is the momentum, x is the position, J is the angular momentum, n is an integer, and h is Planck's constant.

Prior to Heisenberg, however, Einstein (1917) had proposed (1.17) as a quantization condition that was preferable to that developed by Bohr. It was further developed in later years by both Brillouin (1926) and Keller (1958). Today, (1.17) is known as EBK quantization for these three gentlemen.

Heisenberg then pointed out that the derivative of this equation, with respect to the integer n, would yield a value of the angular momentum defined by Planck's constant h. Importantly, as a disciple of Bohr, he wanted nothing to do with the wave picture put forward by de Broglie and firmly supported Bohr's observer requirement for reality. His colleague von Neumann (1932) would publish a textbook describing the Heisenberg approach in terms of sets in a linear vector space and matrix mechanics, which became highly read and cited. However, Heisenberg's kinematics remained somewhat obtuse and confusing to most scientists.

Only a short time later, it would be Erwin Schrödinger (1926) who would give us the equation for the waves:

$$i\hbar\frac{d\psi(x,t)}{dt} = -\frac{\hbar^2}{2m}\nabla^2\psi(x,t) + V(x,t)\psi(x,t). \tag{1.18}$$

It is clear that the solution of this equation depends upon an initial condition, which means that the wave function ψ evolves from this initial condition. In this, and in de Broglie's work, this evolution from the initial condition restores causality to the system's mechanics. Hence, wave mechanics was unacceptable to the Bohr acolytes. With causality and determinism, one does not depend upon the observer to establish reality; reality exists whether the results are measured or not. Moreover, a great many scientists were already familiar with waves, so that the Schrödinger approach would be easy for them to understand and to use. Not only did Schrödinger bring us the wave approach, but he also established that it was fully compatible with matrix mechanics and yielded exactly the same results.

With the work of de Broglie and Schrödinger, the wave picture was fully established, and it was connected to the particle picture. On the other hand, Heisenberg's approach was solely for particle mechanics, although later work would shake this foundation (due to the uncertainty principle). It could be argued that it was the wave and particle view that could explain the duality that existed in the measurements of Taylor, and in the interpretation of electromagnetic waves in terms of photons. One could say that the measurement was important as it determined which behavior, wave or particle, would appear in the experimental results. However, this is independent of reality—it is clear that reality existed for both wave and particle behavior in the measurements of Taylor. In a sense, duality was the new reality!

1.5 GETTING TO THE SCHRÖDINGER EQUATION

If one looks at (1.18) closely, there are certain characteristics that seem to be familiar from classical physics, where one can write the total energy as

$$E = T + V = \frac{p^2}{2m} + V(x,t). \tag{1.19}$$

Certainly, the last term in (1.18) clearly connects to the last term in (1.19). And if we consider that the wave function appearing in (1.18) is a wave such as in (1.7), the left-hand side of (1.18) may be written as

$$i\hbar \frac{d\psi(x,t)}{dt} = i\hbar A \frac{d\exp(-i\omega t)}{dt} = \hbar\omega\psi(x,t) = E\psi(x,t). \tag{1.20}$$

Hence, it is clear that the energy is the eigenvalue of the time derivative in the separation of variables that may be applied to (1.19) (especially beyond one dimension). This means that the last remaining connection is that

$$T\,\psi(x,t) = \frac{p^2}{2m}\psi(x,t) = -\frac{\hbar^2}{2m}\nabla^2\psi(x,t) \tag{1.21}$$

must represent the kinetic energy in quantum mechanics, just as it does in classical mechanics.

Classically, the term p^2 is a product such as $p^2 = p \cdot p$. However, in the last term, there appear differential operators. Normally, these are applied sequentially, so that the natural interpretation of the connection in (1.21) is that

$$p = -i\hbar\nabla, \tag{1.22}$$

so that the momentum becomes a differential operator. This is new to the quantum world—certain variables become differential operators. In fact, the Schrödinger equation (1.18) is an operator equation, and we should interpret both x and p as operators, as will be discussed in the next section. However, the remaining variable t (time) is not an operator. The variable t is simply a parameter which we use to measure the passage of time itself. The two operators are reversible, as each has a positive and negative direction. While we like to think about reversibility in equilibrium physics, time is not reversible. We can never grow younger! While the Schrödinger equation itself is *mathematically* reversible, any time it is applied to the real world of motion, or of currents in electrical circuits, there has to be a transition to the irreversible situation to account for the fact that one cannot really reverse the arrow of time in the real world (Price, 1997; Prigogine, 1981).

1.6 CONNECTIONS TO THE CLASSICAL WORLD

How does the new quantum mechanics connect to the classical world, which is the world of Newtonian (or Hamiltonian) mechanics? If we are to believe the philosophy of Bohr, there is no such connection. And while we may think that the wave view of de Broglie and Schrödinger has no such connection either, this would be wrong. In fact, in both Heisenberg's matrix approach and

Schrödinger's wave approach, there is a strong connection to classical mechanics that runs throughout the theories. In this section, we want to develop this connection further. Our approach is best developed with Hamiltonian mechanics (Goldstein, 1951), in which we can define the total energy as (we work in one dimension for clarity and consider the normal reversible case where the energy is conserved, and therefore, the total energy is not a function of time) it is expressed in (1.18)–(1.21).

That is, the total energy of a particle, or a system of particles if we sum over all the particles, is given by the sum of the kinetic energy and the potential energy. The potential energy V is the same as that appearing in the Schrödinger equation (1.18). In this approach, the time rate of change of the position and momentum gives rise to Newton's equations (or here as Hamilton's version) as

$$\begin{aligned} \frac{dx}{dt} &= v = \frac{\partial H}{\partial p} = \frac{p}{m} \\ \frac{dp}{dt} &= -\frac{\partial H}{\partial x} = -\frac{\partial V}{\partial x} \end{aligned} . \tag{1.23}$$

(It is important to note here that $dp/dt \neq m(dv/dt)$ in the general case, as Newton never considered that there might be cases where the mass is a function of time or energy; for example, in relativity theory or in semiconductors with nonparabolic energy bands.)

The two variables x and p are known as a conjugate pair. Position and momentum, as expressed here, are one type of a conjugate pair. Voltage and current are another conjugate pair; consider that Kirchhoff's current and voltage laws express continuity conditions upon each of the conjugate pairs. Similar conditions are going to appear on the pair of position and momentum as well. These are our common variables for classical motion and will carry over to quantum mechanics, although our understanding of them will change.

1.6.1 Position and Momentum

We begin by looking at how position and momentum are characterized in wave mechanics. The wave function $\psi(x,t)$ is a complex quantity, while the classical position and momentum of the particle are real quantities. Moreover, the wave is a distributed quantity, while we expect the classical particle to be relatively localized in space. This suggests that we relate the *probability* of finding the electron at a position x to the square of the magnitude of the wave (while the wave itself has a complex amplitude, the magnitude squared is a real and positive quantity). Probability was relatively well known in classical physics although it was also new. It appears in statistical physics, which was developed in the last half of the nineteenth century, most famously by Boltzmann. In statistical physics, one dealt with probabilities; e.g., the probability of a particle's position when there were noise and uncertainty in a large ensemble of such particles. In this world, Newton's law was extended to an ensemble, for which we use Langevin's equation (1908)

$$\frac{dp}{dt} = \frac{d(mv)}{dt} = F - \frac{mv}{\tau} + R(t). \tag{1.24}$$

where the acceleration of a selected particle is given by the applied force F (determined from the potential) plus a random force R, and it is decelerated by collisions with a mean time between collisions of τ. This decay makes the equation irreversible; changing the sign on t gives a different equation. The random force has an average value of zero, but it imparts a random motion characterized by the temperature, with the random thermal velocity being given by

$$v_T = \sqrt{\frac{3k_B T}{m}} \tag{1.25}$$

in three dimensions.

The major point of statistical physics is that the N particles could be characterized by a distribution function, such as the Maxwell–Boltzmann distribution, itself characterized by a temperature T. This distribution would tell us the probability that a particle existed at a given energy would vary as

$$P \sim \exp\left(-\frac{E}{k_B T}\right). \tag{1.26}$$

Thus, the concepts of probability and distribution functions and taking ensemble averages already appear in classical physics. What is new in quantum wave mechanics is to say that the probability of finding the particle at position x is given by $|\psi(x,t)|^2$. As a proper probability distribution function, it is required to normalize the probability with

$$\int_{-\infty}^{\infty} |\psi(x,t)|^2 \, dx = 1. \tag{1.27}$$

That is, (1.27) tells us that the particle is somewhere, but we don't know where. It is clear that we are equating the square of the magnitude of the wave function with a probability density function. This allows us to compute immediately the expectation value, or average value, of the position of the particle with the normal definitions introduced in probability theory. That is, the average value of the position is given by (we denote the *expectation* value with the sharp brackets)

$$\langle x \rangle = \int_{-\infty}^{\infty} x |\psi(x,t)|^2 \, dx = \int_{-\infty}^{\infty} \psi^{\dagger}(x,t) x \psi(x,t) dx. \tag{1.28}$$

The angular brackets mean the expectation value. The small superscript "cross" on the first wave function, in the last expression, denotes the adjoint (usually the complex conjugate) of the wave function. The second new point is to recognize that the position itself is an *operator* in quantum mechanics (the momentum has already been shown to be a differential operator, so the conjugate variable must also be an operator). That is, the x inside the integral is an *operator*, and that is why it is placed *between the two wave functions* in the last integral of (1.28). It is important to note that the expectation value of an operator is evaluated with the operator *between* the two wave functions, with the adjoint/complex conjugate form appearing first in the expression.

The wave function described in (1.18) contains only variations in space and time and is not a uniform quantity. In fact, if it is to describe a localized particle, it must vary quite rapidly in space. However, there is no discussion of the momentum in this equation. So, where is the momentum hiding? To investigate this, we take the Fourier transform of this wave function in order to get a representation that describes the spatial frequencies that are involved, and this Fourier transform will then describe a wave function in momentum space (the Fourier transform of one of a pair of conjugate operators will produce a description in the space of the other operator). Then, the wave function can be written in terms of the spatial frequencies as an inverse transform (again, in one spatial dimension for clarity):

$$\psi(x,t) = \frac{1}{\sqrt{2\pi}} \int_{-\infty}^{\infty} \varphi(k,t) e^{ikx} \, dk. \tag{1.29}$$

The inverse transform can be expressed as

$$\varphi(k,t) = \frac{1}{\sqrt{2\pi}} \int_{-\infty}^{\infty} \psi(x,t) e^{-ikx} \, dx. \tag{1.30}$$

The variable k appeared in the wave description (1.7) with the product kx. This latter product is dimensionless, but we have expressed the momentum as the conjugate variable to position. This connection is given by the equation for momentum (1.8), so that we recognize that $p = \hbar k$.

Suppose now that we are using the position representation wave functions. How then are we to interpret the expectation value of the momentum? The wave functions in this representation are functions only of x and t. We use the various Fourier transforms corresponding to the functions φ and ψ. Then, we may write the expectation value of the momentum as

$$\begin{aligned}\langle p\rangle &= \int_{-\infty}^{\infty}\varphi^{\dagger}(k,t)p\varphi(k,t)dk = \int_{-\infty}^{\infty}\varphi^{\dagger}(k,t)\hbar k\varphi(k,t)dk \\ &= \frac{\hbar}{2\pi}\int_{-\infty}^{\infty}\int_{-\infty}^{\infty}\psi^{\dagger}(x',t)e^{ikx'}dx'k\int_{-\infty}^{\infty}\psi(x,t)e^{-ikx}dx\,dk \\ &= \frac{\hbar}{2\pi}\int_{-\infty}^{\infty}\int_{-\infty}^{\infty}\psi^{\dagger}(x',t)e^{ikx'}dx'\int_{-\infty}^{\infty}\psi(x,t)i\frac{\partial}{\partial x}e^{-ikx}dx\,dk \\ &= -i\hbar\int_{-\infty}^{\infty}\psi^{\dagger}(x,t)\frac{\partial}{\partial x}\psi(x,t)dx\end{aligned} \tag{1.31}$$

This is reassuring that the result gives us the expectation value of (1.22) in one dimension. So, a consistency in results has occurred.

An interesting result that will be useful in the future is the integration by parts that was done (the evaluation at the limits is assumed to vanish) after replacing k by the partial derivative. The last line of (1.31) is achieved by recognizing the delta function representation

$$\delta(x-x') = \frac{1}{2\pi}\int_{-\infty}^{\infty}e^{-ik(x-x')}dk. \tag{1.32}$$

1.6.2 Non-Commuting Operators

The description of the momentum operator in the position representation is that of a differential operator. This means that the operators corresponding to the position and to the momentum will not commute, just as suggested by Heisenberg. However, this is a result of the Schrödinger equation as well, since that is how we determined that the momentum was a differential operator. To see this, we define the commutator relation for position and momentum as

$$[x,p] = xp - px. \tag{1.33}$$

The left-hand side of (1.29) defines a quantity that is called the *commutator bracket* and is defined by the right-hand side of the equation. So, the left-hand side is a short-cut notation. However, by itself it only has no implied meaning. The terms contained within the brackets are operators and must actually operate on some wave function.

Classically, the result of the right-hand side would be zero, because position and momentum are just parameters. However, electrical engineers are familiar with things that don't satisfy this zero, and these are matrices. If we have two square matrices A and B, then quite generally the result

$$AB - BA \neq 0. \tag{1.34}$$

And it may be recalled that Heisenberg's quantum mechanics was called matrix mechanics! The fact that we are using Schrödinger's wave mechanics does not take us away from the non-commutation of our operators—it must be remembered that even Schrödinger showed that the two approaches were fully equivalent. So, our operators in quantum mechanics act just as the matrices one learned in circuit theory; they do not commute.

With this recognition, we can proceed to examine the role of the commutator by considering the inner product, or expectation value. This may be written as

$$\begin{aligned}\langle [x,p] \rangle &= -i\hbar \int_{-\infty}^{\infty} \psi^{\dagger}(x,t)\left(x\frac{\partial}{\partial x} - \frac{\partial}{\partial x}x \right)\psi(x,t)dx \\ &= -i\hbar \int_{-\infty}^{\infty} \psi^{\dagger}(x,t)\left[\left(x\frac{\partial}{\partial x} - x\frac{\partial}{\partial x} \right)\psi(x,t) - \psi(x,t)\frac{\partial}{\partial x}x \right]\psi(x,t)dx = i\hbar\end{aligned}, \quad (1.35)$$

where an integration by parts has been performed on the second term within the integral to obtain the second line. If conjugate variables, or operators, do not commute, there is an implication that these quantities cannot be measured simultaneously. Here again, there is another and deeper meaning. In the previous section, we noted that the operation of the position operator x on the wave function in the position representation produced an eigenvalue $\langle x \rangle$, which is actually the expectation value of the position. The momentum operator does not produce this simple result with the wave function of the position representation. Rather, the differential operator produces a more complex result. For example, if the differential operator were to produce a simple eigenvalue, then the wave function would be constrained to be of the form $exp(ipx/\hbar)$ (which can be shown by assuming a simple eigenvalue form with the differential operator and solving the resulting equation). This form is not integrable (it does not fit our requirements on normalization), and thus, the same wave function cannot simultaneously yield eigenvalues for both position and momentum. Since the eigenvalue relates to the expectation value, which corresponds to the most likely result of an experiment, these two quantities cannot be simultaneously measured.

There is a further level of information that can be obtained from the Fourier transform pair of position and momentum wave functions. If the position is known, for example, if we choose the delta function of (1.32), then the Fourier transform has unit amplitude everywhere; that is, the momentum has equal probability of taking on any value. Another way of looking at this is to say that since the position of the particle is completely determined, it is impossible to say anything about the momentum, as any value of the momentum is equally likely. Similarly, if a delta function is used to describe the momentum wave function, which implies that we know the value of the momentum exactly, then the position wave function has equal amplitude everywhere. This means that if the momentum is known, then it is impossible to say anything about the position, as all values of the latter are equally likely. As a consequence, if we want to describe both of these properties of the particle, the position wave function and its Fourier transform must be selected carefully to allow this to occur. Then there will be an uncertainty Δx in position and Δp in momentum.

1.6.3 Uncertainty

The fact that one cannot measure position and momentum simultaneously is called the uncertainty principle, which is attributed to Heisenberg (1927). However, the idea of uncertainty is quite old and familiar in classical physics (as well as in thought processes). The common example is when one is stopped at a railroad crossing for a passing train. Try to figure out how to measure the length of one of the train cars as it passes. Since it is moving, it is difficult to determine its length, but if one tries to time when the two ends pass a single point, it is necessary to know the velocity. In short, it is very difficult to know where the two ends of the car are located at one point in time and to know how fast the train is moving. In classical physics, this is a difficult measurement problem. Heisenberg tells us that it is a fundamental problem in quantum mechanics.

This fundamental problem is inherently tied to the above discussion of the non-commutation of two conjugate variables. Here, we have added a caveat. Uncertainty pertains to the inability to simultaneously measure two non-commuting operators. Since position and momentum do not commute, there will be a fundamental uncertainty in trying to measure them both at the same time. However, there is no uncertainty between energy and time, because time is not an operator (in non-relativistic quantum mechanics) and these two variables do not fit the requirement of non-commuting operators. As remarked above, the variable *t* is merely a parameter marking the passage of time. (Clocks exist to try to give everyone a common value of *t*, hence a common reference time. However, with so many "time zones" around the world, it remains confusing as to what time one refers to, especially with daylight saving time of one form or another in a great many locales around the world.)

It is fair to ask just how classical uncertainty correlates with quantum uncertainty. To do this, one needs a method of computing the uncertainties Δx in position and Δp in momentum, for example, using our two most famous non-commuting operators. For this, we need to progress a little further into the depths of quantum mechanics and the wave function, so this discussion is deferred to Chapter 2.

1.7 ANOTHER VIEW

It has been suggested that the introductory basics of quantum mechanics arise from the changes from classical mechanics that are brought to an observable level by the smallness of some parameter, such as the size scale. This, of course, is not the view held by a number of the founders of quantum mechanics, who insist that this is a fundamentally new science, unrelated to classical physics. To believe that reality arises from the measurement, and did not exist before the measurement, would certainly support the idea that quantum mechanics is fundamentally new and different. However, perhaps it is the prevalence of the latter view that led Richard Feynman (1965) to proclaim that "I think I can safely say that nobody understands quantum mechanics!" In support of the first view, that quantum mechanics has evolved as an extension or improvement or correction to classical mechanics, various analyses were done almost as soon as the first quantum theories arrived. These approaches examined in quite some detail the wave equation developed by Schrödinger and connected it to a deeper understanding of classical mechanics (Kennard, 1928; Madelung, 1926). These latter authors decided to look at the two parts of the Schrödinger equation (the real and the imaginary parts) separately to investigate further how they represent the understanding of quantum physics in relation to classical physics. Some two decades later, Bohm (1952) would rediscover this approach and draw significantly more attention to this understanding.

The most important aspect of quantum theory, and difference from classical physics, is the appearance of operators to replace dynamical variables, and the non-commuting nature of these operators. However, mathematics has always developed and used operators of various forms, and the non-commuting of matrices is common in circuits and systems. So neither of these new views in quantum theory are actually all that new. However, the idea of being either in a position representation or a momentum representation (Fourier space) is different—there is, in the view so far, no mixed phase space (with both position and momentum) that is common in quantum physics (such a mixed phase-space view would be developed by Wigner, 1932). Certainly, the concept of a wave function, either in the position or momentum representation, as introduced by Schrödinger (1926), and relating the squared magnitude to the probability of finding the equivalent particle at a particular point in space, brings statistical physics into the picture. Nevertheless, it is useful to have a look at what the separate real and imaginary parts of the Schrödinger equation tell us about quantum mechanics.

In order to separate the real and imaginary parts, we need to define a wave function in a manner that allows this separation. For this, we chose to let

$$\psi(x,t) = A(x,t)exp\left(iS(x,t)/\hbar\right). \tag{1.36}$$

where *A* is a real amplitude of the wave function and is usually taken to be positive definite. The quantity $S/\hbar$ is the phase of the wave function. *S* actually appears in classical physics and is known as the *action*. Planck's reduced constant $\hbar$ has the units of energy·time. Hence, S must have the same units, as the argument of the exponential must be dimensionless. The action may then be recognized as the integral over time of the energy (or more properly the negative integral). If (1.36) is now introduced to the Schrödinger equation (1.18), the result may be written as

$$-A\frac{\partial S}{\partial t}+i\hbar\frac{\partial A}{\partial t}=\frac{A}{2m}\left(\frac{\partial S}{\partial x}\right)^2-\frac{i\hbar A}{2m}\frac{\partial^2 S}{\partial x^2}-\frac{i\hbar}{m}\frac{\partial S}{\partial x}\frac{\partial A}{\partial x}-\frac{\hbar^2}{2m}\frac{\partial^2 A}{\partial x^2}+VA. \tag{1.37}$$

For this equation to be valid, it is necessary that both the real parts and the imaginary parts balance separately, so this leads to two equations expressed as

$$\frac{\partial S}{\partial t}+\frac{1}{2m}\left(\frac{\partial S}{\partial x}\right)^2+V-\frac{\hbar^2}{2mA}\frac{\partial^2 A}{\partial x^2}=0 \tag{1.38}$$

and

$$\frac{\partial A}{\partial t}+\frac{A}{2m}\frac{\partial^2 S}{\partial x^2}+\frac{1}{m}\frac{\partial S}{\partial x}\frac{\partial A}{\partial x}=0. \tag{1.39}$$

Let us begin by first attacking the second of these equations, (1.39). To proceed, the form of the last term on the left, and its similarity to the second term on the left, suggests that we multiply this equation by the factor $2A$. This leads to

$$2A\frac{\partial A}{\partial t}+\frac{A^2}{m}\frac{\partial^2 S}{\partial x^2}+\frac{2A}{m}\frac{\partial S}{\partial x}\frac{\partial A}{\partial x}=0, \tag{1.40}$$

which may be rearranged to the form

$$\frac{\partial A^2}{\partial t}+\frac{\partial}{\partial x}\left(\frac{A^2}{m}\frac{\partial S}{\partial x}\right)=0. \tag{1.41}$$

An important point to recognize at this point is that

$$A^2=\left|\psi(x,t)\right|^2. \tag{1.42}$$

That is, A^2 is the probability density in position space (the probability of finding something at point x at time t). If we were to write down the 3D version of (1.41), it would look like

$$\frac{\partial A^2}{\partial t}+\nabla\cdot\left(\frac{A^2}{m}\nabla S\right)=0. \tag{1.43}$$

An equivalent equation may be found in electromagnetics as Gauss' law (or even in circuit theory where it is known as Kirchhoff's current conservation law) in the 3D form

$$\frac{\partial \rho}{\partial t}+\nabla\cdot\boldsymbol{J}=0, \tag{1.44}$$

where $\boldsymbol{J}=\rho\boldsymbol{v}$, where $\boldsymbol{v}$ is the velocity. (The two equations differ by the electric charge, which is of course constant.) So, (1.41) is a continuity equation for the probability density, and we may recognize the quantum velocity as

$$v=\frac{1}{m}\nabla S\rightarrow\frac{1}{m}\frac{\partial S}{\partial x}, \tag{1.45}$$

where the last form returns us to one dimension. Obviously, then ∇S is the momentum associated with the probability flow (much like that in hydrodynamics).

Now, let us return to the first equation, (1.38). As discussed above, the first term in this equation is related to the energy, so that this equation may be expressed as

$$E = \frac{1}{2m}\left(\frac{\partial S}{\partial x}\right)^2 + V - \frac{\hbar^2}{2mA}\frac{\partial^2 A}{\partial x^2}. \tag{1.46}$$

We have already discovered that the first term on the right-hand side is $p^2/2m$, or the kinetic energy, and the second term is obviously the potential energy. Only the last term is not recognizable at the moment. Without this last term, (1.46) is merely the definition of the classical energy. The last term is new and is often defined as

$$Q = -\frac{\hbar^2}{2mA}\frac{\partial^2 A}{\partial x^2}. \tag{1.47}$$

Kennard (1928) expressed this as the quantum potential. It gives rise to the fact that each element of the probability would move as a particle according to Newton's laws under the classical forces, those arising from the potentials in (1.19), plus any *quantum force*. Q is one form of the quantum potential that gives rise to the quantum force. We note that, because of the amplitude in the denominator of (1.47), the quantum force depends upon the shape of the wave function and not on the amplitude of it. Different forms of a quantum potential have been discussed by Wigner (1932), Feynman and Hibbs (1965), and Ferry (2000).

However, there is more to say about S. Suppose we put S into (1.17), and we discover an interesting fact (but one that is obvious):

$$\oint \boldsymbol{p}\cdot d\boldsymbol{r} = \oint \nabla S \cdot d\boldsymbol{r} = 2n\pi\hbar. \tag{1.48}$$

where n is an integer. This is precisely Heisenberg's quantization condition (1.17) on momentum. Thus, S is not determined entirely by the local dynamics, as would be the classical case, but must also satisfy a topological condition which follows from its origin as the phase of an independent complex field—the wave function. Another way of expressing this is that, if there are any vortices that form in the quantum flow, their angular momentum must be quantized as well. (It was thought for a long time that vortices could not exist in quantum mechanics, but they are as common in quantum waves as they are in classical hydrodynamics, where they are termed whirlpools.)

Equation (1.48) was not original to Heisenberg. As pointed out earlier in this chapter, it was originally proposed by Einstein (1917) and further developed by Brillouin (1926) and Keller (1958). However, it is important to understand how this simple quantization relates to the topological nature of the wave function. This importance was highlighted in discussions of the geometrical phase associated with the wave function by Berry (1989). We consider motion around a closed path according to (1.48), which has a period of T around this path such that $\mathbf{r}(T) = \mathbf{r}(0)$. The state of the system still evolves according to the Schrödinger equation. While the energy may be slowly varying in space and time (despite the constant energy assumption used previously), quite generally there is no firm relation between the phases at different points in space. If the system evolves adiabatically, then the states will evolve in a manner in which the wave function can be written as

$$\psi(t) \sim e^{-\frac{i}{\hbar}\int_0^t E dt'} e^{i\gamma(t)}\psi(r(t)), \tag{1.49}$$

which differs from one form of the Schrödinger equation only by the second exponential. Berry points out that this phase factor, $\gamma(t)$, is nonintegrable and cannot be written as a function of the position variable, as it is usually not single-valued as one moves around the circuit. This phase factor is determined by the requirement that it satisfies (1.48). The point is that there is more to both the Schrödinger equation and its hydrodynamic interpretation than meets the eye. The factor $\gamma(t)$ is often called the Berry phase, but it is another (and deeper) topic to explore than is necessary at this point. One final point here is that this phase gives rise to the Aharonov–Bohm phase discussed earlier.

1.8 WAVE PACKETS

While we have assumed that the momentum wave function obtained from the Fourier transform of the position wave function in (1.30) was centered at zero momentum, this is not the general case. In the previous section, the momentum of the wave function was found from the phase $S/\hbar$ of the position wave function. This suggests that, quite generally, the wave function is moving, especially if it is supposed to represent a moving particle. The question is just how we relate a moving wave function to the concept of a particle that is localized in a small region of space even when moving. The answer to this is the wave packet. This wave packet is not just for the representation of a particle, but think of a single photon moving through space. This single photon is also localized within a small region of space, as pointed out by Taylor (1909). Such a wave packet has often been described in electromagnetic theory.

The idea is that a wave packet is not just a single plane wave. Such a localized wave packet is shown in Figure 1.4. Rather, it is composed of an entire set of plane waves in a manner such that the superposition of this set leads to interference giving rise to a single localized packet, whose dimensional spread is quite small. However, there are some constraints on this packet which arise from our above discussions.

We assume that the momentum wave function is centered at a displaced value of k, given by k_0. Then, the entire position representation wave function moves with this average momentum and shows an average velocity:

$$v_0 = \hbar k_0 / m. \tag{1.50}$$

We can expect that the peak of the position wave function, x_{peak}, moves, but does it move with this velocity? The position wave function is made up of a sum of a great many Fourier components, each of which arises from a slightly different momentum. Does this affect the uncertainty in position that characterizes the half-width of the position wave function? The answer to both of these questions is yes, but we will try to demonstrate that these are the correct answers.

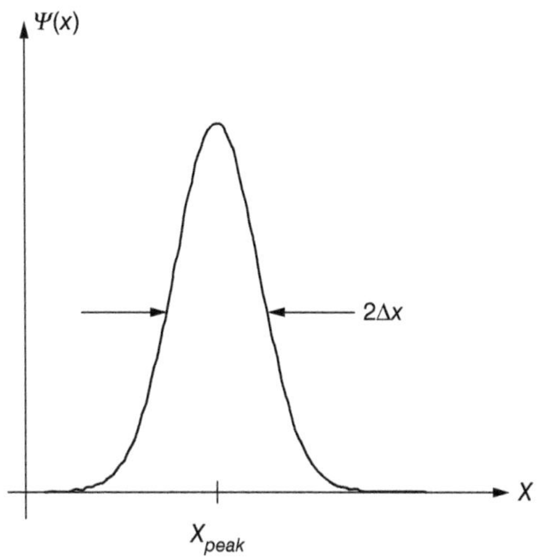

FIGURE 1.4 The positional variation of the square magnitude of a typical wave packet.

As mentioned, the approach is based upon the inverse Fourier transform (1.29). This equation expresses the position wave function $\psi(x)$ as a summation of individual Fourier components, each of whose amplitudes is given by the value of $\phi(k)$ at that particular k. Each of the Fourier terms can be considered to be a plane wave corresponding to that value of k, but with different amplitudes, by introducing the spatial frequency term via

$$\psi(x,t)=\frac{1}{\sqrt{2\pi}}\int_{-\infty}^{\infty}\varphi(k,t)e^{i(kx-\omega t)}\,dk. \tag{1.51}$$

While the frequency term has not been shown with a variation with k, it must be recalled that we might be considering a particle wave. While a light wave may have $\omega = ck_0$, that is with all waves traveling at the speed of light, the particle wave is not constrained in that manner. The particle wave is certainly not traveling at the speed of light, and the dispersion may not be linear (e.g., it does not satisfy $\omega = vk_0$). Hence, each of the Fourier components may actually possess a slightly different frequency and therefore a different velocity. If the main frequency corresponds to the peak of the momentum wave function, then the frequency can be expanded as

$$\omega(k)=\omega(k_0)+(k-k_0)\frac{\partial\omega}{\partial k}\Big|_{k=k_0}+\cdots. \tag{1.52}$$

This is just a normal Taylor series. The interpretation of the position wave function is now that it will be composed of a group of closely related waves, all propagating in the same direction (we assume that $\varphi(k) = 0$ for $k < 0$, but this is merely for convenience and is not critical to the overall discussion). Thus, $\psi(x, t)$ is now defined as a *wave packet*. Equation (1.52) defines the *dispersion* of this wave packet, as it gives the gradual change in frequency for different components of the wave packet. To understand how the dispersion affects the propagation of the wave functions, we insert (1.52) into (1.51) and define the difference variable $u = k - k_0$. Then, (1.51) becomes

$$\psi(x,t)=\frac{1}{\sqrt{2\pi}}e^{i(k_0x-\omega_0t)}\int_{-\infty}^{\infty}\varphi(u+k_0,t)e^{i(ux-\omega' ut)}\,du, \tag{1.53}$$

where ω_0 is the leading term in (1.55), and ω' is the partial derivative in the second term of (1.52). The higher-order terms of (1.52) are neglected, as the first two terms are the most significant. If u is factored out of the argument of the exponential within the integral, it is seen that the position variable varies as $x - \omega' t$. This is our guide as to how to proceed. We will reintroduce k_0 within the exponential but multiplied by this factor, so that

$$\begin{aligned}\psi(x,t)&=\frac{1}{\sqrt{2\pi}}e^{-i(x-\omega' t)}e^{i(k_0x-\omega_0t)}\int_{-\infty}^{\infty}\varphi(u+k_0,t)e^{i(x-\omega' t)}e^{i(ux-\omega' ut)}\,du\\&=\frac{1}{\sqrt{2\pi}}e^{i(\omega_0-\omega' k_0)t}\int_{-\infty}^{\infty}\varphi(u+k_0,t)e^{i(u+k_0)(x-\omega' t)}\,du\\&=e^{i(\omega_0-\omega' k_0)t}\psi(x-\omega' t,t).\end{aligned} \tag{1.54}$$

The leading exponential provides a phase shift in the position wave function. This phase shift has no effect on the square of the magnitude, which represents the expectation value calculations. On the other hand, the entire wave function moves with a velocity given by ω'. This is not surprising. The quantity ω' is the partial derivative of the frequency with respect to the momentum wave vector and hence describes the *group velocity* (a concept familiar in electromagnetics, especially in fibers or waveguides) of the wave packet. Thus, the average velocity of the wave packet in position space is given by the group velocity:

$$v_g = \omega' = \frac{\partial\omega}{\partial k}\Big|_{k=k_0}. \tag{1.55}$$

This answers the first question: the peak of the position wave function remains the peak and moves with an average velocity defined as the group velocity of the wave packet. This group velocity is defined by the frequency variation with respect to the wave vector. Is this related to the average momentum given by k_0? The answer again is affirmative, as we cannot let k_0 take on any arbitrary value. Rather, the peak in the momentum distribution must relate to the average motion of the wave packet in position space. Thus, we must impose a value on k_0 so that it satisfies the condition of actually being the average momentum of the wave packet:

$$v_g = \frac{\hbar k_0}{m} = \frac{\partial \omega}{\partial k}. \tag{1.56}$$

If one is dealing with a light wave, then we have $\omega = ck_0$, and (1.56) yields c, the velocity of light. However, if we are dealing with a matter particle, which does not travel at the speed of light, we can integrate the equation of the last two terms of (1.56) with respect to the wave vector, and recover

$$E = \hbar\omega = \frac{\hbar^2 k^2}{2m} = \frac{p^2}{2m}. \tag{1.57}$$

It is clear that it is the group velocity of the wave packet that describes the average momentum of the wave function and also relates the velocity (and momentum) to the energy of the particle, whether this particle is a photon or an electron.

1.9 WHY QUANTUM MECHANICS

The mystery of quantum mechanics seems to arise from the fact that photons and electrons, normally waves and particles, respectively, do not differ at the fundamental level. Rather, it has been shown that photons can behave as particles or waves, and electrons can behave as particles or waves. It is the dual nature of these respectable entities that has become the primary mystery in this new view of nature's peculiarities. In this sense, quantum mechanics furnishes a formalism for treating the wave–particle duality.

It has become a belief that the main importance of the quantum treatment is for structures and times that are both small, and thus for which the *interference* of the waves can become important. However, this is not always true. Indeed, the sun shines because of quantum tunneling that allows the various nuclei to get close enough to establish nuclear fusion. While the nuclei are small, and the distances involved are also small, it is obvious that the sun is not a small object. Nevertheless, it has been termed quantum tunneling. However, tunneling in the wave world has been around for quite some time—the quantum nature arises, as we shall see, from treating the nuclei as waves and not distinct particles.

The mathematics of quantum mechanics is not new to engineers, as it has often been called applied mathematics. This is the world of various integral transforms, Fourier and Laplace being two such transforms, and of boundary value problems in various coordinate systems (see, for example, Morse and Feshbach, 1953). What has appeared new with quantum mechanics is that the classical dynamical variables have become operators, and that these operators affect the wave function that lies at the basis of probability theory, or statistical physics as it is often called. As operators, these variables often no longer commute, and there is a basic uncertainty relation between non-commuting operators. The non-commuting nature arises from it being no longer possible to generate a wave function that simultaneously yields eigenvalues for *both* of the conjugate operators, representing the fact that they cannot be simultaneously measured. Again, uncertainty in the measurement of conjugate operators is not new to quantum mechanics, but now it is not just a measurement problem but is more fundamental and is tied to the uncertainty arising from the operator formalism.

REFERENCES

Aharonov, Y. and Bohm, D. 1959. Significance of the electromagnetic potentials in the quantum theory. *Phys. Rev.* **115** 485–491.

Ångström, A. J. 1862. On the Fraunhofer-lines visible in the solar spectrum. *Phil. Mag. Ser. 4.* **24** 1–11.

Ångström, A. J. 1868. *Recherches sur le spectre solaire.* (Uppsala: W. Schultz).

Balmer, J. 1885. Notiz über die Spektrallinien des Wasserstoffs. *Verh. Naturfors. Basel.* **7** 548–560.

Becquerel, E. 1839. Mèmoire sur let Effets Électriques Produits sous L'Influence des Rayons Solarires. *Comp. Rend.* **9** 561–567.

Berry, M. V. 1989. Quantum scars of classical closed orbits in phase space. *Proc. Roy. Soc. London.* **423** 219–231.

Bohm, D. 1952. A suggested interpretation of the quantum theory in terms of "hidden" variables. I. *Phys. Rev.* **85** 166–179.

Bohr, N. 1913. On the constitution of atoms and molecules. *Phil. Mag., Ser. 6.* **26** 1–25.

Brillouin, L. 1926. Remarques sur la Mécanique Ondulatoire. *J. Phys. Radium.* **7** 353–368.

Davisson, C. and Kunsman, C. H. 1921. The scattering of electrons by nickel. *Science.* **54** 522–524.

Davisson, C. and Kunsman, C. H. 1923. The scattering of low speed electrons by platinum and magnesium. *Phys. Rev.* **22** 242–258.

de Broglie, L. 1923. Waves and quanta. *Nature.* **112** 540–540.

de Broglie, L. 1924. A tentative theory of light quanta. *Phil. Mag., Ser. 6.* **47** 446–458.

de Broglie, L. 1937. *Matière et Lumière.* (Paris: Albin Michel).

Einstein, A. 1905a. Zur Elektrodynamik bewegter Körper. *Ann. Phys.* **17** 891–921.

Einstein, A. 1905b. Über einen die Erzeugung und Verwandlung des Lichtes betreffenden heuristischen Gesichtspunkt. *Ann. Phys.* **17** 132–148.

Einstein, A. 1917. Zum Quantensatz von Sommerfeld und Epstein. *Verhand. Deut. Phy. Gesell.* **19** 82–92.

Elsasser, W. 1925. Bemerken zur Quantenmechanik freier Elektronen. *Naturwiss.* **13** 711–711.

Ferry, D. K. 2000. Effective potentials and the onset of quantization in ultrasmall MOSFETs. *Superlatt. Microstruc.* **28** 419–423.

Feynman, R. P. 1965. *The Character of Physical Law.* (Cambridge, MA.: The MIT Press) 129.

Feyman, R. P. and Hibbs, A. R. 1965. *Quantum Mechanics and Path Integrals.* (New York: McGraw-Hill) pp. 279–285.

Gell-Mann, M. 1979. *The Nature of the Physical Universe: 1976 Nobel Conference.* Ed. by D. Huff and O. Prewett. (New York: Wiley-Interscience) p. 29.

Goldstein, H. 1951 *Classical Mechanics.* (New York: Addison-Wesley).

Heisenberg, W. 1925. Über quantentheoretische Umdeutung kinematischer und mechanischer Beziehunger. *Z. Phys.* **33** 879–893; Tr. in B. L. van der Warden. 1967. *Sources of Quantum Mechanics.* (Amsterdam, North-Holland) 261–276.

Heisenberg, W. 1927. Uber den anschaulichen inhalt der quantentheorischen Kinematik und Mechanik. *Zeit. Phys.* **43** 172–198.

Huefner, M., Molitor, F., Jacobsen, A., Pioda, A., et al. 2010. The Aharonov-Bohm effect in a side-gated graphene ring. *New J. Phys.* **12** 043054.

Keller, J. B. 1958. Corrected Bohr-Sommerfeld quantum conditions for nonseparable systems. *Ann. Phys. (NY).* **4** 180–188.

Kennard, E. H. 1928. On the quantum mechanics of a system of particles. *Phys. Rev.* **31** 876–890.

Lummer, O. and Pringsheim, E. 1899. Die Verteilung der Energie im Spectrum des Schwarzen Körpers und des Blanken Platins. Verhand. Deut. Phys. *Gesell.* **1899**: 215–230.

Madelung, E. 1926. Quantentheorie in hydrodynamischer form. *Z. Phys.* **40** 322–326.

Morse, P. M. and Feshbach, H. 1953. *Methods in Theoretical Physics.* (New York: McGraw-Hill).

Nagaoka, H. 1904. Kinetic of a system of particles illustrating the line and the band spectrum and the phenomena of radioactivity. *Phil. Mag., Ser. 6.* **7** 445–455.

Newton, I. 1687. *Philosophiæ Naturalis Principia Mathematica.* (London: S. Pepys).

Planck, M. 1900. Zur Theorie des gesetzes der Energievertheilung im Normalspektrum. *Verhand. Deut. Phy. Gesell.* **2** 237–245.

Price, H. 1997. *Time's Arrow and Archimedes' Point: New Directions for the Physics of Time.* (Oxford, UK: Oxford University Press).

Prigogine, I. 1981. *From Being to Becoming: Time and Complexity in the Physical Sciences.* (San Franciso: W. H. Freeman).

Rydberg, J. R. 1889. La constitutions des spectres d'emission des elements chimiques. *Kongl. Swedish Vet.-Akad. Handl.* **23** 1–177.

Schrödinger, E. 1926. Quantizierung als eigenwertprobleme. Ann. Physik. **79** 361–379.
Taylor, G. I. 1909. Interference fringes with feeble light. *Proc. Cambridge Phil. Soc.* **15** 114–115.
Thomson, G. P. 1927. Diffraction of cathode rays by a thin film. *Nature.* **119** 890–890.
Thomson, J. J. 1907. On the ionization of gases by ultra-violet light and on the evidence as to the structure of light afforded by its electrical effects. *Proc. Cambridge Phil. Soc.* **14** 417–424.
von Neumann, J. 1932. *Mathematische Grundlagen der Quantenmechanik.* (Berlin: Springer).
Wien, W. 1896. Über die Energieverteilung im emissionspektrum eines Schwarzen Körpers. *Ann. Phys.* **294** 662–669.
Wigner, E. 1932. On the quantum correction for thermodynamic equilibrium. *Phys. Rev.* **40** 749–759.
Young, T. 1809. *Course of Lectures on Natural Philosophy and Mechanical Arts.* (London: J. Johnson) Lecture XXXIX.

PROBLEMS

1. Calculate the energy density for the plane electromagnetic wave described by the complex field strength:

$$E_c = E_0 e^{i(\omega t - kx)}$$

and show that its average over a temporal period T is

$$\hbar\omega = (E/2)|E_c|^2 1$$

2. What are the de Broglie frequencies and wavelengths of an electron and a proton accelerated to 100 eV? What are the corresponding group and phase velocities?
3. Show that the position operator x is represented by the differential operator:

$$i\hbar\frac{\partial}{\partial \mathrm{p}}$$

in momentum space, when dealing with momentum wave functions.
4. Express the expectation value of the kinetic energy of a Gaussian wave packet in terms of the expectation value and the uncertainty of the momentum wave function. You may express the uncertainty in position from the width of the Gaussian in position (width at half maximum value), as per Figure 1.4. The uncertainty of the momentum wave function comes from the width at half maximum of the resulting Gaussian in momentum space.
5. Consider the amplitude function:

$$A(k) = \begin{cases} \dfrac{1}{\sqrt{\Delta k}}, & k_0 - \dfrac{\Delta k}{2} \le k \le k_0 + \dfrac{\Delta k}{2}, \\ 0, & \textit{elsewhere} . \end{cases}$$

   a. Calculate $\psi(x,t)$.
   b. Calculate the corresponding probability density $P(x,0)$. Plot the probability density $P(x,0)$. How does $P(x,0)$ change when we double Δk? Explain.
6. A particle is represented by a wave packet propagating in a dispersive medium, described by

$$\omega = \frac{A}{\hbar}\left\{\sqrt{1+\frac{\hbar^2 k^2}{mA}} - 1\right\}.$$

What is the group velocity as a function of $\hbar\omega$.

7. The longest wavelength that can cause the emission of electrons from silicon is 296 nm. (1) What is the work function of silicon? (2) If silicon is irradiated with light of 250 nm wavelength, what is the energy and momentum of the emitted electrons? What is their wavelength? (3) If the incident photon flux is 5 mW/cm^{-2}, what is the photoemission current density?
8. For particles which have a thermal velocity, what is the wavelength at 300 K of electrons, helium atoms, and the a-particle (which is ionized ^{4}He)?
9. A wave function has been determined to be given by the spatial variation:

$$\Psi(x)=\begin{cases} 2A, & -a<x<0, \\ 2A(a-x), & 0<x<a, \\ 0 & elsewhere. \end{cases}$$

Determine the value of A, the expectation value of x, x^2, p, and p^2.
10. A wave function has been determined to be given by the spatial variation:

$$\Psi(x)=\begin{cases} 2Asin(\pi/a), & -a<x<a, \\ 0, & elsewhere\ . \end{cases}$$

Determine the value of A, the expectation value of x, x^2, p, and p^2.
11. Consider a Gaussian wave packet (with $k_0 = 0$) of the form:

$$\psi(x,t)=\left(\frac{1}{2\pi}\right)^{1/4}\frac{1}{\sqrt{L\left(1+\frac{i\hbar t}{2mL^2}\right)}}exp\left[-\frac{1}{4L^2}\frac{x^2\left(1-\frac{i\hbar t}{2mL^2}\right)}{1+\left(\frac{\hbar t}{2mL^2}\right)^2}\right].$$

Derive the expression for the probability current density for this wave packet.

2 The Schrödinger Equation

The introductory basics of quantum mechanics were said to arise from the changes from classical mechanics in the first chapter. In some sense, these changes are brought to an observable level by the smallness of some parameter, such as the size scale. In another sense, however, quantum mechanics introduces some fundamental changes, that are not at all present in classical mechanics. Perhaps the most important effect is the appearance of operators in place of dynamical variables and the non-commuting nature of these operators. While these properties are present in normal mathematics, they are not usually associated with discussions of classical mechanics.

In one formulation of quantum mechanics, the so-called wave picture developed by Schrödinger, we determined a wave function, which could appear either in the position or momentum representation, whose squared magnitude is related to the probability of finding the position or momentum of the equivalent particle. More importantly, the wave could be determined from a linear differential equation that had a diffusive nature (only incorporating the first-order time derivative, common in diffusion equations).

In the first few sections of this chapter, we want to gain insight into the quantization process and the effects it causes in normal systems. This will be done with some simple examples. In subsequent sections, we will then apply the Schrödinger equation to a variety of problems to gain further insight. But let us be clear. The quantum aspect of treating, e.g., an electron and its behavior arises from the decision to treat the particle as a wave. It is a wave equation that we are solving. In particular, when we ignore the time variation of the wave function, we arrive at a nearly classical boundary value problem that must be solved. We are doing no more than solving a version of the Sturm–Liouville problem, which differs little from solutions of other boundary value problems in mathematics and electromagnetics. Again, the quantum nature lies entirely in the decision to deal with the wave approach.

2.1 WAVES AND THE DIFFERENTIAL EQUATION

The choice of a wave function–based approach to quantum mechanics is not the only option. As we remarked in the first chapter, there were two separate formulations of the new quantum mechanics that appeared almost simultaneously. One was developed by Werner Heisenberg (1925), at that time a lecturer in Göttingen (Germany) and a postdoctoral fellow in Copenhagen (Denmark). In this approach, a calculus of non-commuting operators was developed. This approach was quite mathematical and required considerable experience to work through in any detail. It remained until later to discover that this calculus was actually representable by normal matrix calculus. The second formulation was worked out by Erwin Schrödinger (1926), at the time a Professor in Vienna. In Schrödinger's formulation, a wave equation was found to provide the basic understanding of quantum mechanics. Although not appreciated at the time, Schrödinger presented the connection between the two approaches in subsequent papers. In a somewhat political environment, Heisenberg received the 1932 Nobel Prize for "discovering" quantum mechanics, while Schrödinger was forced to share the 1933 prize with Paul Dirac for advances in atomic physics. Nevertheless, it is Schrödinger's formulation that is almost universally used today, especially in textbooks. This is especially true for students with a background in electromagnetic fields, as the concept of a wave equation is not foreign to them.

In this chapter, we want to deal now with the wave equation—the Schrödinger equation, from which one version of quantum mechanics—wave mechanics—has evolved. The equation was introduced in Chapter 1 and may be expressed as

$$i\hbar\frac{d\psi(x,t)}{dt} = -\frac{\hbar^2}{2m}\nabla^2\psi(x,t) + V(x)\psi(x,t). \tag{2.1}$$

DOI: 10.1201/9781003717195-2

An important point is that Equation (2.1) is separable into temporal and spatial parts. To see this, let us assume that the wave function can be written as (again in one dimension)

$$\psi(x,t) = \varphi(x)\chi(t). \tag{2.2}$$

Following a standard procedure, this wave function is inserted in (2.1), and the result is divided by the wave function to give

$$\frac{i\hbar}{\chi}\frac{\partial\chi}{\partial t} = \frac{1}{\varphi}\left[-\frac{\hbar^2}{2m}\frac{\partial^2\varphi}{\partial x^2} + V(x)\varphi\right]. \tag{2.3}$$

Now, the left-hand side is a function of t alone, while the right-hand side is a function of x alone. The only way in which this can be true for all x and t is for each side to be equal to the same constant, which is here denoted as the energy E. The left-hand side may then be solved as

$$\chi(t) = \chi_0 exp\left(-\frac{iEt}{\hbar}\right). \tag{2.4}$$

Similarly, the right-hand side leads to what is known as the time-independent Schrödinger equation:

$$-\frac{\hbar^2}{2m}\frac{\partial^2\varphi(x)}{\partial x^2} + V(x)\varphi(x) = E\varphi(x). \tag{2.5}$$

It is this latter equation that is quite often used to solve quantum boundary-value problems.

2.2 SOME SIMPLE EXAMPLES

In this section, we want to examine some simple examples of solutions to the Schrödinger wave equation. Make no mistake, these examples are worked assuming that the particle we are dealing with is, in fact, a wave. Thus, the results all depend upon the quantum nature, requiring the property of duality. The natural state of a classical particle has now become a quantum wave, and it is the solutions to the wave equation that we are choosing in treating the quantum particle. This is, in fact, the heart of the quantization.

2.2.1 The Free Particle

We begin by first considering the situation in which the potential is zero. Then the time-independent equation becomes (in one dimension, and we revert to using $\psi(x)$ rather than φ)

$$-\frac{\hbar^2}{2m}\frac{\partial^2\psi(x)}{\partial x^2} = E\psi(x). \tag{2.6}$$

This may be rewritten as

$$-\frac{\hbar^2}{2m}\left[\frac{\partial^2\psi(x)}{\partial x^2} + k^2\psi(x)\right] = 0, \tag{2.7}$$

where

$$\frac{\hbar^2 k^2}{2m} = E, \quad k = \sqrt{\frac{2mE}{\hbar^2}}. \tag{2.8}$$

Clearly, the prefactor in (2.7) can be ignored and only the term within the square bracket considered. The natural solution of this bracketed term is of the form of a combination of sines and cosines, although the exponentials of imaginary arguments can be used in their place. That is, we may assume (from prior familiarity with equations of this form) that the solution may be written as

$$\psi(x) = Ae^{ikx} + Be^{-ikx} \tag{2.9}$$

The above are just the plane-wave solutions with which we began our treatment of quantum mechanics. The plane-wave form becomes more obvious when the time variation $e^{-i\omega t}$ is reinserted into the total wave function. Then, in looking at the solutions of this equation, and with the time variation restored, so that

$$\psi(x) = Ae^{ikx}e^{-i\omega t} + Be^{-ikx}e^{-i\omega t}. \tag{2.10}$$

We see that the first wave (first term on the right-hand side of the equation) is traveling to the right (toward positive x) while the second term on the right is a wave traveling to the left (toward negative x). Thus, in free space, the solutions to the Schrödinger equation are just a pair of plane waves, one each traveling in the positive and negative directions. We note that each has a positive amplitude and a single value of the wave number k, so that its formulation in momentum space is a delta function at the proper momentum. We cannot determine its position in real space, because the amplitude is uniform everywhere.

2.2.2 A Potential Step

In order to understand the role that the potential plays, the next example investigates a simple potential step, in which the potential is defined as

$$V(x) = \begin{cases} V_0, & x \geq 0, \\ 0, & x < 0. \end{cases} \tag{2.11}$$

The potential is shown in Figure 2.1. The potential has a height of V_0 for positive x and is zero for the negative-x region. This potential creates a barrier to the wave function, and a wave incident from the left (the negative region) will have part (or all) of its amplitude reflected from the barrier. The results that are obtained depend upon the relative energy of the wave. If the energy is less than V_0, the wave cannot propagate in the region of positive x. This is clearly seen from incorporating the potential into the definition of k via

$$\frac{\hbar^2 k^2}{2m} = E - V(x), \quad k = \sqrt{\frac{2m(E - V(x))}{\hbar^2}}. \tag{2.12}$$

The wave vector is imaginary for $E < V_0$ $(x \geq 0)$. Only one exponent can be retained, as we require that the wave function remain finite (but near zero) as $x \to \infty$. So, we will examine two cases, one for which the energy is larger than the potential and one for which the energy is smaller than the potential step. These are illustrated in Figure 2.2.

Let us first consider the low-energy case, where the wave is a non-propagating wave for $x \geq 0$. In the negative half-space, we consider the wave function ψ_1 to be of the form of (2.9), composed of an incident wave (the positive-exponent term) and a reflected wave (the negative-exponent term). That is, we write the wave function for $x < 0$ exactly as (2.9), where the energy and the wave vector k are given by (2.8). For the positive half-space, where the potential exists, we write the wave function as

$$\psi_2(x) = Ce^{-\gamma x}, \tag{2.13}$$

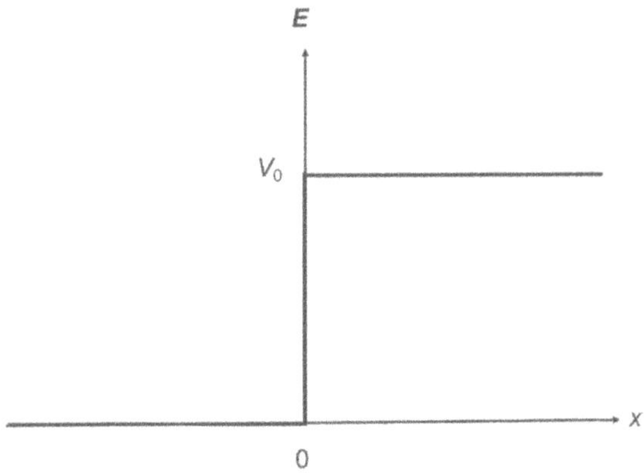

FIGURE 2.1 Schematic view of the potential of (2.11), which is nonzero (and constant) only in the positive half-space.

where

$$\gamma = \sqrt{\frac{2m[V(x)-E]}{\hbar^2}}. \tag{2.14}$$

Here, we have defined a wave function in two separate regions, with the potential constant in each region. These two wave functions must be smoothly joined where the two regions meet.

While three constants are defined (A, B, C), one of these is defined by the resultant normalization of the wave function (we could, e.g., let $A = 1$ without loss of generality). Two boundary conditions are required to evaluate the other two coefficients in terms of A. The boundary conditions can vary with the problem, but one must describe the continuity of the probability across the interface between the two regions. Thus, one boundary condition is that the wave function itself must be continuous at the interface, or

$$\psi_1(0) = \psi_2(0) \rightarrow A + B = C. \tag{2.15}$$

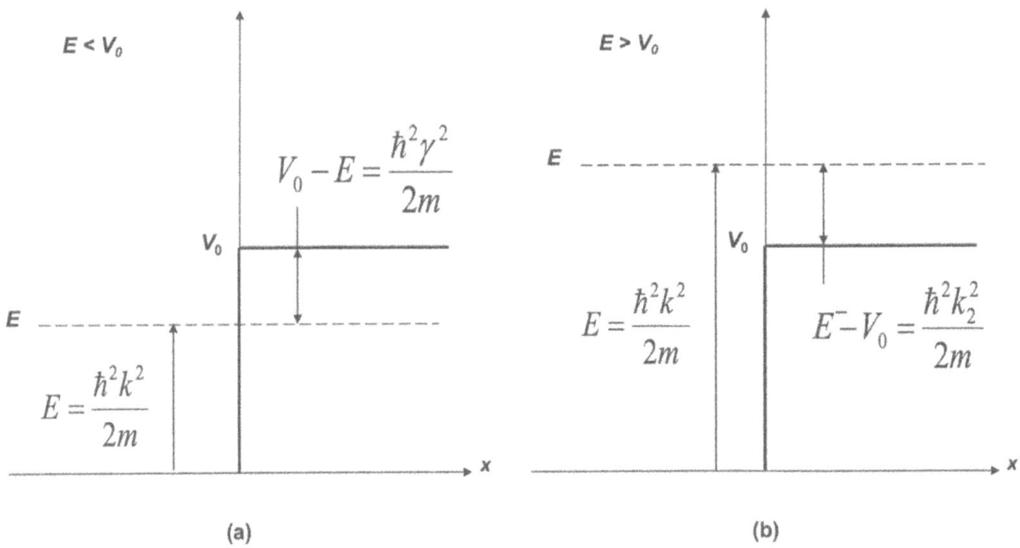

FIGURE 2.2 The various wave vectors are related to the energy of the wave: (a) the case for $E < V_0$; (b) the case for $E > V_0$.

and

$$\begin{aligned}\langle x^2 \rangle &= \frac{1}{a}\int_{-a}^{a} cos^2\left(\frac{n\pi x}{a}\right)x^2 dx \\ &= \frac{1}{2a}\int_{-a}^{a}\left[1+cos\left(\frac{2n\pi x}{a}\right)\right]x^2 dx = \frac{a^2}{3}-\frac{a^2}{2n^2\pi^2}\end{aligned} \quad (2.80b)$$

Similarly, the momentum expectation values are (using the differential operator form for the momentum)

$$\langle p \rangle = \frac{-i\hbar}{a}\int_{-a}^{a} cos\left(\frac{n\pi x}{a}\right)\frac{n\pi}{a} sin\left(\frac{n\pi x}{a}\right)dx = 0 \quad (2.80c)$$

and

$$\langle p^2 \rangle = -\frac{\hbar^2}{2a}\left(\frac{n\pi}{a}\right)^2\int_{-a}^{a} cos^2\left(\frac{n\pi x}{a}\right)dx = -\frac{1}{2}\left(\frac{\hbar n\pi}{a}\right)^2. \quad (2.80d)$$

We note here that the uncertainty in momentum is imaginary due to the negative sign on the last factor. This is because *each wave function has a very precise energy level and therefore a very precise momentum.* We do note, however, that as we move up in the ladder of allowed energy states, the spreading of these leads to an increase in the uncertainty in position. But the conclusion is that this is not a good system in which to discuss the Heisenberg uncertainty principle! There is another important result: the solutions to the wave functions in this infinite potential well give precise energy values, or precise energy eigenvalues. Hence, these wave functions satisfy an eigenvalue equation. Such functions do not lead to uncertainty (eigenvalues derived from an eigenvalue equation have zero uncertainty).

2.7 NUMERICAL SOLUTIONS OF THE SCHRÖDINGER EQUATION

The numerical solutions that we dealt with in the preceding sections were largely carried out by programs available at NanoHub.org. To understand the limitations of these programs and to gain appreciation for the assumptions and limitations inherent in their use, consider the task of simulating the Schrödinger equation itself and in multiple dimensions. If some arbitrary function is known at a series of equally spaced points, one can use a Taylor series to expand the function about these points and evaluate it in the regions between the known points. Consider the set of points shown in Figure 2.15, for example. Here, each point is separated from its neighbors by the distance a (in position). This allows us to develop a finite-difference scheme for the numerical evaluation of the Schrödinger equation. If we first expand the function in a Taylor series about the points on either side of x_0, we get

$$\begin{aligned}f(x_0+a) &= f(x_0)+a\frac{\partial f}{\partial x}\Big|_{x=x_0}+\frac{a^2}{2}\frac{\partial^2 f}{\partial x^2}\Big|_{x=x_0}+\cdots \\ f(x_0-a) &= f(x_0)-a\frac{\partial f}{\partial x}\Big|_{x=x_0}+\frac{a^2}{2}\frac{\partial^2 f}{\partial x^2}\Big|_{x=x_0}+\cdots\end{aligned} \quad (2.81)$$

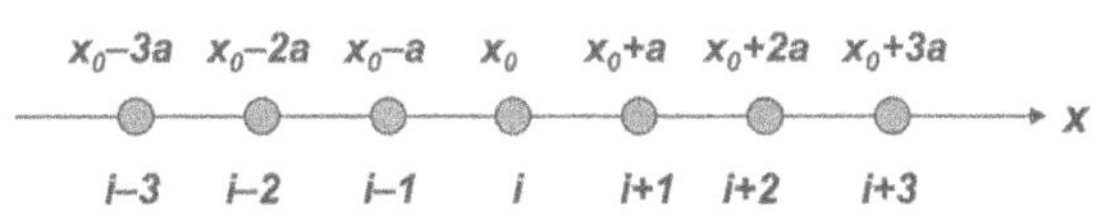

FIGURE 2.15 A one-dimensional, equidistant mesh for the evaluation of the Schrödinger equation by numerical methods.

To obtain a second boundary condition, we require that the derivative of the wave function is also continuous (we can have only two boundary conditions, as we have only one boundary at $x = 0$, and the equation is only of second order in the derivative). In some situations, we cannot specify such a boundary condition, as there may not be a sufficient number of constants to evaluate (this will be the case in the next section). Equating the derivatives of the wave functions at the interface leads to

$$\frac{d\psi_1}{dx}\bigg|_{x=0} = \frac{d\psi_2}{dx}\bigg|_{x=0} \rightarrow ik\left(A-B\right) = -\gamma C. \tag{2.16}$$

This last equation can be rearranged by placing the momentum term in the denominator on the right-hand side. Then adding (2.15) and (2.16) leads to

$$\frac{C}{A} = \frac{2ik}{ik-\gamma}. \tag{2.17}$$

This result can now be used in (2.15) to yield

$$\frac{B}{A} = \frac{ik+\gamma}{ik-\gamma}. \tag{2.18}$$

The amplitude of the reflected wave is unity, so there is no probability amplitude transmitted across the interface. In fact, the only effect of the interface is to phase shift the reflected wave; that is, the wave function is ($x < 0$)

$$\psi_1\left(x\right) = A\left(e^{ikx} + e^{-i(kx+\xi)}\right), \tag{2.19}$$

with

$$\xi = 2arctan\left(\frac{\gamma}{k}\right). \tag{2.20}$$

The probability amplitude is given by

$$\left|\psi_1\left(x\right)\right|^2 = 2A^2, x < 0. \tag{2.21}$$

As may have been expected, this is a *standing-wave* pattern, with the probability oscillating from 0 to twice the value of A^2. The first peak occurs at a distance $x = -\xi/2k$, that is, the distance to the first peak is dependent upon the phase shift at the interface. If the potential amplitude is increased without limit, $V_0 \rightarrow \infty$, the damping coefficient $\gamma \rightarrow \infty$, and the phase shift approaches π. However, the first peak occurs at a value of $kx = \pi/2$, which also leads to the result that *the wave function becomes zero* at $x = 0$. We cannot examine the other limit ($V_0 \rightarrow 0$), as we do not have the proper transmitted wave, but this limit can be probed when the transmission mode is examined. It may also be noted that a calculation of the probability current for $x > 0$ leads immediately to zero as the wave function is real. Thus, no probability current flows into the right half-plane. It is a simple calculation to show that the net probability current in the left half-plane vanishes as well, as the reflected wave carries precisely the same current away from the interface as the incident wave carries toward the interface.

Let us now look at the second case, for which $E > V_0$. In this case, the wave can propagate on both sides of the interface. As above, the wave function in the left half-space is assumed to be of the form of (2.9), which includes both an incident wave and a reflected wave. Similarly, the transmitted wave will be assumed to be of the form

$$\psi_2 = Ce^{ik_2 x}, \tag{2.22}$$

where

$$k_2 = \sqrt{\frac{2m(E - V_0)}{\hbar^2}}. \tag{2.23}$$

Here, k_2 represents the kinetic energy that remains above the potential barrier V_0. The relationships between this wave vector and that for the region $x < 0$ are schematically described in Figure 2.2(b). Again, we will match both the wave function and its derivative at $x = 0$. This leads to

$$\begin{aligned} \psi_1(0) = \psi_2(0) &\rightarrow A + B = C \\ \frac{d\psi_1}{dx}\bigg|_{x=0} = \frac{d\psi_2}{dx}\bigg|_{x=0} &\rightarrow k(A - B) = k_2 C. \end{aligned} \tag{2.24}$$

These equations can now be solved to obtain the constants C and B in terms of A. One difference here from the previous treatment is that these will be real numbers now, rather than complex numbers. Indeed, adding and subtracting the two equations of (2.24) leads to

$$\frac{C}{A} = \frac{2k}{k + k_2}, \quad \frac{B}{A} = \frac{k - k_2}{k + k_2}. \tag{2.25}$$

If $V_0 \rightarrow 0$, $k_2 \rightarrow k$ and the amplitude of the reflected wave vanishes, while the amplitude of the transmitted wave is equal to the incident wave.

2.2.3 Generalities of the Solutions

The solutions that were found above were all in the rectilinear form of general rectangular coordinates. Although carried out for one dimension, it is important to note that even one dimension will vary considerably in different coordinate systems (Morse and Feshbach, 1953). The time-independent Schrödinger equation is a more or less standard form of the general Sturm–Liouville boundary-value problem (Hildebrand, 1962). The form that the equation takes depends significantly on the nature of the coordinate system because of the various scaling factors that appear between the different systems.

The time-independent Schrödinger equation is a second-order differential equation and can therefore accommodate only two boundary conditions. In the case of a free space particle, the two conditions were the values for the constants A and B. They were not actually evaluated because the values depend upon the intensity (amplitude) of the electromagnetic wave. In the second example, the potential step, the two boundary conditions encompassed the continuity of the wave function and its derivative at the edge of the potential step. Also, imposing an integral constraint upon the solution does not add a boundary condition; it only imposes a proper normalization that is required for the Schrödinger wave function. But, we will see future examples in which our application of boundary conditions will be restricted—essentially, we would like to have more boundary conditions even though we are limited to two. To use more than two over-specifies the solution and is unphysical. Hence, one must consider clever ways to create the "proper" boundary conditions (by which one usually means those which have been shown to work).

2.3 DENSITY AND CURRENT

The full Schrödinger equation is a complex diffusion equation. The wave function ψ itself is a complex quantity. The potential energy $V(x,t)$, however, is usually a real quantity. Already in Chapter 1, it was demonstrated that the probability of a particle being at a point was a real quantity, as it was given by the squared magnitude of the wave function (the product of the wave function and its complex conjugate). Thus, the probability "density" of the wave function is just

$$P(x,t)=\psi^*(x,t)\psi(x,t)=|\psi(x,t)|^2. \tag{2.26}$$

This result is referred to as a density, in the sense that it was discussed in Section 1.7 in its connection to the continuity equation. Another point of view is that, if we multiply (2.26) by the electronic charge, we do in fact achieve a charge density function. And this connection is often carried through the field by engineers in particular, because it allows a real connection with the current continuity equation with

$$e\frac{\partial\rho(x,t)}{\partial t}=\nabla\cdot\boldsymbol{J}(x,t). \tag{2.27}$$

While the complex conjugate of the wave function has been mentioned, one must be careful in using it, as it has a subtle difference in the Schrödinger equation. This latter is written as

$$-i\hbar\frac{d\psi^*(x,t)}{dt}=-\frac{\hbar^2}{2m}\nabla^2\psi^*(x,t)+V(x)\psi^*(x,t). \tag{2.28}$$

Our goal here is to justify the form of (2.27) and identify the form of the current density that appears in this equation. Thus, we want to multiply Equation (2.1) by $\psi^*(x,t)$, multiply (2.28) by $\psi(x,t)$, and subtract one from the other. When we do this, the potential term drops out, as it contains just the product of the two wave functions. What remains is

$$\begin{aligned} i\hbar\left[\psi^*(x,t)\frac{\partial\psi(x,t)}{\partial t}+\psi(x,t)\frac{d\psi^*(x,t)}{dt}\right]&= \\ =-\frac{\hbar^2}{2m}\left[\psi^*(x,t)\nabla^2\psi(x,t)-\psi(x,t)\nabla^2\psi^*(x,t)\right]& \end{aligned}. \tag{2.29}$$

The term in the brackets on the left-hand side is just the time derivative of $P(x,t)$. The bracket on the right-hand side can be rewritten with the common resulting term cancelling to yield

$$i\hbar\frac{\partial P}{\partial t}=-\frac{\hbar^2}{2m}\nabla\cdot\left[\psi^*(x,t)\nabla\psi(x,t)-\psi(x,t)\nabla\psi^*(x,t)\right]. \tag{2.30}$$

This now allows us to recognize the probability current as

$$\boldsymbol{J}_\psi=\frac{\hbar}{2mi}\left[\psi^*(x,t)\nabla\psi(x,t)-\psi(x,t)\nabla\psi^*(x,t)\right]. \tag{2.31}$$

Recognizing the momentum operator from (1.22) (and its complex conjugate for use with the conjugate wave function), we see that

$$\boldsymbol{J}_\psi=\frac{1}{2m}2Re\left\{\boldsymbol{p}(x,t)|\psi(x,t)|^2\right\}. \tag{2.32}$$

When the momentum is a "good" operator, which means that it is measurable, the eigenvalue is a real quantity. Then, the imaginary part vanishes, and (2.32) is simply the product of the velocity and the probability, which yields the probability current.

The result (2.32) differs from the earlier form that appears in (1.45). If the expectation of the momentum is real, then the two forms agree, as the gradient of the action just gives the momentum.

On the other hand, if the expectation of the momentum is not real, then the two results differ. For example, if the average momentum were entirely imaginary, then (2.32) would yield zero identically, while (1.45) would give a nonzero result. However, (1.45) was obtained by separating the real and imaginary parts of (1.36), and the result in this latter equation assumed that S was entirely real. An imaginary momentum would require that S be other than purely real. Thus, (1.45) was obtained for a very special form of the wave function. On the other hand, (2.32) results from a quite general wave function, and while the specific result depended upon a plane wave, the approach was not limited to this. If (2.2) is used with the general wave function, then (2.32) is evaluated using the expectation values of the momentum and suggests that, in fact, these eigenvalues should be real *if a real current is to be measured.*

This raises an interesting point. It was stated in the last paragraph that if a quantity is measurable, then the eigenvalue corresponding to that operator must be real. Simply put, a measurement in the classical world yields a real value (we don't do imaginary numbers in the real world, which is why they are called imaginary). Consider an operator which shall be defined by $A(x,t)$, so that when it operates on the wave function, it produces a number a according to

$$A(x,t)\psi(x,t) = a\psi(x,t). \tag{2.33}$$

This last expression is an equation since $A(x,t)$ is an operator. The number a is the eigenvalue for this operator. Suppose that a were not real; so that

$$a^* = \langle A^\dagger(x,t)\rangle = \int \psi^* A^\dagger \psi\, dx = \int A\psi^* \psi\, dx, \tag{2.34}$$

where the latter result arises since $\psi^*\psi$ is a real number, and therefore, the simple integral over A must give a real number. In comparison,

$$a = \langle A(x,t)\rangle = \int \psi^* A\psi\, dx = a\int \psi^* \psi\, dx. \tag{2.35}$$

So, if a number is equal to its own complex value, then the number has to be real. Along this line, the fact that an operator has a real eigenvalue also implies that it is a Hermitian operator for which $A^\dagger = A$ (one may recall that a Hermitian matrix is equal to its own transpose-complex conjugate).

Let us examine one of the previous cases using the continuity of current. We consider the transport over the barrier in the case for which $E > V_0$. The velocity in the two regions (for waves moving to the right) is $\hbar k/m$ for the wave denoted by A, and $\hbar k_2/m$ for the wave denoted by C. The magnitude squared for wave A was unity, while for wave C, it was $[2k/(k+k_2)]^2$. The ratio of the transmitted current to the incident current is the transmission

$$T = \frac{J_C}{J_A} = \frac{4kk_2}{(k+k_2)^2}, \tag{2.36}$$

and is reciprocal; it is the same in either direction left-to-right or right-to-left. It is easy enough to demonstrate that

$$T + R = 1, \tag{2.37}$$

which tells us that what is not transmitted is reflected! This is, of course, what was learned in electromagnetics for waves traversing a dielectric interface.

2.4 THE POTENTIAL WELL

If we now put two barriers, such as those in Figure 2.2, adjacent to one another, we have a choice of making a potential in which there is a barrier between two points in space, or a well between two points in space. The former case, in which we will be interested in transmission through or over

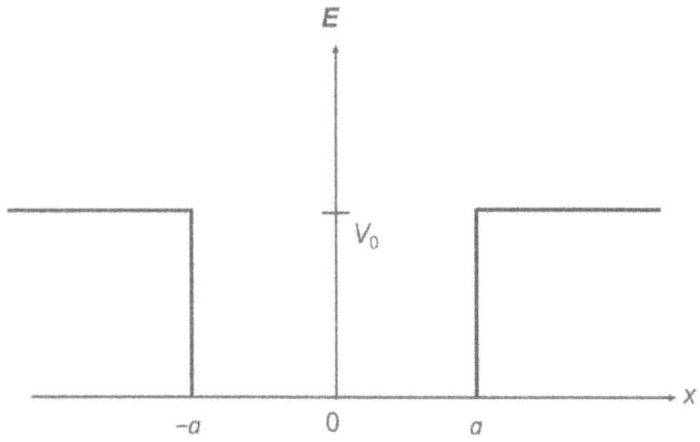

FIGURE 2.3 A potential well is formed by two barriers located at $|x| = a$.

the barrier, is the subject of Chapter 3. Here, we want to consider the latter case in which particles can be captured and constrained to lie within a quantum well, as shown in Figure 2.3. In this case, the two barriers are taken to be located at $|x| = a$. In general, the wave function will penetrate into the barriers a distance given roughly by the decay constant γ. As we know from (2.14), the size of the decay constant will depend upon the height of the barrier and the energy of the particle confined between the two barriers. One special case, however, is when the barrier is infinite in height, so that $\gamma \to \infty$, so that the wave does not penetrate the barrier. We deal with this case first, as it is considerably easier.

2.4.1 The Infinite Potential Well

From the results obtained in Section 2.1, it is clear that the wave function decays infinitely rapidly under an infinite barrier. This leads to a boundary condition that requires the wave function to vanish at the barrier interfaces, that is, $\Psi = 0$ at $|x| = a$. Within the central region, the potential vanishes, and the Schrödinger equation becomes just (2.5), with the wave vector defined by (2.8). The solution is now given, just as in the free-particle case, by

$$\psi(x) = Ae^{ikx} + Be^{-ikx}. \tag{2.38}$$

Here, we have two equations with two unknowns, apparently. Using the boundary conditions at $x = \pm a$, we have

$$\begin{aligned} &Ae^{ika} + Be^{-ika}, \quad x = a, \\ &Ae^{-ika} + Be^{ika}, \quad x = -a. \end{aligned} \tag{2.39}$$

However, one of the constants must be determined by normalization, so only A or B can be treated as unknown constants. The apparent dilemma is resolved by recognizing that the wave vector k cannot take just any value, and the allowed values of k are recognized as the second unknown. Since the two equations cannot give two solutions, they must be *degenerate*, and the determinant of the coefficients must vanish, that is

$$\begin{vmatrix} e^{ika} & e^{-ika} \\ e^{-ika} & e^{ika} \end{vmatrix} = sin(2ka) = 0. \tag{2.40}$$

This leads to discrete values for the wave vector k, which are given as

$$k_n = \frac{n\pi}{2a}, \; E_n = \frac{n^2\pi^2\hbar^2}{8ma^2}, \tag{2.41}$$

where the energy is given by using (2.8). Thus, for the infinite potential well, there are an infinite number of energy levels that are spaced quadratically in n.

Since the Schrödinger equation is of second order, it is expected that there will be two families of solutions. These can be found by putting each value of k_n back into (2.38) and examining the resulting form for the wave function. This is a simple enough task, and the result is a relation between B and A that may be expressed as

$$\frac{B}{A} = -e^{in\pi} = (-1)^{n+1}. \tag{2.42}$$

The astute reader has already recognized that these two possible signs lead to families of cosines and sines, as

$$\psi(x) = \begin{cases} A sin\left(\frac{n\pi x}{a}\right), & n\ even, \\ A cos\left(\frac{n\pi x}{a}\right), & n\ odd. \end{cases} \tag{2.43}$$

It is important to note that n cannot be 0. This would lead to a constant wave function, and the boundary conditions force this constant to be zero.

Let us now change the potential well, so that we shift the $x = 0$ point to minus a, and replace the length with $L = 2a$. Now, the condition (2.40) becomes

$$sin\left(\frac{ka}{L}\right) = 0, \tag{2.44}$$

so that

$$k_n = \frac{n\pi}{L},\ E_n = \frac{n^2\pi^2\hbar^2}{2mL^2}. \tag{2.45}$$

The wave functions (2.42) now become

$$\psi(x) = A sin\left(\frac{n\pi x}{a}\right), n\ integer. \tag{2.46}$$

The cosinusoidal wave functions can no longer satisfy the boundary conditions, and only sine waves are involved. The lowest wave function is now $n = 1$, and the wave function has only a single peak in the range $0 \le x \le L$. Thus, the squared magnitude of the wave function has a single peak. The second wave function, for $n = 2$, is $sin(2n\pi/L)$ and the squared magnitude exhibits two peaks. Thus, while the first version (with $2a$) clearly exhibited the even and odd symmetry of the probability density and wave function, this formulation clearly links the number of peaks in the probability density with the index n. Each version provides a different form of clarity. By even and odd symmetry, we refer to what happens when we make the change $x \to -x$. (The cosine function is then even, while the sine function is odd.) This is a general result when the potential is an even function; that is, $V(x) = V(-x)$. In the Schrödinger equation, the equation itself is unchanged when the substitution $x \to -x$ is made, provided that the potential is an even function. Thus, for a bounded wave function, $\psi(-x)$ can differ from $\psi(x)$ by no more than a constant, say α. Repeated application of this variable replacement shows that $\alpha^2 = 1$, so α can only take on the values ± 1, which means that the wave function is either even or odd under the variable change. Unfortunately, this nice symmetry only occurs when the potential has even symmetry, which it did in the first solution form, but not in the second.

We still have to normalize the wave functions. To do this, we can use either of the two forms of the solutions. Using the second version (running from 0 to L)

$$1 = A^2 \int_0^L sin^2\left(\frac{n\pi x}{L}\right)dx = \frac{A^2}{2}\int_0^L \left[1 - cos\left(\frac{2n\pi x}{L}\right)\right]dx$$
$$= \frac{A^2}{2}\left[x - \frac{L}{2n\pi} sin\left(\frac{2n\pi x}{L}\right)\right]_0^L = \frac{A^2 L}{2} \quad . \tag{2.47}$$

Thus, we find that

$$A = \sqrt{\frac{2}{L}} = \frac{1}{\sqrt{a}}. \tag{2.48}$$

If the particle resides exactly in a single energy level, we say that it is in a *pure* state. The more usual case is that it moves around between the levels, and on average, many different levels contribute to the total wave function. Then the total wave function is a sum over the Fourier series, with coefficients related to the probability that each level is occupied. We will examine this further in Chapter 6.

2.4.2 The Finite Potential Well

Now consider the situation in which the potential is not infinite in amplitude so that the wave function penetrates into the regions outside the quantum well and under the barriers. The potential remains a symmetric potential centered about the point $x = 0$ and extending to $x = \pm a$. However, it is clear that the treatment will need to be split into two cases: one for energies that lie above the top of the barriers, where the wave function can propagate over the barriers, and a second case for energies where the wave function decays outside the well, simulating a particle confined in the potential well. In this regard, the system is precisely like the single finite barrier that was discussed in Section 2.2.2. When the energy is below the height of the barrier, the wave must decay into the region where the barrier exists, as indicated in Figure 2.4. On the other hand, when the energy is greater than the barrier height, propagating waves exist in all regions, but there is a mismatch in the wave vectors, which leads to quasi-bound states and reflections from the interface. We begin with the case for the energy below the barrier height, which is the case shown in Figure 2.4.

For energies below the potential, the particle has freely propagating characteristics only for the range $|x| \le a$, for which the Schrödinger equation becomes the same as (2.5) with the energy and

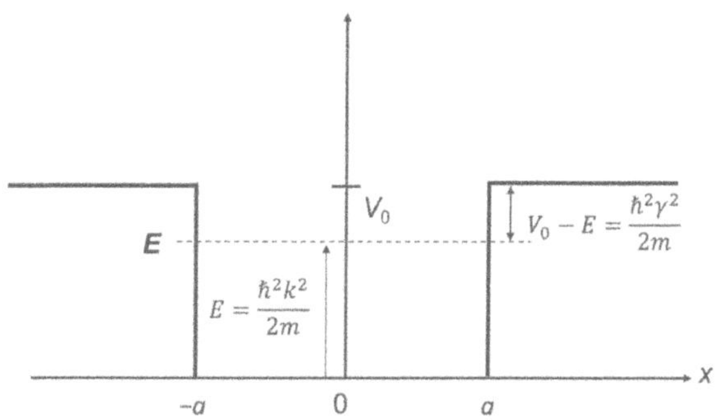

FIGURE 2.4 A potential well described by the two potential barriers for $|x| > a$. The parameters k and γ are appropriate wave vectors for the propagating (in the well) wave function and decaying (under the barrier) wave function.

wave vector given in (2.8). In this central region, the potential is zero just as for the previous case. In fact, the solutions are going to be of the same form. The difference will be that the boundary conditions are going to be different because of the penetration of the wave function into the barrier regions. In these latter regions, the barrier height V_0 is the magnitude of the potential barrier and is a positive quantity. For this region, the range $|x| > a$, the Schrödinger equation becomes

$$\frac{d^2\psi}{dx^2} - \gamma^2\psi = 0, \quad \gamma^2 = \frac{2m(V_0 - E)}{\hbar^2}. \tag{2.49}$$

In fact, this is just the same situation as penetration into the same barrier in Section 2.2.2.

It was pointed out at the end of the previous section that, with the potential being a symmetric quantity, the solutions for the Schrödinger equation would have either even or odd symmetry. The basic properties of the last section will carry over to the present case, and we expect the solutions in the well region to be either sines or cosines. Of course, these solutions have the desired symmetry properties and will allow us to solve for the allowed energy levels somewhat more simply.

As a result, the even and odd solutions can be treated separately. In either case, the solutions of (2.49) for the penetrating wave under the barrier will be of the form $Ce^{\mp\gamma x}, |x| > a$. This will be matched to the proper sine or cosine function at the boundaries. However, in Section 2.2, both the wave function and its derivative were matched at each boundary. If we attempt to do the same here, this will provide four equations. However, there are only two unknowns—the amplitude of C relative to that of either the sine or cosine wave and the allowed values of the wave vector k (and hence γ, since it is not independent of k) for the bound-state energy levels. We cannot over-specify the problem, so we must resort to just two boundary conditions. One common approach to finesse the problem is to make the ratio of the derivative to the wave function itself continuous. That is, we make the logarithmic derivative ψ'/ψ continuous. (This is obviously called the logarithmic derivative since it is the derivative of the logarithm of ψ.) Of course, if we choose the solutions to have even or odd symmetry, the boundary condition at $-a$ is redundant, as it is the same as that at a by these symmetry relations.

To begin, consider the even-symmetry wave function (the cosine states) for which the logarithmic derivative lies at $x = a$. Now, selecting the logarithmic derivative of the well and barrier wave functions, the boundary condition is given by

$$\frac{-k\sin(ka)}{\cos(ka)} = -k\tan(ka) = -\gamma. \tag{2.50}$$

On the far right-hand side, the logarithmic derivative of the damped function arises because of the magnitude in the argument of the exponent. We note that we can match the boundary condition at either a or $-a$, and the result is the same, a fact that gives rise to the even function that we are using. This transcendental equation now determines the allowed values of the energy for the bound states. If we define a new, reduced variable $\xi = ka$, then this equation becomes

$$\tan(\xi) = \frac{\gamma}{k} = \sqrt{\left(\frac{\beta}{\xi}\right)^2 - 1}, \quad \beta^2 = \frac{2mV_0a^2}{\hbar^2}. \tag{2.51}$$

The right-hand side of the transcendental equation is a decreasing function of ξ, while the left-hand side is an increasing function of ξ. Thus, there will be a series of points where the curves cross. These are the solutions, but only those values for which the energy lies in the range $(0, V_0)$ constitute bound states. In general, the solution is usually found graphically. This is shown in Figure 2.5, in which we plot the left-hand side and the right-hand side of (2.51) separately. The crossings are allowed energy levels.

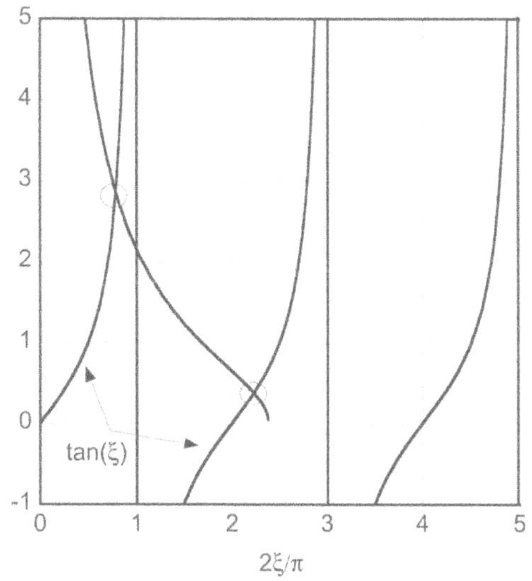

FIGURE 2.5 The graphical solution of (2.51) is indicated by the circled crossings. Here, we have used the values of $a = 5$ nm, $V_0 = 0.3$ eV, and $m = 0.067m_0$, appropriate to a GaAs quantum well between two layers of GaAlAs. The two circled crossings indicate that there are two even-symmetry solutions.

As the potential amplitude is made smaller, or as the well width is made narrower, the value of β is reduced, and there is a smaller range of ξ that can be accommodated before the argument of the square root becomes negative. Variations in the width affect both parameters, so we should prefer to think of variations in the amplitude, which affects only β. We note, however, that the right-hand side varies from infinity (for $\xi = 0$) to zero (for $\xi = \beta$), regardless of the value of the potential. A similar variation, in inverse range, occurs for the tangent function (that is, the tangent function goes to zero for $\xi = 0$ or $n\pi$, and the tangent diverges for ξ taking on odd values of $\pi/2$). Thus, there is always at least one crossing. However, there may only be one. As the potential amplitude is reduced, the intercept β of the decreasing curve in Figure 2.5 moves toward the origin. Thus, the solution point approaches $\xi = 0$, or $k = 0$. By expanding the tangent function for small ξ, it is found that the solution is approximately $\beta\xi$. However, this requires $E \simeq V_0$, which means that the energy level is just at the top of the well. Thus, there is at least one crossing of the curves for $\xi < \pi/2$. For larger values of the amplitude of the potential, the zero point (β) moves to the right, and more allowed energy levels appear for the even functions. It is clear from the construction of Figure 2.5 that at least one solution must occur, even if the width is the parameter made smaller, as the ξ-axis intersection cannot be reduced to a point where it does not cross the tan(ξ) axis at least once. The various allowed energy levels may be identified with the integers 1,3, 5, …, just as is the case for the infinite well (it is a peculiarity that the even-symmetry wave functions have the odd integers), although the levels no longer involve exact integers.

Let us now turn to the odd-symmetry wave functions in (2.43). Again, the continuity of the two logarithmic derivatives of the propagating waves for $|x| < a$, and the decaying wave under the barrier, may be equated as

$$\frac{k\cos(ka)}{\sin(ka)} = k\cot(ka) = -\gamma. \tag{2.52}$$

Using the same change of variables and parameters, this becomes

$$\cot(ka) - \frac{\gamma}{k} = -\sqrt{\left(\frac{\beta}{\xi}\right)^2 - 1}. \tag{2.53}$$

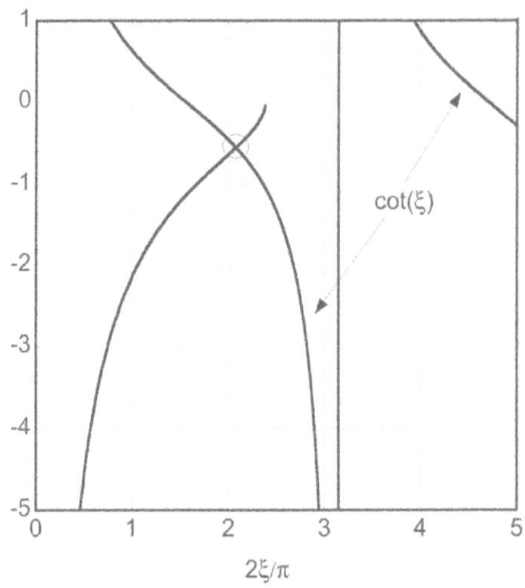

FIGURE 2.6 The graphical solution of (2.53). The parameters here are the same as those used in Figure 2.5. Only a single solution (circled) is found for the antisymmetric solution, and this energy lies between the two solutions found in Figure 2.5.

Again, a graphical solution is usually used to find the bound states. This is shown in Figure 2.6. The difference between this case and that for the even wave functions is that the left-hand side of (2.52) starts on the opposite side of the ξ-axis from the right-hand side, and we are not guaranteed to have even one solution point. On the other hand, it may be seen by comparing Figures 2.5 and 2.6 that the solution point that does occur lies between those that occur for the even-symmetry wave functions. Thus, this may be identified with the integers 2, 4, …, even though the solutions no longer involve exact integers.

We can summarize these results by saying that for small amplitudes of the potential, or for small widths, there is at least one bound state lying just below the top of the well. As the potential, or width, increases, additional bound states become possible. The first (and, perhaps, only) bound state has an even-symmetry wave function. The next level that becomes bound will have odd symmetry. Then a second even-symmetry wave function will be allowed, then another odd-symmetry one, and so on.

Now, all this plotting and graphing to find the solutions is rather slow given the technology available today. In fact, a site known as NanoHub.org has brought together a large set of tools for computational science and given them all a common graphical user interface (accounts are free). We can examine the solutions here with the "bound-state laboratory" available in the AQME set of tools for quantum mechanics (Klimeck *et al.*, 2019). In Figure 2.7, part of the entry panel is shown for this tool, along with the parameters that are entered for the simulation. These parameters are those for GaAs and are listed in the caption of Figure 2.5 (a 10 nm width well has $a = 5$ nm). In Figure 2.8, the solutions for the energy levels are shown overlaid on the potential well itself. The three energy levels lie at 34.2, 132, and 271 meV. In Figure 2.9, the wave functions for these three energy levels are plotted, and the clear even-odd-even parity can be seen.

Now, attention is turned to the completely propagating waves that exist for energies above the potential well. It might be thought that these waves will show no effect of the quantum well, but this is not the case. Each interface is equivalent to a dielectric interface in electromagnetics, and the thin layer is equivalent to a thin dielectric layer in which interference phenomena can occur. The same is expected to occur here. Calculations for these phenomena involve determining the transmission coefficient for waves propagating from the left (negative x) to the right (positive x).

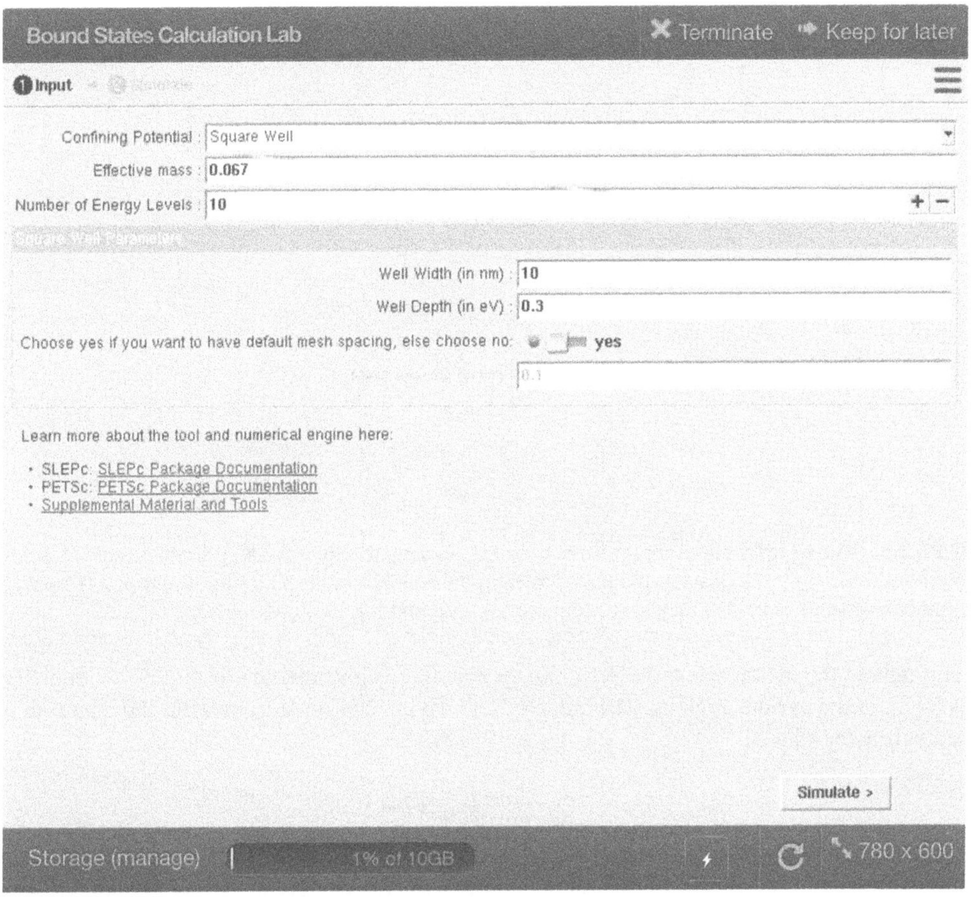

FIGURE 2.7 The front panel of AQME, at NanoHub.org, showing that the "bound states lab" has been selected. The parameters from Figure 2.5 have been added at the appropriate spots.

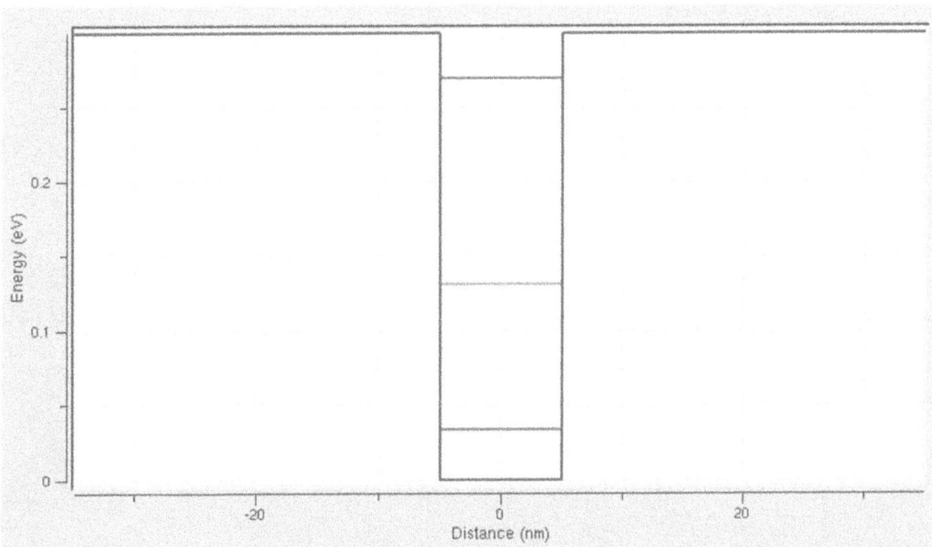

FIGURE 2.8 The potential well with the three bound States found from the "bound States lab" is shown.

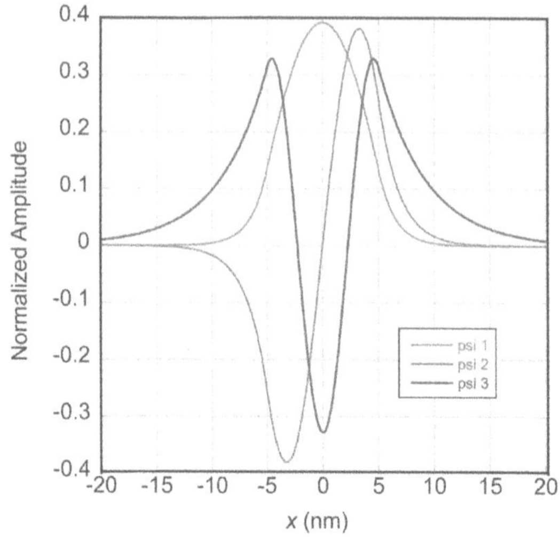

FIGURE 2.9 The wave functions for the three bound states shown in Figure 2.8: (psi 1) state at 34.2 meV (red in the ebook version), (psi 2) state at 132 meV (green in the ebook version), and (psi 3) state at 271 meV (blue in the ebook version). Note that they have progressively 1, 2, and 3 peaks.

Throughout the entire space, the Schrödinger equation is given by the form (2.5), with different values of k in the various regions. The value of k given in (2.8) remains valid in the quantum well region, while for $|x| > a$,

$$k_0^2 = \frac{2m(E - V_0)}{\hbar^2}, \tag{2.54}$$

just as in the case of the single barrier, although the situation is more complicated here because of the two barriers, which more than doubles the difficulty, as was found in the confined solutions.

As remarked above, it is assumed that a wave is arriving from the left ($x < -a$), and is partially reflected at the left boundary ($x = -a$), so that this wave may be written.

$$\psi_L = e^{ik_0x} + Be^{-ik_0x}, \tag{2.55}$$

while for the outgoing wave ($x > a$), we can take the form

$$\psi_R = Ee^{ik_0x}. \tag{2.56}$$

In the central region, which is the quantum well, we use the normal approach with

$$\psi_C = Ce^{ikx} + De^{-ikx}. \tag{2.57}$$

Now, there are four coefficients that need to be determined, and two boundaries for this process. This means we can write a boundary for the continuity of both the wave functions and their derivative. This gives four equations, as

$$\begin{aligned} e^{-ik_0a} + Be^{ik_0a} &= Ce^{-ika} + De^{ika}, \\ ik_0\left(e^{-ik_0a} - Be^{ik_0a}\right) &= ik\left(Ce^{-ika} + De^{ika}\right), \\ Ce^{ika} + De^{-ika} &= Ee^{ik_0a}, \\ ik\left(Ce^{ika} - De^{-ika}\right) &= ik_0Ee^{ik_0a}. \end{aligned} \tag{2.58}$$

While these are the four equations, it is easiest to put them into a matrix equation, which is

$$\begin{bmatrix} -e^{ik_0a} & e^{-ika} & e^{ika} & 0 \\ e^{ik_0a} & \frac{k}{k_0}e^{-ika} & -\frac{k}{k_0}e^{ika} & 0 \\ 0 & e^{ika} & e^{-ika} & -e^{ik_0a} \\ 0 & e^{ika} & -e^{-ika} & -\frac{k_0}{k}e^{ik_0a} \end{bmatrix}\begin{bmatrix} B \\ C \\ D \\ E \end{bmatrix} = \begin{bmatrix} e^{-ik_0a} \\ e^{-ik_0a} \\ 0 \\ 0 \end{bmatrix}. \tag{2.59}$$

This can now be solved for E in terms of the unit input (right-hand side of this last equation) to give

$$E = \frac{e^{-2ik_0a}}{\cos(2ka) - i\frac{k^2+k_0^2}{2kk_0}\sin(2ka)}. \tag{2.60}$$

This last result is symmetrical in the two wave numbers, which is what was also found for the single barrier. The transmission through the structure, from input to output in either direction, is thus found to be

$$T = |E|^2 = \frac{1}{1+\left[\frac{k^2+k_0^2}{2kk_0}\right]^2 sin^2(2ka)}. \tag{2.61}$$

It is interesting that, when $2ka$ is a multiple of π, the sine term is zero and the transmission becomes unity. These are the "over the barrier resonances" or sometimes referred to as quasi-bound states. A deeper potential well causes a greater discontinuity in the wave vector, and this leads to a larger modulation of the transmission coefficient. An example of the transmission modulation is shown in Figure 2.10.

It seems strange that a wave function that lies above the quantum well should not be perfectly transmitted. It is simple enough to explain this via the idea of "dielectric discontinuity," but is this really telling the whole truth of physics? Yes and no. It explains the physics with the mathematics,

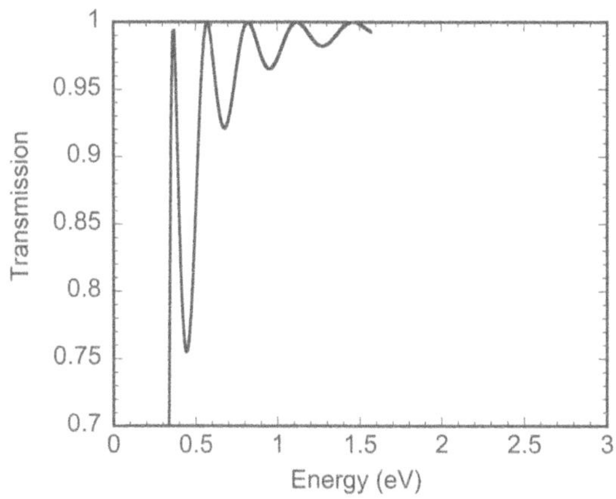

FIGURE 2.10 The transmission, given by (2.61), for a finite potential well. The parameters are those of Figure 2.5, appropriate to a GaAs quantum well situated between two GaAlAs layers.

but it does not convey the understanding of what is happening. In fact, it is perhaps easier to think about the incident wave as a particle. When it impinges upon the region where the potential well exists, it cannot be trapped there, as its energy lies above the top of the well. However, the well potential can *scatter* the particle as it arrives. In the present case, the particle is scattered back into the direction from which it came with a probability given by the reflection coefficient. It can then also be scattered back and forth by the two interfaces. It proceeds in an un-scattered state with a probability given by the transmission coefficient. In general, the scattering probability is nonzero, but at special values of the incident energy, the scattering vanishes. In this case, the particle is transmitted with unit probability. This type of potential scattering is quite special, because only the direction of the momentum (this is a one-dimensional problem) is changed, and the energy of the particle remains unchanged. This type of scattering is termed *elastic*, as in elastic scattering of a billiard ball from a "cushion" in three dimensions. We will see other examples of this in the following chapters.

2.5 THE TRIANGULAR POTENTIAL WELL

Another type of potential well is quite common in everyday semiconductor devices, such as the common metal–oxide–semiconductor (MOS) transistor (Figure 2.11(a)). The MOS field-effect transistor is the workhorse in nearly all microprocessors and computers today, yet the presence of quantization has not really been highlighted in the operation of these devices. These devices depend upon capacitive control of the charge at the interface between the oxide and the semiconductor. If we consider a parallel-plate capacitor made of a metal plate, with an insulator made of silicon dioxide, and a second plate composed of the semiconductor silicon, we essentially have the MOS transistor. Voltage applied across the capacitor varies the amount of charge accumulated in the metal and in the semiconductor, in both cases at the interface with the insulator. On the semiconductor side, contacts (made of n-type regions embedded in a normally p-type material) allow one to pass current through the channel in which the charge resides in the semiconductor. Variation of the current, through variation of the charge via the capacitor voltage, is the heart of the transistor operation.

Consider the case in which the semiconductor is p-type, and hence, the surface is biased by the positive gate voltage into an "inverted" condition (more electrons than holes). Here, mobile electrons are drawn to the interface by the positive voltage on the metal plate (the channel region is isolated from the bulk of the semiconductor by the inversion process). The surface charge in the semiconductor is composed of two parts: (1) the surface electrons, and (2) ionized acceptors from which the holes have been pushed into the interior of the semiconductor. In both cases, the charge that results is negative and serves to balance the positive charge on the metal gate. The electron charge is localized right at the interface with the insulator, while the ionized acceptor charge is distributed over a large region. In fact, it is the localized electron charge that is mobile in the direction along the interface, and that is quantized in the resulting potential well. The field in the oxide is then given by the total surface charge through Gauss's law (we use the approximation of an infinite two-dimensional plane) as (we use F for the field here to avoid confusion with the energy E)

$$F_s = \frac{e}{\varepsilon_{ox}}\left(N_a W + n_s\right), \tag{2.62}$$

where W is the thickness of the layer of ionized acceptors N_a (normal to the surface), the surface electron density n_s is assumed to be a two-dimensional sheet of charge, and the permittivity is that of the oxide. On the semiconductor side of the interface, the normal component of D is continuous, which means that E in (2.62) is discontinuous by the dielectric constant ratio. Thus, just inside the interface, (2.62) represents the field if the oxide permittivity is replaced by that of the

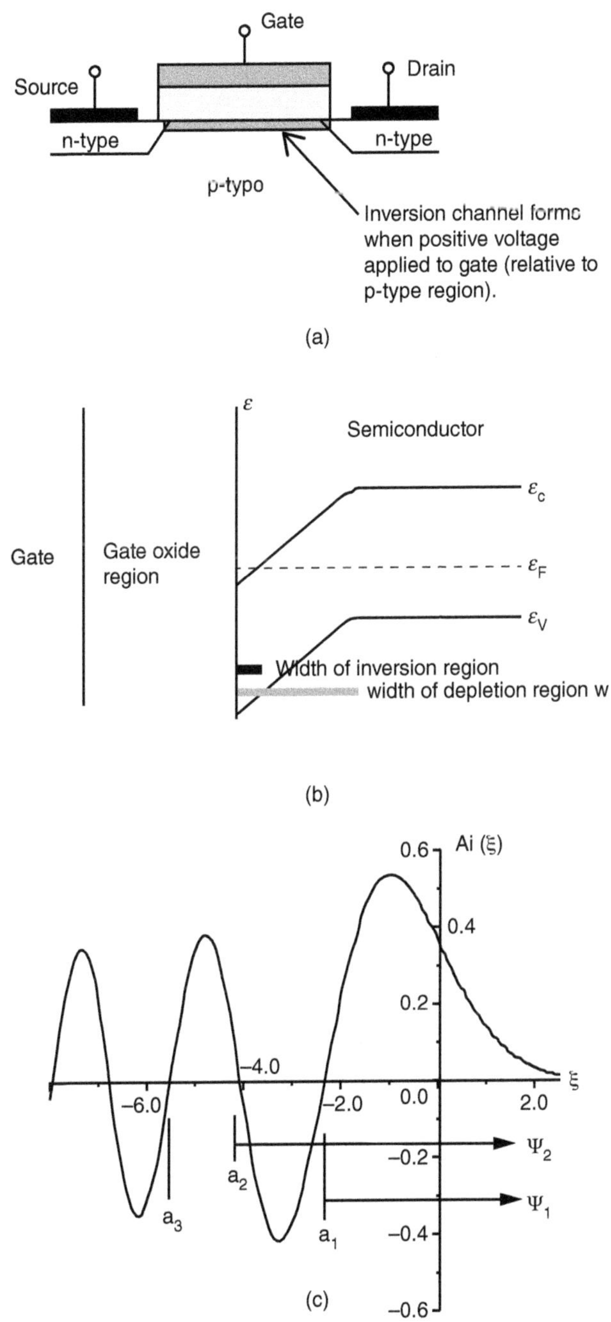

FIGURE 2.11 (a) A MOS field-effect transistor, (b) the triangular potential, and (c) the Airy function and the use of the zeros to match the boundary conditions.

semiconductor. However, just a short distance further into the semiconductor, the field drops by the amount produced by the surface electron density. Thus, the average field in the semiconductor, in the region where the electrons are located, is approximately

$$F_s = \frac{e}{\varepsilon_{ox}}\left(N_a W + \frac{n_s}{2}\right). \tag{2.63}$$

In this approximation, a constant electric field in this region gives rise to a linear potential in the Schrödinger equation (Figure 2.11(b)). We want to solve for just the region inside the semiconductor, near the oxide interface. Here, we can write the Schrödinger equation in the form

$$-\frac{\hbar^2}{2m}\frac{\partial^2}{\partial x^2}\psi + eF_s x\psi = E\psi, \quad x > 0. \tag{2.64}$$

It may be assumed that the potential barrier at the interface is infinitely high, so no electrons can get into the oxide, which leads to the boundary condition that $\psi(0) = 0$. The other boundary condition is merely that the wave function must remain finite, which means that it also tends to zero at large values of x.

While the above discussion has been for the case of a MOS transistor, such as found in silicon integrated circuits, quite similar behavior arises in the GaAs–AlGaAs heterojunction high-electron-mobility transistor (HEMT). In this case, the AlGaAs plays a role similar to the oxide in the MOS transistor, with the exception that the dopant atoms are placed in this layer. The dopants near the interface are ionized, with the electrons falling into a potential well on the GaAs side of the interface, a process that is facilitated by the barrier created by the difference in the two conduction band edges, as shown in Figure 2.12. There are very few ionized dopants in the GaAs, although the interface electric field still satisfies (2.62). This field is created by the positively charged, ionized donors and the negatively charged electrons in the potential well. By placing a metal gate on the surface of the AlGaAs layer, bias applied to the gate can affect the density of the carriers in the quantum well, reducing them to zero. This is a *depletion* device, as opposed to the inversion device of the MOS transistor. That is, the gate *removes* carriers from the channel in the HEMT while the gate pulls carriers into the channel in the MOS transistor. Since the impurities are removed from the region where the carriers reside, the mobility is higher in the HEMT, hence its name. Such devices have found extensive use as analog microwave amplifiers, either as power amplifiers or as low-noise amplifiers in receivers. Nevertheless, the potential variation near the interface still appears as approximately a triangular potential well, just as in the MOS transistor.

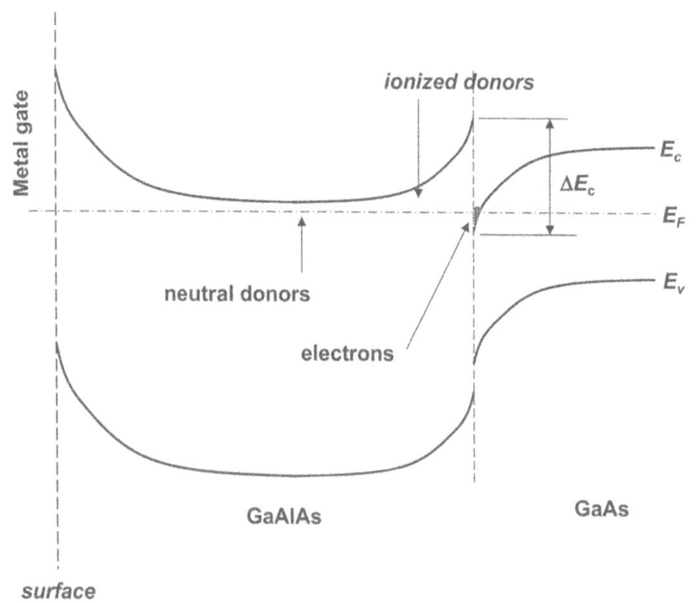

FIGURE 2.12 Band alignment for the AlGaAs–GaAs high-electron-mobility transistor. Ionized donors in the GaAlAs produce the electrons that reside in the triangular quantum well on the GaAs side of the interface.

The goal is to solve (2.64) for the triangular potential that arises from the nearly constant electric field. To simplify the approach, a change of variables will be made in (2.64), which will put the equation into a standard form. One may then recognize (2.64) as the Airy equation, which is one standard form in the Sturm–Liouville problem, and it guides our choice for the change of variables. For this, we redefine the position and energy variables as

$$z=\left(\frac{2meF_s}{\hbar^2}\right)^{1/3}x, \quad z_0=\frac{2mE}{\hbar^2 z^?}. \tag{2.65}$$

A shift $\xi = z - z_0$ in the coordinates is also made, so (2.64) becomes

$$\frac{\partial^2}{\partial\xi^2}\psi-\xi\psi=0. \tag{2.66}$$

This is now the final reduced form that is usually used for the Airy equation (Abramowitz and Stegun, 1964).

Airy functions are combinations of Bessel functions and modified Bessel functions. It is not important here to discuss their properties in excruciating detail. The important facts for us are that: (1) the Airy function $\mathrm{Ai}(-\xi)$ decays as an exponential for positive ξ; and (2) $\mathrm{Ai}(\xi)$ behaves as a damped sinusoid with a period that also varies as ξ, as shown in Figure 2.11(c). For our purposes, this is all we need. The second solution of (2.65), the Airy functions $\mathrm{Bi}(\xi)$, diverges in each direction and must be discarded in order to keep the probability function finite. The problem is in meeting the desired boundary conditions. The requirement that the wave function decay for large x is easy. This converts readily into the requirement that the wave function decay for large ξ, which is the case for $\mathrm{Ai}(-\xi)$. However, the requirement that the wave function vanish at $x = 0$ is not arbitrarily satisfied for the Airy functions. On the other hand, the Airy functions are oscillatory. In the simple quantum well of the last two sections, we noted that the lowest bound state had a single peak in the wave function, while the second state had two, and so on. This suggests that we associate the vanishing of the wave function at $x = 0$ with the intrinsic zeros of the Airy function, which we will call a_s. Thus, choosing a wave function that puts the first zero a_1 at the point $x = 0$ would fit all the boundary conditions for the lowest energy level (Figure 2.11(c)). Similarly, putting the second zero, a_2, at $x = 0$ fits the boundary conditions for the next level, corresponding to $n = 2$. By this technique, we build the set of wave functions and also the energy levels for the bound states in the wells.

Consider the lowest bound state, which will be taken as an example. For this, one requires the first zero of the Airy function. Because the numerical evaluation of the Airy functions yields a complicated series, we cannot give exact values for the zeros. However, they are given approximately by the relation

$$a_s=-\left[\frac{3\pi(4s-1)}{8}\right]^{2/3}. \tag{2.67}$$

Thus, the first zero appears at approximately $-(9\pi/8)^{2/3}$. Now, this may be related to the required boundary condition at $x = z = 0$ through

$$\xi=-\left(\frac{9\pi}{8}\right)^{\frac{2}{3}}\rightarrow -z_0=\frac{2mE}{\hbar^2}\left(\frac{\hbar^2}{2mF_s}\right)^{2/3}, \tag{2.68}$$

or

$$E_1=\frac{\hbar^2}{2m}\left(\frac{9\pi meE_s}{4\hbar^2}\right)^{\frac{2}{3}}. \tag{2.69}$$

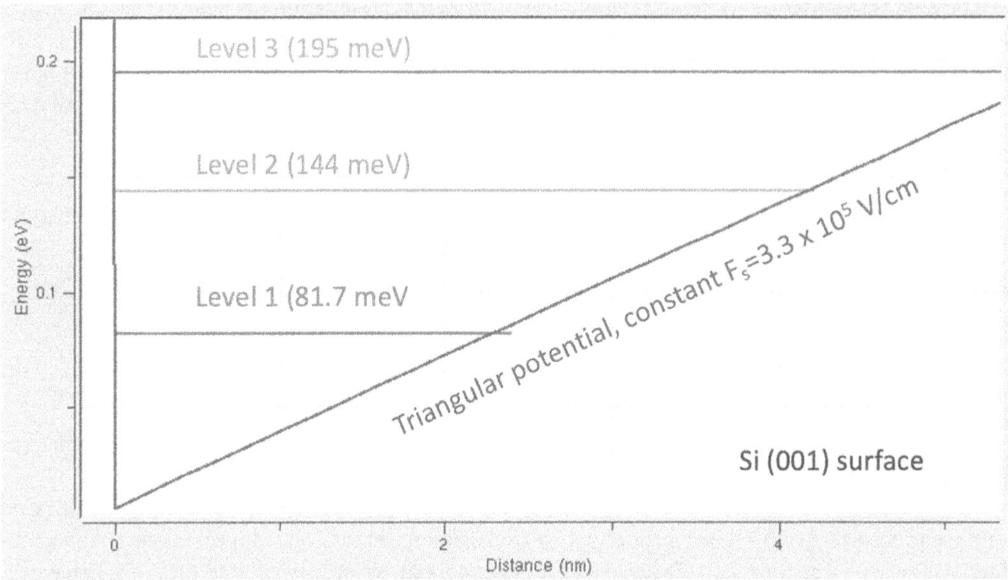

FIGURE 2.13 The triangular potential well and the ladder of bound states determined from the bound-state tool at Nanohub.org.

It must be remembered, however, that this is an approximate value since the value for the zero of the Airy function is only an approximate one.

In fact, we can return to NanoHub.org and once again use our AQME tool to find the solutions. In the "bound states lab," one can select a *triangular well*. Then, we select the Si heavy mass ($0.91m_0$) and a field of 0.33 MV/cm. (If we consider that we have 1 V on the gate of the MOS, and an oxide thickness of 10 nm, we find that the oxide field is 1 MV/cm.) The field in the semiconductor is smaller by the ratio of the permittivities. The bound-state lab gives the lowest energy level as 83.43 meV. In comparison, the lowest energy level in the Airy function approach, Equation (2.69), gives 86.4 meV. The second and third levels are at 147 and 198 meV, respectively. The energy levels for this potential, as determined by this tool, are plotted in Figure 2.13. In Figure 2.14, the magnitude squared of the first three wave functions is plotted.

2.6 UNCERTAINTY

The description of the momentum operator in the position representation is that of a differential operator, as we discussed in Section 1.5. In that treatment, we found that the position and momentum do not commute, but satisfy a relation

$$\left[x, p\right] = xp - px \rightarrow i\hbar, \tag{2.70}$$

where this relationship yields the result on the right-hand side *when it operates upon a wave function*. If variables, or operators, do not commute, there is an implication that these quantities cannot be measured simultaneously. Here again, there is another and deeper meaning.

In the previous chapter, we noted that the operation of the position operator x on the wave function in the position representation produced an expectation value $<x>$. The momentum operator does not produce this simple result with the wave function of the position representation. Rather, the differential operator produces a more complex result. For example, if the differential operator were to produce a simple eigenvalue, then the wave function would be constrained to be of the form

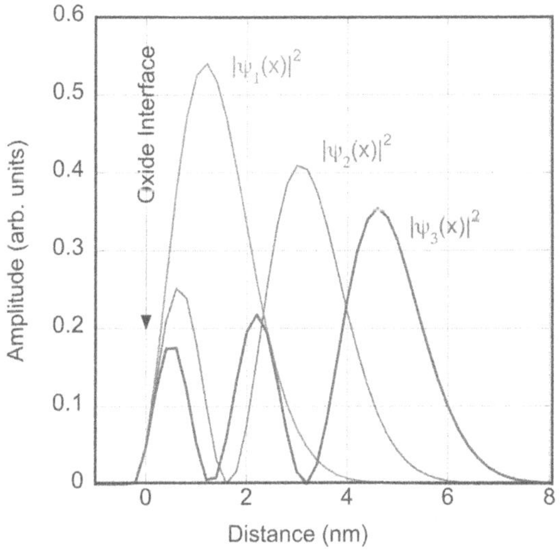

FIGURE 2.14 The squared magnitude of the wave function for the first three energy levels in the triangular potential.

$\exp(ipx/\hbar)$ (which can be shown by assuming a simple eigenvalue form with the differential operator and solving the resulting equation). This form is not integrable (it does not fit our requirements on normalization), and thus, the same wave function cannot simultaneously yield eigenvalues for both position and momentum. Since the eigenvalue relates to the expectation value, which corresponds to the most likely result of an experiment, these two quantities cannot be simultaneously measured. Thus, we will find that this leads to an uncertainty in the values of the two non-commuting operators. If the position is known, for example, if we choose a delta function in space, then the Fourier transform has unit amplitude everywhere; that is, the momentum has equal probability of taking on any value. Another way of looking at this is to say that since the position of the particle is completely determined, it is impossible to say anything about the momentum, as any value of the momentum is equally likely. Similarly, if a delta function is used to describe the momentum wave function, which implies that we know the value of the momentum exactly, then the position wave function has equal amplitude everywhere. Another way to look at this is that if the momentum is completely known, then it corresponds to a single value of k, and this means a plane wave, whose amplitude is constant throughout space. This means that if the momentum is known, then it is impossible to say anything about the position, as all values of the latter are equally likely.

As a consequence, if we want to describe both of the above properties of the particle, the position wave function and its Fourier transform must be selected carefully to allow this to occur. Then there will be an uncertainty Δx in position, and there will be a corresponding uncertainty Δp in momentum. Heisenberg (1927) gives an explicit value for the uncertainty in any two non-commuting operators in the form of

$$\Delta x \Delta p \geq \frac{\hbar}{2}. \tag{2.71}$$

But, given this, how are we to define Δx and Δp? The answer is that they are defined in exactly the same manner as the effective value of an a.c. signal is defined—the root-mean square of the signal. That is, the above two quantities are defined as

$$\begin{aligned} \Delta x &= \sqrt{\langle x^2 \rangle - \langle x \rangle^2}, \\ \Delta p &= \sqrt{\langle p^2 \rangle - \langle p \rangle^2}. \end{aligned} \tag{2.72}$$

The above can be examined further with the Gaussian wave function shown in Figure 1.4. This wave function may be expressed in common form as

$$\psi(x)=\frac{1}{(2\pi)^{1/4}\sqrt{\sigma}}exp\left(-\frac{x^2}{4\sigma^2}\right), \tag{2.73}$$

where the prefactor assures the normalization of the probability and σ is the standard deviation of the probability. This wave function has its peak at $x = 0$, and is symmetrical about this point. This assures us of the first result

$$\langle x\rangle = \frac{1}{\sqrt{2\pi\sigma}}\int_{-\infty}^{\infty} exp\left(-\frac{x^2}{2\sigma^2}\right)xdx = 0. \tag{2.74}$$

The next number we need is the expectation value of x^2, which is

$$\langle x^2\rangle = \frac{1}{\sqrt{2\pi\sigma}}\int_{-\infty}^{\infty} exp\left(-\frac{x^2}{2\sigma^2}\right)x^2dx = \sigma^2 , \tag{2.75}$$

which is a well-known property of a Gaussian, and in fact defines the meaning of the standard deviation. Hence, evaluating the first line of (2.72) now leads us to $\Delta x = \sqrt{\langle x^2\rangle} = \sigma$.

The next step is to determine the appropriate momentum wave function, which is found by Fourier transforming this position wave function (2.73). Using (1.30), this gives

$$\phi(k)=\left(\frac{2}{\pi}\right)^{1/4}\sqrt{\sigma}exp\left(-\sigma^2k^2\right). \tag{2.76}$$

This is a Gaussian in its own right. Here, we proceed exactly as above to find

$$\langle \mathrm{p}\rangle = \left(\frac{2}{\pi}\right)^{1/4}\sqrt{\sigma}\int_{-\infty}^{\infty} exp\left(-2\sigma^2k^2\right)\hbar kdk = 0, \tag{2.77}$$

as this Gaussian is also centered about zero momentum, and

$$\langle \mathrm{p}^2\rangle = \left(\frac{2}{\pi}\right)^{1/4}\sqrt{\sigma}\int_{-\infty}^{\infty} exp\left(-2\sigma^2k^2\right)\hbar^2k^2dk = \frac{\hbar^2}{4\sigma^2}. \tag{2.78}$$

Thus, just as above, we find that $\Delta p = \sqrt{\langle p^2\rangle} = \hbar/2\sigma$.

With these two values for the variances, the Heisenberg uncertainty relation (2.71) can be evaluated to show that, for the Gaussian wave packet, the result is

$$\Delta x\Delta p = \frac{\hbar}{2}. \tag{2.79}$$

Here, we have achieved the minimum uncertainty expected by Heisenberg. Thus, the Gaussian wave packet is often referred to as a minimum uncertainty wave packet. Let us now examine what this says about our infinite square well wave functions (2.43). Using these wave functions, we can examine the various quantities we need. Let us begin with the even-symmetry wave functions, for which the position expectation values are

$$\langle x\rangle = \frac{1}{a}\int_{-a}^{a} cos^2\left(\frac{n\pi x}{a}\right)xdx = \frac{1}{2a}\int_{-a}^{a}\left[1+cos\left(\frac{2n\pi x}{a}\right)\right]xdx = 0 \tag{2.80a}$$

By adding the two equations together, we arrive at

$$\frac{\partial^2 f}{\partial x^2}\Big|_{x=x_0} \approx \frac{1}{a^2}\left[f(x_0+a)+f(x_0-a)-2f(x_0)\right]. \tag{2.82}$$

Let us now introduce a short-hand notation. We can arbitrarily set $x_0 = x_i$, and then denote $f(x_0)$ as f_i. Then, we can write the time-independent Schrödinger equation as

$$-\frac{\hbar^2}{2ma^2}\left(\psi_{i+1}+\psi_{i-1}-2\psi_i\right)+V_i\psi_i = E\psi_i. \tag{2.83}$$

Of course, the trouble with an eigenvalue equation such as the Schrödinger equation is that solutions are only found for certain values of the energy E. This means that the energies (eigenvalues) must be found initially before the wave functions can be determined. However, the form (2.83) easily admits to this problem, given a reasonable set of computational routines on a modern computer. Equation (2.83) can be rewritten as a set of equations, one for each value of i. This then gives us a matrix equation that can be written as

$$[S][\psi]=0. \tag{2.84}$$

If there are N grid points in Figure 2.15, then $[\psi]$ is a $N\times 1$ column matrix; for example, a single column with N rows. The boundary conditions are imposed on the zeroth node (the supposed value of the node just below the first node) and the N+1 node. The matrix $[S]$ is now a tridiagonal matrix of size $N \times N$. The diagonal elements, and first off-diagonal elements, are given by

$$S_{ii} = \frac{\hbar^2}{ma^2}+V_i-E, \quad S_{i,i+1} = S_{i,i-1} = -\frac{\hbar^2}{2ma^2}. \tag{2.85}$$

The last term on the right is often called the "hopping energy" ($t=\hbar^2/2ma^2$) as it connects two adjacent sites on the grid. Now, immediately from (2.85), we know that the S matrix is singular; that is, its determinant must be zero if we are to have nonzero results, and

$$det[S]=0. \tag{2.86}$$

Expanding the determinant gives an Nth order equation in energy whose solutions are the N values of the energy E. This latter equation is known as the eigenvalue equation, and the solutions for the energy are the eigenvalues. For N nodes, we can find N solutions, each of which is representative of the various combinations of amplitudes of the wave function at the nodes. These are the eigen*functions*.

Let us take as an example an infinite potential well, whose width L is 20 nm (we use free electrons). The exact solutions are given in (2.45) in terms of the total width of the well (we set $2a = L$). Suppose we renormalize the energy such that $\pi^2\hbar^2/2mL^2=1$, then the energies are given by n^2. To simulate the wave functions, we take the discretization variable (in Figure 2.15) as a = 1.0 nm, so that N = 20. Since the end points i = 0 and i = 20 both have the wave function zero because of the infinite height of the barrier, we only need to consider the N – 1 interior grid points and $[S]$ is a 19 × 19 matrix. We compare the results with the numerically determined values of the lowest six energies in Table 2.1. It may be seen that while the results are close, the numerical estimates are somewhat below the actual values. However, this is mainly true only for the low-energy eigenvalues. They will deviate from the n^2 values at high energies due to the periodicity that has been imposed upon the system (this will be discussed more in Chapter 4).

TABLE 2.1
Comparison of Numerically Computed Energies with the Actual Values

n	Energy	Calculated
1	1	0.99795
2	4	3.96795
3	9	8.83468
4	16	15.4805
5	25	23.744103
6	36	33.41287

We now turn to computing the wave functions. Since the wave function has only been evaluated on the nodes, using the energy values E_n in the determinant of $[S]$ gives the set of values the node functions take for each energy. Not all values can be found for each energy level, with the last value needing to be determined from the normalization condition. It is important to note that the boundary conditions at $i = 0$ and $i = N$ must be imposed at the beginning in establishing (2.86). In Figure 2.16, we plot the first and fourth wave functions for this example, showing the exact solutions, with the numerically computed discrete eigenfunctions. The solid curves are the exact values from (2.45), while the solid points are the numerically computed values at the grid points. It may be seen that these appear to be fairly accurate, given the large scale of the figure.

In Figure 2.17, the 20 values of energy are plotted as a function of the index (level number). Note that they do not continue to follow the square law of (2.45) at higher energies. This deviation arises for several reasons, as mentioned above. First, the well is not infinite on the computer, as computers don't deal with infinities (at least not on purpose). Second, the discretization of the space into a grid with uniform spacing creates a *lattice*, just as in any crystalline structure. Because of the propagating wave nature of the particle in this simulated well, the energies tend to form into an arrangement that appears to be points on a cosinusoidal curve, which they actually are. In fact, the bandwidth (the range of energies) is determined by the hopping energy t. In one dimension, the bandwidth is $2t$. In two dimensions, it will be twice this value.

When the problem is time-varying, the numerical approach is more complicated. The Schrödinger equation is a diffusion equation, but with complex coefficients. This becomes more apparent if we rewrite (2.1) as

$$\frac{\partial \psi}{\partial t} = \frac{i\hbar}{2m}\frac{\partial^2 \psi}{\partial x^2} + \frac{1}{i\hbar}V(x,t)\psi. \tag{2.87}$$

The quantity $\hbar/2m$ has the units of a diffusion constant ($cm^2\ s^{-1}$). Nevertheless, we can use a Taylor series, exactly as we did earlier, for the time derivative as

$$\psi(t+\Delta t) = \psi(t) + \Delta t \frac{\partial \psi}{\partial t}\Big|_{\Delta t=0} + \cdots. \tag{2.88}$$

We still use the subscript i to denote the position at which the function is evaluated and denote the time evolution by $t = n\Delta t$. Hence, the explicit first-order evaluation of (2.87) is given by

$$\psi_i^{n+1} = \psi_i^n + \frac{i\hbar\Delta t}{2ma^2}\left(\psi_{i+1}^n + \psi_{i-1}^n - 2\psi_i^n\right) - \frac{i\Delta t}{\hbar}V_i\psi_i^n. \tag{2.89}$$

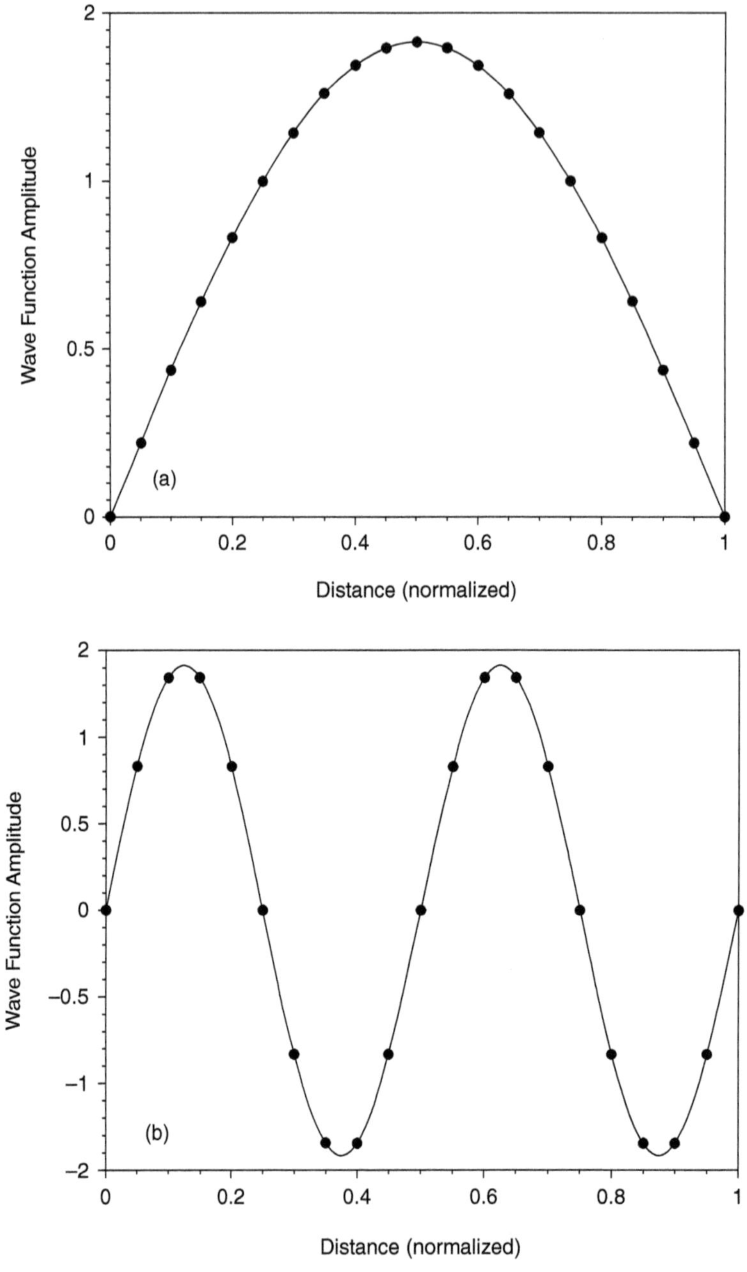

FIGURE 2.16 A comparison of the exact wave function with a numerically computed estimate for an infinite quantum well. In (a), ψ_1 is shown, while in (b), ψ_4 is plotted.

This is basically the simplest approach and develops the value at grid point i and new time $t + \Delta t$ in terms of the function evaluated at the preceding time step.

There are clearly errors associated with the time step Δt and the space step a. In fact, if these are too big, then an instability develops in the solution. We can check the linear stability by assuming that the wave function is composed of a set of Fourier modes

$$\psi = \tilde{\psi} e^{iqx}, \tag{2.90}$$

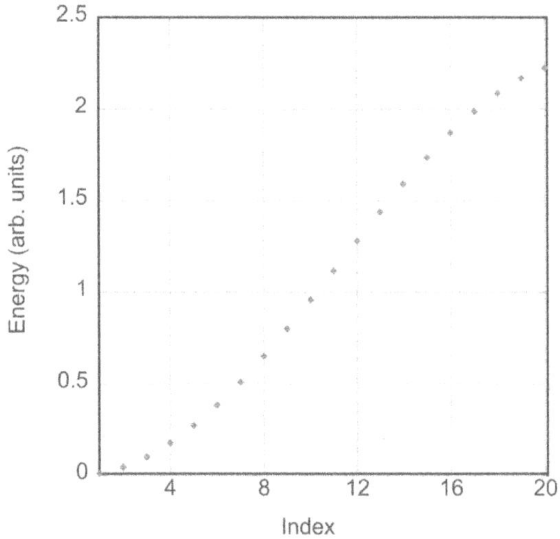

FIGURE 2.17 The 20 energy values for the simulation of the wave functions in Figures 2.15 and 2.16 are plotted as a function of the level index (level 1 is the lowest energy state). The cosinusoidal arrangement is related to the periodicity imposed by the grid.

for which (2.89) becomes (we ignore the potential for the moment)

$$\tilde{\psi}^{n+1}e^{iqx} = \tilde{\psi}^{n}e^{iqx} + \frac{i\hbar\Delta t}{2ma^2}\left(\tilde{\psi}^{n}e^{iq(x+a)} + \tilde{\psi}^{n}e^{iq(x-a)} - 2\tilde{\psi}^{n}e^{iqx}\right), \tag{2.91}$$

or

$$\tilde{\psi}^{n+1} = \tilde{\psi}^{n}\left[1 - \frac{2i\hbar\Delta t}{ma^2}sin^2\left(\frac{qa}{2}\right)\right]. \tag{2.92}$$

For stability, the second term in brackets must be less than 1 for all values of q. This can only occur if

$$\frac{2i\hbar\Delta t}{ma^2} \leq 1 \tag{2.93}$$

or

$$\Delta t \leq \frac{ma^2}{2\hbar}. \tag{2.94}$$

For example, for the previous example, in which a = 1 nm, the time step must be smaller than 4.3×10^{-15} s for a free electron and less than 2.9×10^{-16} s for an electron in GaAs. This is a severe limitation for many applications.

An improvement is to go to a second-order method, due to Crank and Nicholson (1947) (see also Richtmyer and Morton, 1967). In this, a two-step approach is utilized in which an estimate for ψ_i^{n+1} is obtained using (2.89). Once this estimate is obtained, an improved value is found by including this in the update of the final value of the wave function via

$$\begin{aligned}\psi_i^{n=1} &= \psi_i^n + \frac{i\hbar\Delta t}{4ma^2}\left(\psi_{i+1}^n + \psi_{i-1}^n - 2\psi_i^n\right) - \frac{i\Delta t}{2\hbar}V_i\psi_i^n \\ &+ \frac{i\hbar\Delta t}{4ma^2}\left(\psi_{i+1}^{n+1} + \psi_{i-1}^{n+1} - 2\psi_i^{n+1}\right) - \frac{i\Delta t}{2\hbar}V_i\psi_i^{n+1}.\end{aligned} \tag{2.95}$$

In essence, this is a form of a predictor–corrector algorithm to estimate the time variation. The intrinsic advantage is that the growth factor (the bracketed term in (2.92)) becomes

$$\left[\psi_i^{n=1}\right] \to \frac{1 - \dfrac{2i\hbar\Delta t}{ma^2} sin^2\left(\dfrac{qa}{2}\right)}{1 + \dfrac{2i\hbar\Delta t}{ma^2} sin^2\left(\dfrac{qa}{2}\right)} \tag{2.96}$$

and the magnitude is less than unity for all values of the coefficients. This means that any combination of a and Δt can be used, and the solution is, in principle, stable. This is, of course, an oversimplification, and the solutions must be checked for stability (convergence) with variable values of, for example, Δt for a given value of a.

To go further, we now want to consider several methods of computing the wave functions for more than a single dimension. This will be done in the following paragraphs.

2.7.1 Transfer Matrices

Determining a solution to the Schrödinger equation begins to become more difficult when the problem lies in two or more dimensions. It is true that one can actually write a single large matrix that contains the entire problem, even in three dimensions. However, this matrix tends to become rather large, and the solution time tends to depend upon the dimension of the matrix as N^3, or so most people believe. So, a number of iterative methods have been constructed to solve the problems more efficiently. One of these methods is the transfer matrix method. In fact, this approach has already been used in Section 2.2.2 to solve the potential step problem. Here, we will formalize this approach so as to apply it to many other problems.

In principle, there are two situations in which a transfer matrix needs to be described. The first situation is for free wave propagation in a medium with a constant potential. This is just going to be like the free particle. The second situation is at a potential discontinuity, which is the situation treated in the previous section mentioned above. Once these two cases are determined, then any arbitrary potential, even one varying in space, can be broken into slices, in which each slice is assumed to have a constant potential within that slice. Then, the two transfer matrices described for the interface between slices and for propagation through a slice are used to describe the role of that slice. At the end of the problem, the total transmission is computed by the product of the individual transfer matrices as

$$\psi_{final} = M_N M_{N-1} M_{N-2} \ldots M_1 M_0 \psi_{initial}, \tag{2.97}$$

where each M_i is one of the two types of matrices as required. In fact, as done in the earlier section, the wave function will contain both forward and backward propagating components.

The first matrix is for the simple propagation of the wave function in a region where the potential is constant. Then, we may write the wave function exactly as (2.38) with k defined by (2.12). For convenience, we take the position $x = a$ as the left-hand side of the propagation region, and $x = b$ as the right-hand side of this region. This means the length is $L = b - a$. We may write the wave function as a matrix on each side as

$$\psi_a = \begin{bmatrix} Ae^{ika} \\ Be^{-ika} \end{bmatrix}, \psi_b = \begin{bmatrix} Ae^{ikb} \\ Be^{-ikb} \end{bmatrix}. \tag{2.98}$$

Each component of the right-hand side just involves the propagator and the length, so one can ignore the coefficients for the present time. Then, the transfer matrix for the simple propagator is

$$M_P = \begin{bmatrix} e^{ikL} & 0 \\ 0 & e^{-ikL} \end{bmatrix}. \tag{2.99}$$

That is, each of the forward and backward propagating parts of the wave function evolves just as the free particle. There are no interactions between the two for a region of constant potential, and the value of the potential only affects the wave number *k*.

The second matrix is for a point where the potential is discontinuous, which is just the situation treated in Section 2.2.2. The discontinuity is treated simply as a step between two distinct values, with the discontinuity appearing at $x = x_0$, as

$$V(x) = \begin{cases} V_2, & x > x_0, \\ V_1, & x < x_0. \end{cases} \tag{2.100}$$

The wave numbers are dependent upon these two potentials, and

$$k = \begin{cases} k_2 = \sqrt{\dfrac{2m(E-V_2)}{\hbar^2}}, & x > x_0, \\ k_1 = \sqrt{\dfrac{2m(E-V_1)}{\hbar^2}}, & x < x_0. \end{cases} \tag{2.101}$$

To proceed, we go beyond the earlier treatment with forward and backward waves in each region, which we write as

$$\psi_1^{(\pm)} \sim e^{\pm ik_1x}, \psi_2^{(\pm)} \sim e^{\pm ik_2x}. \tag{2.102}$$

Requiring both the wave function and its derivative to be continuous across the interface, we can write these as the matrix equation (following Section 2.2.2) as

$$\begin{bmatrix} 1 & 1 \\ k_1 & -k_1 \end{bmatrix}\begin{bmatrix} \psi_1^{(+)}(x_0) \\ \psi_1^{(-)}(x_0) \end{bmatrix} = \begin{bmatrix} 1 & 1 \\ k_2 & -k_2 \end{bmatrix}\begin{bmatrix} \psi_2^{(+)}(x_0) \\ \psi_2^{(-)}(x_0) \end{bmatrix}. \tag{2.103}$$

These two equations can now be solved for the wave function in region 2 as a function of the wave function in region 1, and this defines the transfer matrix as

$$M_I(k_1,k_2) = \frac{1}{2}\begin{bmatrix} \left(\dfrac{k_2+k_1}{k_2}\right) & \left(\dfrac{k_2-k_1}{k_2}\right) \\ \left(\dfrac{k_2-k_1}{k_2}\right) & \left(\dfrac{k_2+k_1}{k_2}\right) \end{bmatrix}. \tag{2.104}$$

This now gives us the so-called propagators for a potential discontinuity and, previously, for a region of constant potential. This can be sequentially applied as in (2.97) to compute the transmission through any arbitrary set of slices.

2.7.2 Usuki's Method for Two-Dimensional Systems

As commented above, there are approaches that can be applied in two or more dimensions. We begin by considering a transverse *slice*, which is the wave function (2.83) solved for the initial slice. Let us think about a two-dimensional system with *N* longitudinal slices and each slice having *M* grid points in the transverse dimension, as shown in Figure 2.18. The longitudinal dimension is oriented along the *x*-axis, and the transverse dimension is oriented in the *y*-direction. This leads to an *i*-index along *x* and a *j*-index along *y*. Each grid point is spaced by a distance *a* from its neighbors

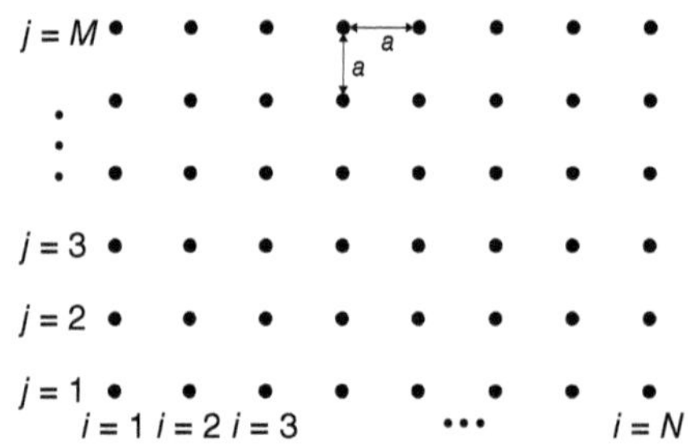

FIGURE 2.18 A two-dimensional grid. This is considered to have N slices along the length (horizontal), with each slice having M grid points (along the width).

in the two primary directions. As above, the value of the wave function at $j = 0$ and $j = M + 1$ will have a zero value for the wave function. It turns out that this does not have to be specified; it will occur naturally due to an intrinsic assumption that the potential will be infinite on these points (if not otherwise specified in the formulation, just as it was in the earlier one-dimensional treatment). The overall wave function for a given slice can be defined as

$$\bar{\psi}_i = \begin{bmatrix} \psi_{i,1} \\ \psi_{i,2} \\ \vdots \\ \vdots \\ \psi_{i,M} \end{bmatrix}. \tag{2.105}$$

Then, the Schrödinger equation may be written as

$$\boldsymbol{H}_{0,i}\bar{\psi}_i - \boldsymbol{t}\bar{\psi}_{i+1} - \boldsymbol{t}\bar{\psi}_{i-1} = E\boldsymbol{I}\bar{\psi}_i, \tag{2.106}$$

where $\boldsymbol{t}$ is the $M \times M$ diagonal matrix whose diagonal elements are the hopping energy (2.85), $\boldsymbol{I}$ is a unit matrix of the same dimension, and $\boldsymbol{H}_{0,i}$ is the tridiagonal Hamiltonian matrix

$$\boldsymbol{H}_{0,i} = \begin{bmatrix} (V_{i,M} + 4t) & -t & 0 & \cdots & 0 & & \\ -t & (V_{i,M-1} + 4t) & -t & & & & \\ & 0 & & \ddots & 0 & & \\ & & & & -t & (V_{i,2} + 4t) & -t \\ & 0 & & \cdots & 0 & -t & (V_{i,1} + 4t) \end{bmatrix}. \tag{2.107}$$

That is, each slice has a solution just as given for (2.83). However, we find this solution only for the zero slice, which becomes a boundary condition for the iteration indicated in (2.106). Thus, the M grid points in the initial slice mean that there are going to be M values of the energy for this slice, just as we found 20 values for the energy in Figure 2.17. But, there are going to be $2M$ values for the momentum, because there are going to be waves propagating from left to right (in Figure 2.18) and waves propagating from right to left just as were used in the transfer matrix (2.101). There are also potentials shown in (2.107), so that the values for the momentum can be either real or imaginary depending upon whether the energy is above or below the local potential. All of this complicates the problem.

So, let's try to walk ourselves into understanding the problem. The tridiagonal Hamiltonian (2.107) arises for any given slice denoted by i. The coupling between slices is handled by the $\boldsymbol{t}$ matrices. Looking at the iteration Equation (2.106) suggests that we follow a form of transfer matrix by writing the equation in terms of two adjacent slices as

$$\begin{bmatrix} \bar{\psi}_i \\ \bar{\psi}_{i+1} \end{bmatrix} = \begin{bmatrix} 0 & \boldsymbol{I} \\ \boldsymbol{I} & \boldsymbol{t}^{-1}\left(\boldsymbol{H}_{0,i} - E\boldsymbol{I}\right) \end{bmatrix} \begin{bmatrix} \bar{\psi}_{i-1} \\ \bar{\psi}_i \end{bmatrix} \equiv \boldsymbol{T}_i \begin{bmatrix} \bar{\psi}_{i-1} \\ \bar{\psi}_i \end{bmatrix}. \tag{2.108}$$

Since the structure of Figure 2.18 looks like a waveguide that controls waves moving horizontally, it is worthwhile to begin the procedure by solving the eigenvalue problem for the transfer matrix for zero and first slices. This begins by rewriting the form in (2.108) to have

$$\boldsymbol{T}_1 \begin{bmatrix} \bar{\psi}_1 \\ \bar{\psi}_0 \end{bmatrix} = \begin{bmatrix} T_{11} & T_{12} \\ T_{21} & T_{22} \end{bmatrix} \begin{bmatrix} \bar{\psi}_1 \\ \bar{\psi}_0 \end{bmatrix} = \lambda \begin{bmatrix} \bar{\psi}_1 \\ \bar{\psi}_0 \end{bmatrix}. \tag{2.109}$$

Now, the normal eigenvalue problem was expressed in (2.84), where S would become an $M \times M$ matrix here for the zero slice. Why have we doubled the size, because the $\boldsymbol{T}$ matrix is $2M \times 2M$? The reason has to do with those two values of momentum at each energy. That is, the problem posed in (2.109) is degenerate, as finding the wave function for the zero slice is just not sufficient, since the same wave function has two exponential propagators, as in (2.38) and (2.98). So, our wave function actually has twice as many values, having one entry for each of the two values of momentum at each grid point. Thus, we need a double-size matrix, and this leads to the size of $\boldsymbol{T}$. In addition, however, when the eigenvalue Equation (2.84) is solved, this is usually done by diagonalizing the S matrix, and this creates a transformation matrix U, which itself is quite interesting. First, in the final form, the columns of the U matrix represent the wave functions for particular energies. That is, column i gives the wave function for E_i on the M grid points. These are called the *mode* wave functions. We recall the Fourier transform pair that appeared in (1.29) and (1.30) of Chapter 1. Here, each column represents a mode, and this mode is a Fourier wave function in momentum space, in that it has a momentum (energy) associated with that mode. For example, the lowest energy mode will have the same sinusoidal shape as in Figure 2.16 all along the slices, so long as the potential is homogeneous. But, inhomogeneities in size or potential will mix the modes.

Each row of U gives the contribution from each momentum state to the wave function at that site. So, the columns represent the wave functions in the *mode* representation, and the rows represent the wave functions in the *site* representation. That is, each row gives the value for the wave function at that site by summing over all Fourier contributions. The mode representation is useful in waveguides, since the mode is either propagating or not, depending upon the mode's energy relative to the local potential. The site representation is useful when one is computing the probability density at each site. These eigenvectors of (2.84) will appear here but must be supplemented to double their number. Thus, the eigenvectors of (2.109) will have the form

$$\begin{bmatrix} \bar{u}_m(\pm) \\ \lambda_m(\pm)\bar{u}_m(\pm) \end{bmatrix}. \tag{2.110}$$

In general, depending upon the size of the potential terms in (2.107), there will be q propagating modes and $M - q$ evanescent modes. The eigenvalues can be expressed in terms of these mode momenta as

$$\begin{aligned} \lambda_m(\pm) &= e^{\pm ik_m a}, m = 1, 2, \ldots, q, \\ \lambda_m(\pm) &= e^{\mp \gamma_m a}, m = q+1, \ldots, M. \end{aligned} \tag{2.111}$$

Here, the "+" sign denotes left-to-right propagating (or decaying) modes, while the "–" sign denotes right-to-left propagating (or decaying) modes. It is useful to now collect the wave functions and the eigenvalues into a useful zero-transmission matrix

$$\boldsymbol{T}_0 = \begin{bmatrix} \boldsymbol{U}_+ & \boldsymbol{U}_- \\ \lambda_+ \boldsymbol{U}_+ & \lambda_- \boldsymbol{U}_- \end{bmatrix}, \tag{2.112}$$

where

$$\boldsymbol{U}_\pm = \left[\bar{u}_1(\pm) \quad \cdots \quad \bar{u}_M(\pm) \right], \tag{2.113}$$

and

$$\lambda_+ = diag\left[\lambda_1(\pm) \quad \cdots \quad \lambda_M(\pm) \right]. \tag{2.114}$$

The matrix $\boldsymbol{T}_0$ represents a global transformation matrix that takes the results from the mode basis to the site basis, while its inverse does the opposite transformation.

This latter transmission matrix, as well as the others defined above, can be used to process the iteration through the various slices just as in (2.97). But, the complex wave numbers that can appear through (2.111) make this a hazardous process, as some of these evanescent waves will have exponentially growing amplitudes through the mathematical iteration. This has to be avoided. Generally, for this reason, one switches from the *transmission* matrices to the so-called *scattering* matrices, which are written in terms of the incoming and outgoing waves at each side of a slice matrix (Ando, 1991). A particularly useful form of this has been developed by Usuki et al. (1995), in which the incoming waves from the left are given the amplitude $\boldsymbol{C}_1^i$ and the incoming waves from the right are given the amplitude $\boldsymbol{C}_2^i$, both for slice i. Then, Usuki demonstrated that the scattering matrices can be used in the iteration as

$$\begin{bmatrix} \boldsymbol{C}_1^{i+1} & \boldsymbol{C}_2^{i+1} \\ \boldsymbol{0} & \boldsymbol{I} \end{bmatrix} = \boldsymbol{T}_i \begin{bmatrix} \boldsymbol{C}_1^i & \boldsymbol{C}_2^i \\ \boldsymbol{0} & \boldsymbol{I} \end{bmatrix} \boldsymbol{P}_i, \tag{2.115}$$

where

$$\boldsymbol{P}_i = \begin{bmatrix} \boldsymbol{I} & \boldsymbol{0} \\ \boldsymbol{P}_{i1} & \boldsymbol{P}_{i2} \end{bmatrix} \tag{2.116}$$

and

$$\begin{aligned} \boldsymbol{P}_{i1} &= -\boldsymbol{P}_{i2} \boldsymbol{T}_{i,21} \boldsymbol{C}_1^i, \\ \boldsymbol{P}_{i2} &= \left[\boldsymbol{T}_{i,21} \boldsymbol{C}_2^i + \boldsymbol{T}_{i,22} \right]^{-1}. \end{aligned} \tag{2.117}$$

The iteration is started by assuming that $\boldsymbol{C}_1^0 = \boldsymbol{I}$ and $\boldsymbol{C}_2^0 = 0$. This implies a situation in which the modes start off as incident from the left with unit amplitude (they can be later given an amplitude determined by the Fermi–Dirac distribution with no loss of ease in computation), and there are no waves coming from the right. In symmetric systems, the waves coming from the right can then be determined by a symmetrical rotation of those coming from the left. In non-symmetrical systems, the left propagating waves are solved separately (remember that transmission is generally symmetrical). Once the waves have been iterated completely to the right, the final transmission matrix $\boldsymbol{t}$ can be found from

$$\boldsymbol{t} = -\left(\boldsymbol{U}_+ \lambda_+\right)^{-1} \left[\boldsymbol{C}_1^{N+1} + \boldsymbol{U}_+ \left(\boldsymbol{U}_+ \lambda_+\right)^{-1} \right]^{-1}. \tag{2.118}$$

The stability of the Usuki method in large part is provided by the fact that the iteration in (2.117) and (2.116) involves products of matrices and inverses of matrices so that the diffusive growth of evanescent waves is canceled out.

It can be shown that the wave function for the final slice can be written in terms of the parameters in (2.116) as

$$\bar{\psi}_N = \boldsymbol{P}_{N2}, \tag{2.119}$$

and the remaining wave functions can. be found by back propagating from this result, as

$$\bar{\psi}_i = \boldsymbol{P}_{i1} + \boldsymbol{P}_{i2}\bar{\psi}_{i+1}. \tag{2.120}$$

The density is then determined by

$$n(i,j) = \sum_k \left[\psi_{i,j,k}\right]^2, \tag{2.121}$$

where k is the mode number. This summation runs over all the modes, not just the propagating modes, as the evanescent modes can still contribute to the density. A more extensive review is given by Akis et al. (2003).

Let us consider a simple example here. We take the material parameters for GaAs and use a grid spacing of $a = 5nm$. The structure is 120 grid points wide and 72 grid points long, although a quantum wire of 20 grid points (100 nm) is defined by inserting a potential $V = 4t \sim 91.2\,meV$. A sweep of the energy E in (2.108) is equivalent to sweeping the Fermi energy in the real material, and as this value is raised, first one, then two, and so on, modes are excited and can propagate. There is a minimum energy for the first mode because of the quantization energy of the transverse size of the wire. In Figure 2.19, the transmission in units of $2e^2/h$ is plotted in panel (a). Panels (b) and (c) give a cross-section of the squared magnitude of the wave function for the situations in which one, and two, modes are propagating, respectively.

2.7.3 The Landauer Formula

Quite some time ago, Rolf Landauer (1957, 1970) presented an approach to transport and the calculation of conductance that was dramatically different from the microscopic kinetic theory based on the Boltzmann equation that is usually used in transport theory and semiconductor textbooks. His approach built upon the idea of modes that were introduced in the previous section. He considered that each mode was effectively a one-dimensional "wire" going through the system and that scattering could induce transitions of particles from one mode to another. Thus, if one could compute the transition from the set of input modes to the set of output modes, determining the conductance would be easy. But this determination of the transmission is precisely what was determined in the previous section by (2.118).

The only real constraint on Landauer's computation was that of lateral confinement so that the two input and output regions could be discussed in terms of their transverse modes. These two reservoirs were required to be in equilibrium with any applied potentials at the terminals. In essence, this treats these terminals as metals with well-defined Fermi levels, and in which the Fermi level at one terminal would be shifted relative to the value at the other terminal by any applied potentials. While he originally considered that the transport was ballistic, this is not required. Rather, the requirement is that we can assign a definitive mode to the electron when it is in either of the two reservoirs, which means that if scattering is present, it must be described specifically as a transfer of the electron from one internal mode to another. This is exactly what is done in transport theory, where an electron is scattered from one well-defined momentum (and energy) state to another. Now, these states lie in one mode or another.

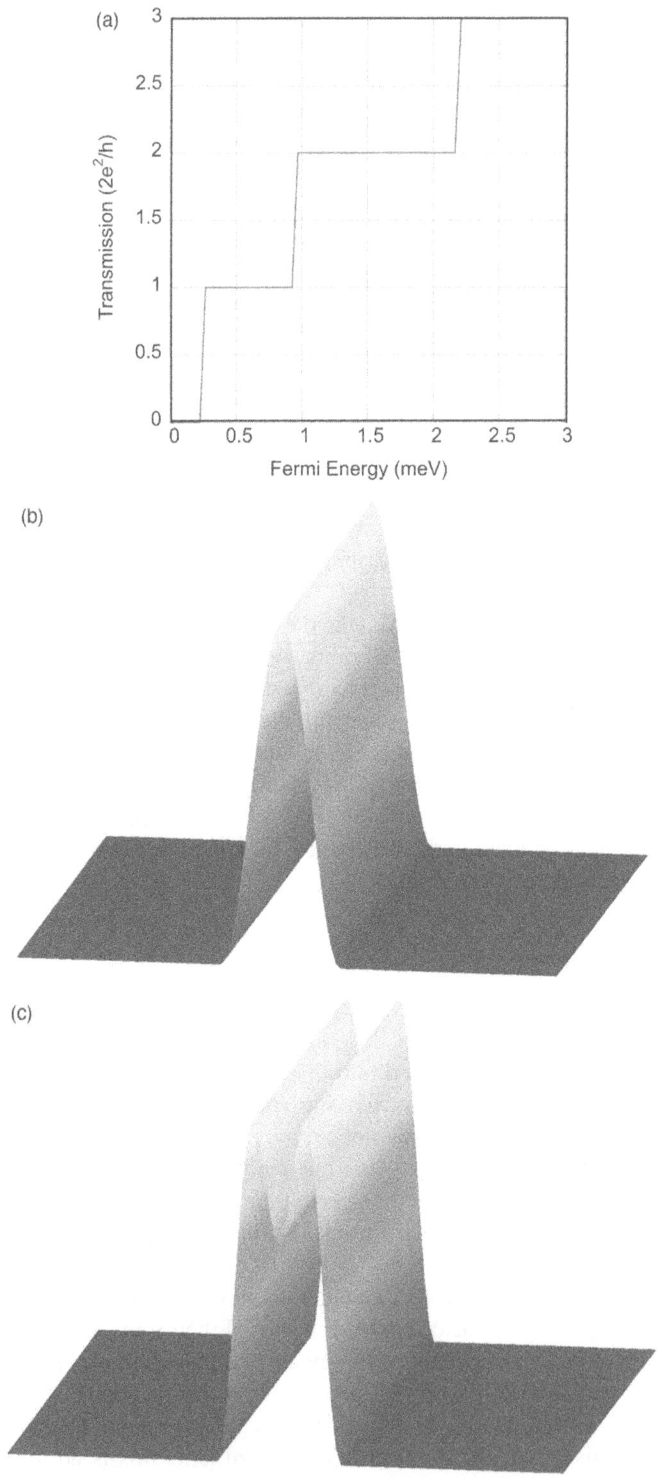

FIGURE 2.19 Simulation of a GaAs nanowire within a larger grid scheme. The details are discussed in the text. (a) Transmission, in units of $2e^2/h$, as a function of the Fermi energy in the conduction band. (b) Cross-section of the squared magnitude of the wave function when only a single mode can propagate. (c) Cross-section of the squared magnitude of the wave function when only two modes can propagate. The value does not go to zero at the center since this is where the peak of the first mode appears.

By adding a potential to one end or the other, it is possible for the carriers to gain energy (and momentum and velocity) as they move down the channel (the simulated region of Figure 2.18). Consider Figure 2.19. In the previous section, the slice zero gave rise to the transverse energies of possible wave functions, each of which represents a mode along the length of the structure. These are values on the central vertical line, where the parabolas cross this line. These are the *transverse* mode energies. But the particles can gain energy from the electric field, and this field makes the Fermi level different at the two ends. Applying a positive voltage at the right end lowers the energy at that end. Modes traveling from left to right have a positive momentum and are shown as the parabolas on the right side of the figure. Modes traveling from right to left have a negative momentum and are shown on the left side of the figure, where the Fermi level is lower due to the positive voltage. Electrons fill the resulting bands up to the Fermi levels, and it is clear that there are more electrons traveling to the right than traveling to the left. This creates a current.

In order to compute the current through the constriction, we need to first evaluate the charge that is occupied in each channel in the energy range between E_F and $E_F + eV$. As we remarked above, if the applied bias is small, then we can estimate the excess charge as

$$\delta Q \approx \frac{e^2 V}{2}\rho_{1d}\left(E_F\right) = e^2 V\sqrt{\frac{m^*}{2\pi^2\hbar^2 E_F}}, \tag{2.122}$$

where $\rho_{1d}(E_F)$ is the one-dimensional density of states at the Fermi energy (on the input side), assuming the potential is quite small. Only one-half of the density of states is used, as we are interested only in those states traveling to the right (those with positive momentum). The one-dimensional density of states can be found by considering that the momentum axis is filled from $-k_F$ to k_F, so that this gives $2k_F$ of momentum. The number of states in this momentum range is set by the factor $2\pi/L$, where L is the length of the wire. Thus, the number of states needed to contain $N/2$ electrons (2 is for spin degeneracy) is then $k_F L/\pi = \sqrt{2mE_F/\hbar^2}\,L/\pi$, and the density of states is (Ferry, 2020)

$$\rho_{1d}\left(E_F\right) = \frac{dN}{dE_2} = \sqrt{\frac{m^*}{2\pi^2\hbar^2 E_F}}. \tag{2.123}$$

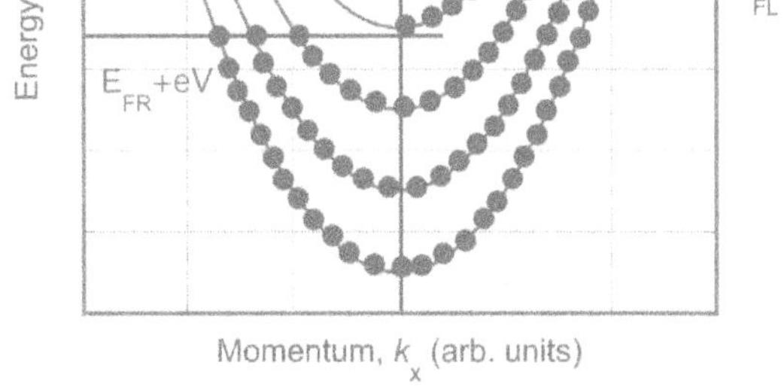

FIGURE 2.20 Energy bands, as a function of momentum along the transport direction, for the different modes. Four modes are assumed to be filled at the input end. The difference in the two Fermi levels reflects the applied voltage.

In order to compute the current, we need only the velocity

$$v = \frac{1}{\hbar}\frac{dE}{sk}\bigg|_{E=E_F} = \frac{\hbar k_F}{m^*} = \sqrt{\frac{2E_F}{m^*}}, \tag{2.124}$$

and the current is then given by

$$I_c = \delta Q v_g = e^2 V\sqrt{\frac{m^*}{2\pi^2\hbar^2 E_F}}\sqrt{\frac{2E_F}{m^*}} = \frac{2e^2}{h}V. \tag{2.125}$$

It is interesting that the energy has dropped out of the equation. Hence, the current is completely independent of the energy, and this is true for each, and every, one-dimensional channel (that is, each and every mode) that flows through the structure. The result has no dependence on the channel index; the current in each channel is identical. Landauer refers to this as the equipartition of the current. Such a partition seems to be unique to quasi-one-dimensional current flow.

But let us take a deeper look at the formulation of the Landauer formula. For this, we can again refer to Figure 2.19. The states on the right-hand side of the figure carry current from the input (left side) to the output. We can write this current as

$$I_+ = -e\int_0^\infty dkv(k)\rho(k)T(E)P_{FD,L}(E) = -e\int_0^\infty dEv(E)\rho_+(E)T(E)P_{FD,L}(E), \tag{2.126}$$

where the "+" on the density of states (as a function of energy) refers only to the positive-going states, and the "L" on the Fermi–Dirac distribution refers to the left (input) end of the device. Similarly, states moving to the left provide a current

$$I_- = -e\int_0^\infty dkv(k)\rho(k)T(E)P_{FD,R}(E) = -e\int_0^\infty dEv(E)\rho_-(E)T(E)P_{FD,R}(\mathrm{E}). \tag{2.127}$$

The total current can now be written as

$$I = -e\int_0^\infty dEv(E)\rho(E)T(E)\left[P_{FD}(E) - P_{FD}(E-eV)\right]. \tag{2.128}$$

Here, the density of states and velocity are simply functions of the energy, and the Fermi–Dirac distribution on the right has been shifted by the applied voltage in the figure. Using the properties of the density of states and the velocity given above, (2.128) simplifies to

$$I = -\frac{2e}{h}\int_0^\infty dET(E)\left[P_{FD}(E) - P_{FD}(E-eV)\right]. \tag{2.129}$$

For a small applied voltage, one can expand the last term in square brackets as

$$P_{FD}(E-eV) \approx P_{FD}(E) - eV\frac{\partial P_{FD}}{\partial E} + \cdots. \tag{2.130}$$

This gives us

$$I = -\frac{2e^2}{h}V\int_0^\infty dET(E)\frac{\partial P_{FD}}{\partial E}. \tag{2.131}$$

Generally, one considers the case of sufficiently low temperature where the derivative of the Fermi–Dirac distribution is a negative delta function at the Fermi energy, or

$$I = \frac{2e^2}{h} V \sum T(E)\delta(E - E_F). \tag{2.132}$$

Since the current in each mode is equal, the total current carried is simply the number of such modes that are occupied at the Fermi energy and the current per mode. Hence, the total current is

$$I_{total} = NI_c = N\frac{2e^2}{h}V, \tag{2.133}$$

and the conductance in our structure becomes

$$G = \frac{I_{total}}{V} = N\frac{2e^2}{h}. \tag{2.134}$$

However, because we have allowed scattering between the various modes, one has to extend the concept of N modes to a summation over all the input and output modes, with

$$N \to T = \sum_{m=1}^{M} \sum_{n=1}^{N} t_{mn}, \tag{2.135}$$

where t_{mn} is the transmission from input mode m to output mode n.

It is interesting that the conductance per mode is a ratio of fundamental constants and has the approximate value of 77.28 μS (microSiemen in SI units). This fact was discovered in 1980 (von Klitzing et al., 1980) and has become the fundamental basis for a new definition of international standard units introduced a few years ago (Bureau International des Poids et Mesures (BIPM), 2019).

2.8 BOUNDARY CONDITIONS ONCE AGAIN

It might seem that we are beating the boundary conditions to death, but in fact, we have not yet discussed just what the boundary conditions should be. In Section 2.2.2, the interface was evaluated by keeping the wave function and its derivative continuous. In Section 2.4.1, we set the wave function to zero at the point at which the potential step occurred. But, in Section 2.4.2, we made the logarithmic derivative of the wave function continuous across the interface. In Section 2.5, with a different set of functions, we imposed the zeroes of the Airy functions to lie at the oxide-semiconductor interface. However, in Section 2.7.1, we returned to using continuity of the wave function and its derivative. These are all interesting choices for boundary conditions and are all commonly used in quantum mechanics. But, are they correct? In fact, we know that physics imposes certain conservation rules, so what do these say about boundary conditions?

2.8.1 Continuity Equations

Electromagnetic field theory, whether in a classical or quantum field formulation, is one of the most well-developed areas of physics. It has been affected by two major revolutions since the advent of quantum mechanics. These are the introduction of relativity theory into electromagnetics (Einstein, 1905) and then the introduction of relativistic quantum theory (Dirac, 1928). Ultimately, it turns out that there are two fundamental quantities that do not depend upon their frame of reference: charge and flux. It is charge, and particularly the charge on the electron, that is of interest here. Classically, the electron is a particle, but in quantum theory, it can be either a wave or a particle. Normally, one thinks that the theory of electromagnetic fields deals with waves, but it also incorporates particles through the consideration of discrete charge.

Of interest here is the fact that simple operations upon Maxwell's (1865) equations lead to a charge continuity equation

$$\nabla \cdot \boldsymbol{J} + \frac{\partial \rho}{\partial t} = 0, \tag{2.136}$$

where $\boldsymbol{J}$ is the current density and ρ is the charge density, both at a point in space. Here, the relationship $\nabla \cdot \boldsymbol{D} = \rho$ has been used as well. This equation also leads to Gauss' Law, in which the charge enclosed at a point must be balanced by the fields passing through a closed surface surrounding that point, although, since it is a mathematical relationship, it was known somewhat earlier in this latter field. Critical to us is the fact that it is the basis of the Kirchhoff relationships in time varying situations (Graneau and Assis, 1994; Kirchhoff, 1857), in which the charge at a circuit node must evolve according to the sum of the currents entering the node. This continuity equation appeared in Chapter 1, Equation (1.44), where it arose from an expanding view of the Schrödinger equation. There it was written in the same form as (2.136), but the current density was explicitly written as the product of the charge density and its velocity $\boldsymbol{J} = \rho \boldsymbol{v}$. It is important to note the use of velocity and not momentum. In particular, this velocity was determined from the phase of the wave function as

$$v = \frac{1}{m} \nabla S \rightarrow \frac{1}{m} \frac{\partial S}{\partial x}. \tag{2.137}$$

This now brings material parameters explicitly into the discussion of boundary conditions. Nevertheless, such boundary conditions apply to probability current as well as normal current.

The important point that the above discussion brings forward for the quantum examples treated in this chapter is that, since no charge is accumulating (or dissipating) in any of these examples, the current must be continuous across any interface where boundary conditions are applied. This point will guide our discussion. If a single charge crosses an interface, the current crossing that interface must be constant, and have the value ev, where v is the velocity, whether it is a wave or particle crossing the interface. This charge can be an actual electrical charge or just the imaginary charge of the probability current (dividing the real current on either side of the interface by the charge e still leaves a constant probability, or density, current crossing the interface).

2.8.2 Dissimilar Materials

The discussion around the Landauer formula in Section 2.7.3 seemed to have avoided the importance of the velocity. But this was because of the one-dimensional nature of this approach. A mode is basically a quasi-one-dimensional entity, and travels along a basically one-dimensional wave guide. In one dimension, the velocity that entered the problem was exactly canceled by a large part of the density of states, leaving just the transmission coefficient in the formula. It is important to note, however, that transmission infers the movement of charge (or probability) from one side of the circuit to the other. This cancellation will not be there in other dimensionalities.

Consider the first example of a potential discontinuity. Suppose we have material 1 for $x < 0$ and material 2 for $x > 0$ as in (2.11). Then the boundary condition that arises from (2.137), when we use the momentum operators, is

$$\frac{1}{m_1} \frac{\partial \psi_1}{\partial x} = \frac{1}{m_2} \frac{\partial \psi_2}{\partial x}. \tag{2.138}$$

If the masses are equal, then this is simply continuity of the derivative of the wave function. What about the case where S itself is imaginary, and we have evanescent waves whose energy lies below the potential? Then (2.138) doesn't apply. Since there is now no real current on one side of the boundary, there can be no real current on the other side of the boundary. Yet, in electromagnetics, one has displacement current to account for the buildup (or decay) of charge as a time variation. The problem with our example was the assumption of no time variation when we used the time-invariant version of Schrödinger's equation. This assumption leads naturally to the conclusion here that any current arriving at the boundary is fully reflected, so that no net current exists. Then, one is free to use other boundary conditions, such as the assumption that both the wave function and its derivative are continuous even if that imposes a phase shift and complex behavior.

2.8.3 Stationary States

The above discussion leads naturally into the stationary states found in the potential well examples. Stationary states have real wave functions, and real wave functions have no associated velocity since $S = 0$ in this case. Even though the total wave function has an $exp(-iEt/\hbar)$ due to the energy of the state involved, the argument about zero velocity remains valid as there is no overt motion involved even with the total wave function.

2.9 THE WKB APPROXIMATION

So far, the barriers that we have been treating are simple barriers in the sense that the potential *V*(*x*) has always been piece-wise constant. The reason for this lies in the fact that if the barrier height is a function of position, then the Schrödinger equation is a complicated equation that has solutions that are special functions. The example treated in Section 2.5 merely had a linear variation of the potential—a constant electric field—and the solutions were identified as Airy functions, which already are quite complicated. What are we to do with more complicated potential variations? In some cases, the solutions can be achieved as well-known special functions—we treat Hermite polynomials in Chapter 5. On the other hand, nearly all of the solution techniques that we have used involve propagating waves or decaying waves, and the rest of the problem lies in matching boundary conditions. This observation suggests an approximation technique to find solutions, the Wentzel–Kramers–Brillouin (WKB) approach (Brillouin, 1926; Kramers, 1926; Wentzel, 1926).

Consider Figure 2.21, in which we illustrate a general spatially varying potential. At any particular energy, there is a position (shown as *a*) at which the wave changes from propagating to decaying. This position is known as a *turning point,* because a classical particle will be reflected here (thus, turned around in direction). In fact, we can generally extend the earlier arguments and definitions of this chapter to say that

$$k(x) = \sqrt{\frac{2m}{\hbar^2}\left[E - V(x)\right]}, \quad E > V(x), \tag{2.139}$$

and

$$\gamma(x) = \sqrt{\frac{2m}{\hbar^2}\left[V(x) - E\right]}, \quad E < V(x). \tag{2.140}$$

These assumptions suggest that, at least to zero order, the solutions can be taken as simple exponentials that correspond either to propagating waves or to decaying waves. We then adopt these forms as the lowest approximation, and we seek higher approximations. To proceed, let us assume that the wave function is generically definable as

$$\psi(x) \sim e^{iu(x)}. \tag{2.141}$$

The task now is to seek a form for the function $u(x)$. This form is then inserted into (2.7), which then gives the differential equation

$$i\frac{\partial^2 u}{\partial x^2} - \left(\frac{\partial u}{\partial x}\right)^2 + k^2(x) = 0. \tag{2.142}$$

Here, the propagating wave has been assumed; the decaying one will follow easily via a sign change. If we had a true free particle, the last two terms would cancel ($u = kx$), and we would be left with

$$\frac{\partial^2 u}{\partial x^2} = 0. \tag{2.143}$$

This suggests that $u(x)$ should be approximated by making this latter form the initial choice as the lowest-order approximation to $u(x)$. To carry this further, the ith iteration can be written as the solution of

$$\left(\frac{\partial u_i}{\partial x}\right)^2 = k^2(x) + i\frac{\partial^2 u_{i-1}}{\partial x^2}. \tag{2.144}$$

Only the first-order correction and approximation will be of concern here. The insertion of the zero-order approximation into the equation for the first-order approximation leads to

$$\frac{\partial u_1}{\partial x} = \sqrt{k^2(x) + i\frac{\partial k}{\partial x}} = \pm k(x) + i\frac{1}{2k(x)}\frac{\partial k}{\partial x}. \tag{2.145}$$

In arriving at this last expression, it has been assumed, in keeping with the approximations discussed, that the second term on the right-hand side in (2.145) is much smaller than the first term on the right. This implies that, in keeping with the discussion of Section 2.1, the potential is slowly varying on the scale of the wavelength of the wave packet.

The result (2.145) can now be integrated over the position, with an arbitrary initial position as the reference point. This gives

$$u_1 \approx \pm\int^x k(x')dx' + \frac{1}{2}ln\left[k(x)\right] + lnC_1. \tag{2.146}$$

This can now be used in (2.141) to give

$$\psi(x) \sim \frac{C_1}{\sqrt{k(x)}} exp\left[\pm i\int^x k(x')dx'\right]. \tag{2.147}$$

The equivalent solution for the decaying wave function is

$$\psi(x) \sim \frac{C_1}{\sqrt{\gamma(x)}} exp\left[\pm\int^x \gamma(x')dx'\right]. \tag{2.148}$$

It may be noted that these results automatically are equivalent to the requirement of making the current continuous at the turning point, which is achieved via the square-root prefactors. This continuity of current $|J| = |p\psi^2|$ was discussed in the last section, as well as the discussion above, to determine the transmission coefficient. Now, if we are to infer a connection formula for the wave function, then we must generalize the connection formula to say that $\frac{1}{\sqrt{|p|}}\psi$ must be continuous. It is the recognition of this fact that has led to the forms (2.147) and (2.148) for the wave functions on either side of the turning point of Figure 2.21. These connection formulas are the heart of the WKB technique.

The remaining problem lies in connecting the waves of one type with those of the other at the turning point. The way this is done is through a method called the method of stationary phase. The details are beyond the present treatment but are actually quite intuitive. In general, the connection formulas are written in terms of sines and cosines, rather than as propagating exponentials, and this will insert a factor of two, but only in the even functions of the propagating waves. In addition, the cosine waves

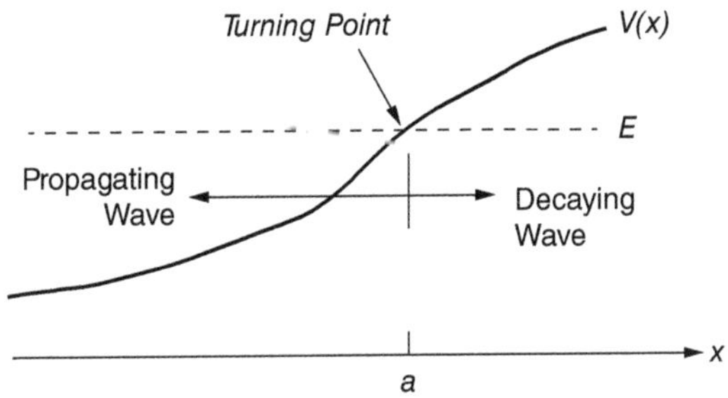

FIGURE 2.21 A simple variation of potential and the corresponding energy surface.

always couple to the decaying solution, and a factor of $\pi/4$ is always subtracted from the phase of the propagating waves (this is a result of the details of the stationary phase relationship and arises from the need to include a factor that is the square root of i). In Figure 2.21, the turning point is to the right of the classical region (where $E > V$). For this case, the connection formulas are given by

$$\frac{2}{\sqrt{k}}cos\left(\int_x^a kdx' - \frac{\pi}{4}\right) \leftrightarrow \frac{1}{\sqrt{\gamma}}exp\left(-\int_a^x \gamma dx'\right) \tag{2.149}$$

and

$$\frac{2}{\sqrt{k}}sin\left(\int_x^a kdx' - \frac{\pi}{4}\right) \leftrightarrow \frac{1}{\sqrt{\gamma}}exp\left(\int_a^x \gamma dx'\right). \tag{2.150}$$

The alternative case is for the mirror image of Figure 2.21, in which the turning point is to the left of the classical region (in which the potential would be a decreasing function of x rather than an increasing function). For this case, the matching formulas are given as (the turning point is taken as $x = b$ in this case)

$$\frac{1}{\sqrt{\gamma}}exp\left(-\int_x^b \gamma dx'\right) \leftrightarrow \frac{2}{\sqrt{k}}cos\left(\int_b^x kdx' - \frac{\pi}{4}\right) \tag{2.151}$$

and

$$-\frac{1}{\sqrt{\gamma}}exp\left(\int_x^b \gamma dx'\right) \leftrightarrow \frac{2}{\sqrt{k}}sin\left(\int_b^x kdx' - \frac{\pi}{4}\right). \tag{2.152}$$

As a first example of the WKB technique and the matching formulas, let us consider the general potential shown in Figure 2.22. Our aim is to find the bound states, or the energy levels, to be more exact. It is assumed that the energy level of interest is such that the turning points are as indicated; that is, the points $x = a$ and $x = b$ correspond to the turning points. Now, in region 1, to the left of $x = b$, we know that the solution has to be a decaying exponential as we move away from b. This means that we require that

$$\psi_1(x) = \frac{1}{\sqrt{\gamma}}exp\left(-\int_x^b \gamma dx'\right), \quad x < b. \tag{2.153}$$

At $x = b$, this must match the cosine wave according to (2.151). Then, in region 2, the wave function is given by

$$\psi_2(x) = \frac{2}{\sqrt{k}}cos\left(\int_b^x kdx' - \frac{\pi}{4}\right), \quad b < x < a. \tag{2.154}$$

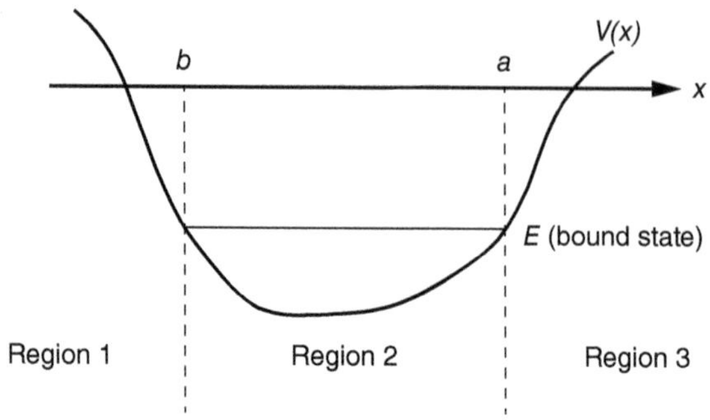

FIGURE 2.22 An arbitrary potential well in which to apply the WKB method.

Now, one proceeds across to $x = a$, and this is done quite simply with simple manipulations of (2.152), as

$$\begin{aligned}\psi_2(x) &= \frac{2}{\sqrt{k}} cos\left(\int_b^x kdx' + \frac{\pi}{4} - \frac{\pi}{2}\right) = \frac{2}{\sqrt{k}} sin\left(\int_b^x kdx' + \frac{\pi}{4}\right) \\ &= \frac{2}{\sqrt{k}} sin\left(\int_b^a kdx' - \int_x^a kdx' + \frac{\pi}{4}\right).\end{aligned} \tag{2.155}$$

The sine function is expanded to give

$$\begin{aligned}\psi_2(x) = &-\frac{2}{\sqrt{k}} cos\left(\int_b^a kdx'\right) sin\left(\int_x^a kdx' - \frac{\pi}{4}\right) \\ &+ \frac{2}{\sqrt{k}} sin\left(\int_b^a kdx'\right) cos\left(\int_x^a kdx' - \frac{\pi}{4}\right).\end{aligned} \tag{2.156}$$

The solution for the matching at the interface $x = a$ must satisfy (2.149), as the wave function in region 3 must be a decaying wave function. This means that at this interface, $\Psi_2(a)$ must be given *only* by the second term of (2.156). In this second term, the sine function is just a constant. The desired result can *only* be achieved by requiring that the first term must vanish, and this requires that

$$cos\left(\int_b^a kdx'\right) = 0, \tag{2.157}$$

or

$$\int_b^a kdx' = (2n+1)\frac{\pi}{2}. \tag{2.158}$$

This equation now determines the energy eigenvalues of the potential well, at least within the WKB approximation.

If we compare (2.158) with the result for a sharp potential such as the infinite quantum well found previously, with $b = -a$, we see that there is an additional phase shift of $\pi/2$ on the right-hand side of the last equation. While one might think that this is an error inherent in the WKB approach, we note that the sharp potentials of the last chapter violate the assumptions of the WKB approach (slowly varying potentials). The extra factor of $\pi/2$ arises from the soft variation of the potentials. Without exactly solving the true potential case, one cannot say whether or not this extra factor is an error, but this factor is a general result of the WKB approach.

REFERENCES

Abramowitz, M. and Stegun, I. A. 1964. *Handbook of Mathematical Function with Formulas, Graphs, and Mathematical Tables.* (Washington, DC: National Bureau of Standards) Sec. 10.4

Akis, R., Bird, J. P., Vasileska, D., Ferry, D. K., de Moura, A. P. S. and Lai, Y.-C. 2003. On the Influence of Resonant States on Ballistic Transport in Open Quantum Dots: Spectroscopy and Tunneling in the Presence of Multiple Conducting Channels. *Electron Transport in Quantum Dots.* Bird, J. P. Ed. (Boston: Kluwer Academic Publishers) 209–276

Ando, T. 1991. Quantum point contacts in magnetic fields. *Phys. Rev. B* **44** 8017–8027

Brillouin, L. 1926. The ondulatory mechanics of Schrödinger; A general method of Resolution by successive approximations. *C. R. Acad. Sci.* **183** 24–27

Bureau International des Poids et Mesures (BIPM). 2019. *The International System of Units (SI), 9th* ed.

Crank, D. C. and Nicholson, P. 1947. *Proc. Cambridge Phil. Soc.* **43** 50–67

Dirac, P. A. M. 1928. The quantum theory of the electron. *Proc. Roy. Soc. A* **117** 610–624

Einstein, A. 1905. Zur Elektrodynamik bewegter Körper. *Ann. Phys.* **17** 891–921

Ferry, D. K. 2020. *Transport in Mesoscopic Devices, 2nd Ed.* (Bristol: IOP Publishing).

Graneau, P. and Assis, A. K. T. 1994 Kirchhoff on the motion of electricity in wires. *Apeiron* **19** 19–25

Heisenberg, W. 1925. Über quantentheoreticsche undeutung Kinematicsher und mechanischer bezeihungen. *Z. Physik* **33** 879–893

Hildebrand, F. B. 1962. *Advanced Calculus for Applications.* (Englewood Cliffs, N.J.: Prentice-Hall)

Kirchhoff, G. 1857. Uber die Bewegung der Elektricität in Leitern. *Ann. Phys.* **102** 529–542

Klimeck, G., Wang, X., and Vasileska, D. 2019. *AQME—Advancing Quantum Mechanics for Engineers.* Nanohub.org. https://nanohub.org/resources/aqme.

Kramers, H. A. 1926. Wellenmechnik und halbzahlige Quantisierung. *Z. Phys.* **39** 828–840

Landauer, R. 1957. Spatial variations in currents and fields due to localized scatterers in metallic conduction. *IBM J. Res. Develop.* **1** 223–231

Landauer, R. 1970. Electrical resistance of disordered one-dimensional lattices. *Phil. Mag.* **21** 863–877

Maxwell, J. C. 1865. A dynamical theory of the electromagnetic field. *Phil. Trans. Roy. Soc.* **155** 459–512

Morse, P. M. and Feshbach, H. 1953. *Methods in Theoretical Physics.* (New York: McGraw-Hill)

Richtmyer, R. D. and Morton, K. W. 1967. *Difference Methods for Initial-Value Problems.* (New York: Interscience)

Schrödinger, E. 1926. Quantizierung als eigenwertprobleme. Ann. Physik **79** 361–379

Usuki, T., Saito, M., Takatsu, M., Kiehl, R. A., and Yokoyama, N. 1995. Numerical analysis of ballistic-electron transport in magnetic fields by using a quantum point contact and a quantum wire. *Phys. Rev. B* **52** 8244–8255

von Klitzing, K., Dorda, G., and Pepper, M. 1980. New method for high accuracy measurement of the fine-structure constant based on quantized hall resistance. *Phys. Rev. Lett.* **45** 494–497

Wentzel, G. 1926. Eine Verallgemeinerung der Quantenbedingungen für die Zwecke der Wellenmechanik. *Z. Phys.* **38** 518–52

PROBLEMS

1. If a Gaussian wave packet approaches a potential step ($V > 0$ for $x > 0$, $k0 > 0$), it is found that it becomes broader for the region $x > 0$. Why?
2. Assume that $\psi(x)$ are the eigenfunctions in an infinite square well ($V \to \infty$ for $|x| > d/2$), as given in the text. Calculate the overlap integrals

$$\int_{-d/2}^{d/2} \psi_n(x)\psi_m(x)dx.$$

3. Suppose that electrons are confined in an infinite potential well of width 0.5 nm. What spectral frequencies will result from transitions between the lowest four energy levels? Use the free-electron mass in your computations.
4. A particle confined to an infinite potential well has an uncertainty that is of the order of the well width, $\Delta x \simeq a$. The momentum can be estimated as its uncertainty value as well. Using these simple assumptions, estimate the energy of the lowest level. Compare with the actual value.

5. Derive expressions for the momentum wave functions $\varphi_n(x)$ for the stationary states of the infinite square well in terms of *k, L, n,* and the fundamental constant. Consider both even and odd values of n separately.
6. Using the graphical solution method, calculate the bound states in a finite potential well whose width is 15 nm and the barrier heights are 0.5 eV, using an effective mass of $0.067m_0$.
7. For a finite potential well of width 20 nm and height 0.3 eV (use an effective mass appropriate to GaAs electrons, $m = 0.067m_0$), compute the energy needed to ionize an electron (move it from the lowest energy level to the top of the well). If a 0.4 eV photon excites the electron out of the well, what is its kinetic energy? What is its wavelength?
8. For the situation in which a linear potential is imposed on a system, compute the momentum wave functions. Show that these wave functions form a normalized set.
9. For a finite potential well of width 7.5 nm and height 0.3 eV (use an effective mass appropriate to GaAs electrons, $m = 0.067m_0$), carry out a numerical evaluation of the energy levels. Use enough grid points in the "forbidden" region to ensure that the wave function is thoroughly damped. How many bound states are contained in the well?
10. The expectation value $\langle Q \rangle$ of an observable Q can be measured in the laboratory as the mean value of a set of measurements on the observable. This quantity must be real if it is measurable. Consider the classical measurable *xp*, the product of position and momentum. Classically, one gets no difference between *xp* and *px*. (a) For a general wave function, show that determining the expectation values $\langle xp \rangle$ and $\langle px \rangle$ gives different results, neither of which is real. (b) Compute the expectation value $\langle C \rangle = \frac{1}{2}\langle xp + px \rangle$. Is this an acceptable representation of the observable *xp*?
11. Consider a potential well with $V_0 \to \infty$ for $x < 0$, and $V(x) = Fx$ for $x > 0$, with $F = 0.02$ eV/nm. Using a numerical procedure from NanoHub.org, compute the ten lowest energy levels in the quantum well.
12. SCHRED is a numerical routine found at NanoHub.org that solves the one-dimensional Schrodinger equation and the Poisson equation simultaneously. Consider the MOS capacitor with a 4-nm-thick SiO_2 insulator and a p-type gallium arsenide substrate doped to 10^{17} cm^{-3}, and a metal gate at 300 K. The potential applied to this metal gate is $V_G = 1$ V, with the substrate grounded. Compute: (a) the conduction band profile, (b) the wave functions for the five lowest bound states in the surface inversion layer, (c) the differences in the electron charge density between a quantum and a classical charge density.
13. A particle is contained within a potential well defined by $V(x) \to \infty$ for $x < 0$ and $V(x) = ax$ for $x > 0$. Using the WKB formula, compute the bound-state energies.
14. A particle is confined in the potential well $V(x) = ax^4$. Calculate the bound states with the WKB approximation.

3 Tunneling

When we dealt with the finite potential well in the last chapter, it was observed that the wave function penetrated into the barriers. This process does not occur in classical mechanics, since a particle will actually bounce off the barrier. However, when we treat the particle as a wave, then the wave nature of barrier penetration can occur. This is familiar in electromagnetic waves, where the decaying wave (as opposed to a propagating wave) is termed an *evanescent* wave. For energies below the top of the barrier, the wave is attenuated, and it decays exponentially. Yet, it takes a significant distance for this decay to eliminate the wave completely. If the barrier is thinner than this critical distance, the evanescent wave can excite a propagating wave in the region beyond the barrier. In optics, this effect is termed *frustrated total internal reflection*. We are familiar with total internal reflection just from a swimming pool. Water has a higher dielectric constant than free space, so when we are under water, there is critical angle beyond which we can no longer see out (from under the water) of the pool. In optics, we can use a prism, and there will be reflection from the back surface for angles greater than a critical angle. The wave in back of the prism is attenuated and decaying. But, if we place another high dielectric constant material close to the prism, the decaying wave can excite a wave in this new material.[1] Thus, the wave can penetrate the barrier and continue to propagate, with an attenuated amplitude, in the trans-barrier region. For particles, this process is termed *tunneling*, with analogy to the miners who burrow through a mountain in order to get to the other side! In fact, it is known that the nuclear fusion process by which our sun produces its power would not occur were there no tunneling. The intra-nuclear repulsion is just too strong and tunneling is required for the nuclei to find one another.

This process is also quite important in modern semiconductor devices, and Leo Esaki (1958) received the Nobel prize for first recognizing that tunneling was important in degenerately doped *p–n* junction diodes. Nevertheless, the optical process described above is a classical tunneling event. For semiconductors, tunneling arises *when the particle is treated as a wave*. The behavior is fully consistent with optics in that regard. However, it is clear that quantum tunneling of particles is a result of the wave–particle duality discussed in Chapter 1. Since Esaki's discovery of the tunnel diode, tunneling has become important in a wider variety of situations. For example, in reverse-biased *p–n* junctions, Zener breakdown occurs when electrons in the valence band can tunnel across the gap into the conduction band, provided a sufficiently high electric field is applied to bring these two bands to the same energy levels (on opposite sides of the junction). Similarly, resonant-tunneling diodes have been fabricated in heterostructures such as GaAs–AlGaAs, and we will discuss these in some detail in a later section. Finally, as semiconductor devices become smaller, particularly the metal–oxide-semiconductor (MOS) field-effect transistor (FET) (MOSFET), where a thin layer of insulator (usually SiO_2 or, more recently, HfO_2) is used as the gate insulator, this thin oxide becomes susceptible to leakage currents via tunneling through the oxide.

In this chapter, we will address this tunneling process. First, we will treat those few cases in which the tunneling probability can be obtained exactly. Then we will discuss its use in solid-state electronics. Following this, we will move to approximate treatments suitable for those cases in which the solution is not readily obtainable in an exact manner.

3.1 THE TUNNELING BARRIER

As discussed above, when we ask how a particle can tunnel through a barrier quantum mechanically, the solution is to invoke wave–particle duality and treat the particle as a wave. This is the same approach used in Chapter 2 for a particle encountering a potential barrier. The difference here is that

DOI: 10.1201/9781003717195-3

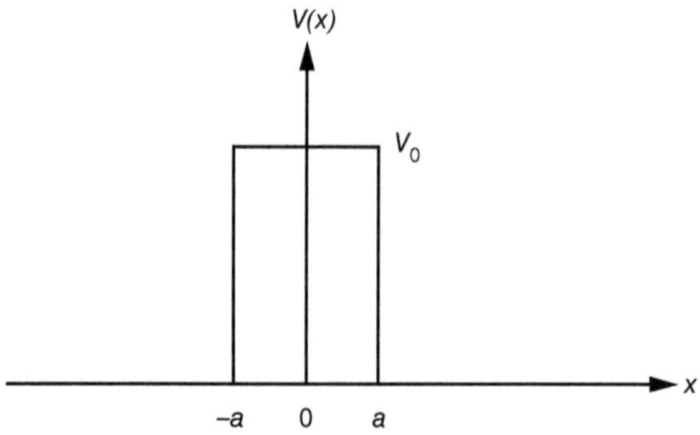

FIGURE 3.1 The simple rectangular tunneling barrier. The important parameters are the height of the barrier V_0 and the thickness $2a$ for which $E < V_0$.

the barrier has a fixed thickness, and this thickness, along with the potential height, will determine the probability that the particle can penetrate through the barrier. The general problem is that posed in Figure 3.1. Here, we have a barrier, whose height is taken to be V_0, that exists in the region $|x| < a$. To the left and to the right of this barrier, the particle can exist as a freely propagating wave, but, in the region $|x| < a$, and for energies $E < V_0$, the wave is heavily attenuated and is characterized by a decaying exponential "wave", due to its evanescent behavior. Our interest is in determining just what the transmission probability through the barrier is for an incident particle of a given energy $E < V_0$. We are also interested in the transmission behavior for energies above the top of the barrier. To solve for these factors, we proceed in precisely the same fashion as we did for the examples of Chapter 2. That is, we assume waves with the appropriate propagation characteristics in each of the regions of interest, but with unknown coefficients. We then apply boundary conditions, in this case the continuity of the wave function and its derivative at each interface, in order to evaluate the unknown coefficients.

3.1.1 The Simple Rectangular Barrier

The simple rectangular barrier is shown in Figure 3.1. Here the potential is defined to exist only between $-a$ and a, and the zero of potential for the propagating waves on either side is the same. We can therefore define the wave vector k in the region $|x| > a$, and the decaying wave vector γ in the region $|x| < a$, by the equations ($E < V_0$) as

$$k = \sqrt{\frac{2mE}{\hbar^2}}, \quad \gamma = \sqrt{\frac{2m(V_0 - E)}{\hbar^2}}, \tag{3.1}$$

respectively. These equations, of course, are exactly (2.8) and (2.12), as they should be. To the right and left of the barrier, the wave is described by propagating waves, while in the barrier region, the wave is attenuated. Thus, we can write the wave function quite generally as was done in the last chapter, via

$$\psi(x) = \begin{cases} Ae^{ikx} + Be^{-ikx}, & x < -a, \\ Ce^{\gamma x} + De^{-\gamma x}, & |x| < a, \\ Fe^{ikx}, & x > a. \end{cases} \tag{3.2}$$

This assumes that the wave is incident from the left (the term with coefficient A). The term with coefficient B is a reflected wave from the barrier. The terms with coefficients C and D are the right-going and left-going, respectively, evanescent waves under the barrier, while the last term with coefficient F is the transmitted wave.

We now have five unknown coefficients to evaluate. However, we can get only four equations from the two boundary conditions when we invoke continuity of the wave function and its derivative at each boundary. The evaluation of the fifth coefficient comes from normalizing the incoming wave (A) from one side or the other. In fact, this won't be necessary as the transition probability will be found by dividing the transmitted wave F with the value of the incoming wave A.

The boundary conditions are applied sequentially at each of the two interfaces. Thus, at the interface $x = -a$, continuity of the wave function and its derivative give rise to

$$\begin{aligned} Ae^{ika} + Be^{-ika} &= Ce^{-\gamma a} + De^{\gamma a}, \\ ik\left(Ae^{ika} - Be^{-ika}\right) &= -\gamma\left(Ce^{-\gamma a} - De^{\gamma a}\right). \end{aligned} \tag{3.3}$$

As in Chapter 2, we can solve for two of these coefficients in terms of the other two. In order to apply matrix multiplication a little later, we solve for A and B in terms of C and D. This gives the matrix equation

$$\begin{bmatrix} A \\ B \end{bmatrix} = \begin{bmatrix} \left(\dfrac{ik+\gamma}{2ik}\right)e^{(ik-\gamma)a} & \left(\dfrac{ik-\gamma}{2ik}\right)e^{(ik+\gamma)a} \\ \left(\dfrac{ik-\gamma}{2ik}\right)e^{-(ik+\gamma)a} & \left(\dfrac{ik+\gamma}{2ik}\right)e^{-(ik-\gamma)a} \end{bmatrix} \begin{bmatrix} C \\ D \end{bmatrix}. \tag{3.4}$$

Now, we turn to the other boundary, $x = a$. The continuity of the wave function and its derivative here give us the two equations

$$\begin{aligned} Ce^{\gamma a} + De^{-\gamma a} &= Fe^{ika} \\ \gamma\left(Ce^{\gamma a} - De^{-\gamma a}\right) &= ikFe^{ika} \end{aligned} . \tag{3.5}$$

Again, we can solve for two of these coefficients in terms of the one on the right. Here, we seek to find C and D in terms of F. This leads to the matrix equation

$$\begin{bmatrix} C \\ D \end{bmatrix} = \begin{bmatrix} \left(\dfrac{ik+\gamma}{2\gamma}\right)e^{(ik-\gamma)a} & -\left(\dfrac{ik-\gamma}{2\gamma}\right)e^{-(ik+\gamma)a} \\ -\left(\dfrac{ik-\gamma}{2\gamma}\right)e^{(ik+\gamma)a} & \left(\dfrac{ik+\gamma}{2\gamma}\right)e^{-(ik-\gamma)a} \end{bmatrix} \begin{bmatrix} F \\ 0 \end{bmatrix}. \tag{3.6}$$

From the pair of equations, (3.4) and (3.6), the two propagating coefficients on the left of the barrier, A and B, can be related directly to those on the right of the barrier, here only F, with the two under the barrier dropping out of consideration. That is, we replace the last matrix in (3.4) with the right-hand side of (3.6) and simply multiply the two 2×2 matrices together. This leads to the matrix equation

$$\begin{bmatrix} A \\ B \end{bmatrix} = \begin{bmatrix} M_{11} & M_{12} \\ M_{21} & M_{22} \end{bmatrix} \begin{bmatrix} F \\ 0 \end{bmatrix}, \tag{3.7}$$

with

$$\begin{aligned} M_{11} = M_{22}^* &= \left[cosh(2\gamma a) - \frac{i}{2}\left(\frac{k^2 - \gamma^2}{k\gamma}\right)sinh(2\gamma a)\right]e^{2ika}, \\ M_{21} = M_{12}^* &= -\frac{i}{2}\left(\frac{k^2 + \gamma^2}{k\gamma}\right)sinh(2\gamma a). \end{aligned} \tag{3.8}$$

It is a simple algebraic exercise to show that for the present case, the determinant of the matrix M is unity, so this matrix has quite interesting properties. It is *not* a unitary matrix, because the diagonal elements are complex. In the general case, the transmission coefficient is simply given by the reciprocal of $|M_{11}|^2$, since the momentum is the same on either side of the barrier and hence the current does not involve any momentum adjustments on the two sides. The transmission probability is the ratio of the currents on the two sides of the barrier, directed in the same direction of course, so

$$T = \frac{1}{|M_{11}|^2} = \left[cosh^2(2\gamma a) + \left(\frac{k^2 - \gamma^2}{k\gamma} \right)^2 sinh^2(2\gamma a) \right]^{-1} = \left[1 + \left(\frac{k^2 + \gamma^2}{2k\gamma} \right)^2 sinh^2(2\gamma a) \right]^{-1} . \tag{3.9}$$

There are a number of limiting cases that are of interest. First, for a very weak barrier, in which $2\gamma a \ll 1$, the transmission coefficient becomes

$$T \to \frac{1}{1 + (ka)^2} . \tag{3.10}$$

On the other hand, when the potential is very strong, where $2\gamma a \gg 1$, the transmission coefficient falls off exponentially as

$$T \to \left(\frac{2k\gamma}{k^2 + \gamma^2} \right)^2 e^{-4\gamma a} . \tag{3.11}$$

It is important to understand that these approximations have existed for some time, but they are not very useful. In Figure 3.2, we plot the above three equations by varying the parameter γa and set the energy to be one-half the barrier height ($E = V_0/2$). It is clear that neither approximation is very good, although the small γa approximation (3.10) is not too bad. However, the large γa approximation (3.11) is very bad, being often orders of magnitude away from the proper value!

It is important to note that the result (3.10) is valid only for a weak potential for which the energy is actually *just below the top of the barrier*. If we consider an incident energy *above* the barrier, we expect the barrier region to act as a thin dielectric and cause interference fringes. We can see this by making the simple substitution suggested by (3.1) through $\gamma \to -ik'$. This changes (3.9) into

$$T = \frac{1}{[M_{11}]^2} = \left[1 + \left(\frac{k^2 - k'^2}{2kk'} \right)^2 sin^2(2k'a) \right]^{-1} , \quad E > V_0, \tag{3.12}$$

which would follow precisely from the results obtained in the last chapter (with a suitable change in the definition of the wave function in the barrier region) if we made the barrier only a finite thickness as we have done here. Thus, above the barrier, the transmission has oscillatory behavior as a function of energy, with resonances that occur for $2k'a = n\pi$. The overall behavior of the tunneling coefficient is shown in Figure 3.3.

3.1.2 A More Complex Barrier

In the previous section, the calculations were quite simple as the wave momentum was the same on either side of the barrier. Now, we want to consider a somewhat more realistic barrier in which the

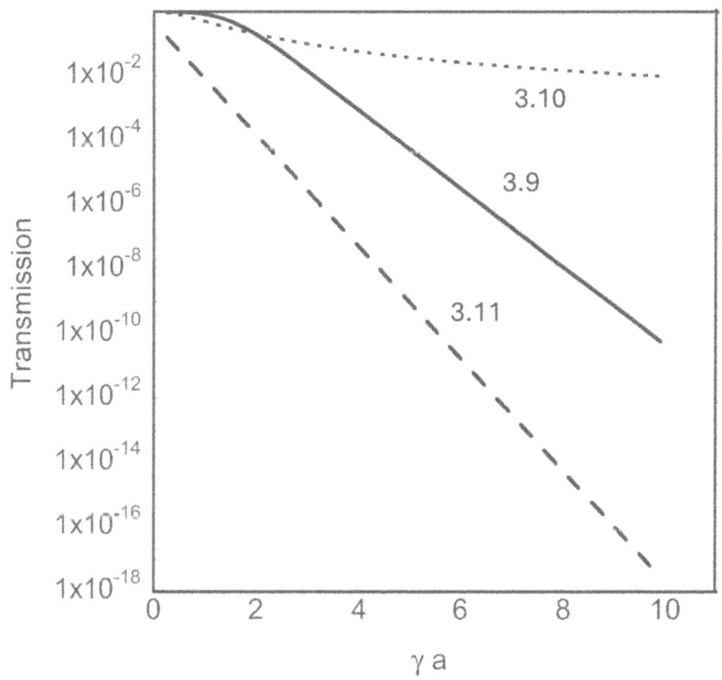

FIGURE 3.2 Comparison of the exact tunneling coefficient (3.9) with two approximations for a weak barrier (3.10) and a very strong barrier (3.11).

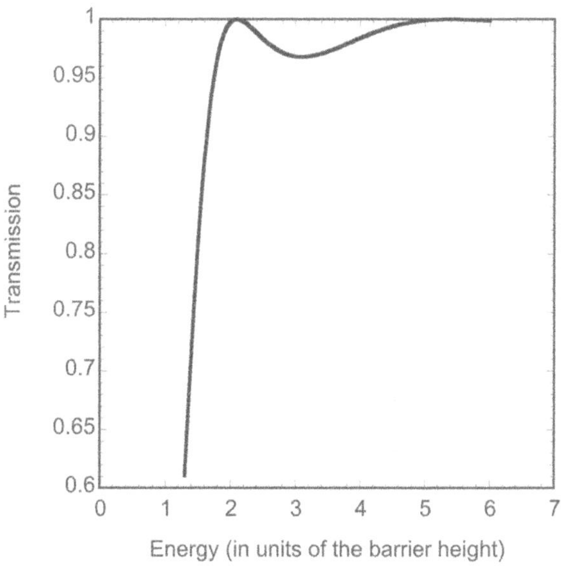

FIGURE 3.3 Tunneling (transmission) probability for a simple barrier. The energy is normalized to the barrier height and the quantity $2\sqrt{2mV_0/\hbar^2}a = 3$. The points at which the transmission goes to unity are where the sine term in (3.12) goes to zero.

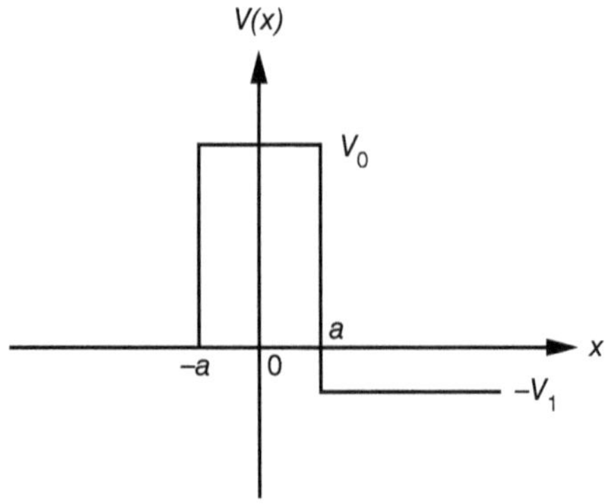

FIGURE 3.4 A more complex tunneling barrier, in which the wave number is different on the two sides of the barrier.

momentum differs on the two sides of the barrier. Consider the barrier shown in Figure 3.4. The interface at $x = -a$ is the same as treated previously, and the results of (3.4) are directly used in the present problem. However, the propagating wave on the right-hand side of the barrier $(x > a)$ is characterized by a different wave vector through

$$k_1 = \sqrt{\frac{2m(E+V_1)}{\hbar^2}}. \tag{3.13}$$

Matching the wave function and its derivative at $x = a$ leads to

$$\begin{bmatrix} C \\ D \end{bmatrix} = \begin{bmatrix} \left(\frac{ik_1+\gamma}{2\gamma}\right)e^{(ik_1-\gamma)a} & -\left(\frac{ik_1-\gamma}{2\gamma}\right)e^{-(ik_1+\gamma)a} \\ -\left(\frac{ik_1-\gamma}{2\gamma}\right)e^{(ik_1+\gamma)a} & -\left(\frac{ik_1-\gamma}{2\gamma}\right)e^{(ik_1+\gamma)a} \end{bmatrix} \begin{bmatrix} F \\ 0 \end{bmatrix}. \tag{3.14}$$

This, of course, differs from (3.6) only in the replacement of k by k_1. But now we find that the entries in the M matrix are changed. These now become

$$\begin{aligned} M_{11} = M_{22}^* &= \left[\frac{1}{2}\left(1+\frac{k_1}{k}\right)\cosh(2\gamma a) - \frac{i}{2}\left(\frac{kk_1-\gamma^2}{k\gamma}\right)\sinh(2\gamma a)\right]e^{i(k+k_1)a}, \\ M_{21} = M_{12}^* &= -\left[\frac{i}{2}\left(\frac{kk_1+\gamma^2}{k\gamma}\right)\sinh(2\gamma a) + \frac{1}{2}\left(\frac{k_1}{k}-1\right)\cosh(2\gamma a)\right]e^{i(k-k_1)a}. \end{aligned} \tag{3.15}$$

The determinant of the matrix M is also no longer unity, but is given by the ratio k_1/k. This determinant also reminds us that we must be careful in calculating the transmission coefficient as well, due to the differences in the momenta, at a given energy, on the two sides of the barrier. We proceed as in the previous section to compute the transmission coefficient. The actual transmission coefficient relates the currents, and we find that

$$T = \frac{k_1}{k}\frac{1}{\left[M_{11}\right]^2} = \frac{4k_1k}{(k_1+k)^2}\left[1+\frac{(\gamma^2+k^2)(\gamma^2+k_1^2)}{\gamma^2(k_1+k)^2}\sinh^2(2\gamma a)\right]^{-1}. \tag{3.16}$$

The first factor on the right-hand side describes the discontinuity between the propagation constants in the two regions to the left and to the right of the barrier. The second factor is the bracketed term which is the denominator and is the actual tunneling coefficient describing the *transparency* of the barrier. It is these two factors together that describe the total transmission of waves from one side to the other. It should be noted that if we take the limiting case of $k_1 = k$, we recover the previous result (3.12).

There is an important relationship that is apparent in (3.16). The result represented by (3.16) is symmetrical in the two wave vectors. They appear symmetrical in the transmission coefficient T. This is a natural and important result of the symmetry. Even though the barrier and energy structure of Figure 3.4 does not appear symmetrical, the barrier is a linear structure that is passive (there is no active gain in the system). Therefore, the electrical properties should satisfy the principle of reciprocity, and the transmission should be the same regardless of the direction from which one approaches the barrier. This is evident in the form of the transmission coefficient (3.16) that is obtained from these calculations.

3.1.3 Approximate Solutions

In many cases, it is not necessary to work out the complete tunneling problem in detail, and a general approximate solution is sufficient. This is certainly the case when we are interested only in the decay of the wave function from one side of the barrier to the other (recall that the input wave was always normalized to unity). It suffices to say that the spirit of the Wentzel–Kramers–Brillouin (WKB) approximation, developed in Section 2.9 can give an acceptable answer. This is true when the result is mainly concerned with the propagation (or decaying) wave vector, and the computation of the argument of the exponential decay function. The result (2.153) is that it is only the combination of forward and reverse waves that matter. For a barrier in which the attenuation is relatively large, only the decaying forward wave is important, and the tunneling probability may be expressed approximately as

$$T \sim exp\left[-2\int_b^a \gamma(x')dx'\right], \tag{3.17}$$

which implies that it is only the numerical coefficients (the prefactor in (3.16), and the term leading the sine term in the denominator, which involves the propagating and decaying wave vectors) that are lost in the WKB method. This tells us that we can use the limiting form of (3.11) $(b = -a)$, or the equivalent limit of (3.17).

3.1.4 Fowler–Nordheim Tunneling

As an example of approximate tunneling through a more complex barrier, consider the structures shown in Figure 3.5. These are appropriate to a MOSFET. We don't want to discuss the full operation of such a device here, but the interested parties can find a discussion in Ferry (2026). The structures are a gate, shown on the left-hand side of both panels. Originally, this gate was a metal, hence the name of the structure, but this metal was superseded by a heavily doped polysilicon layer. However, as materials such as HfO_2 replaced the SiO_2, a return to metal gates followed. The central part is the oxide mentioned in the last sentence. The right-hand side of each panel is a p-type Si layer. The application of a gate voltage (V_G) bends the bands in this latter layer and creates an n-type inversion layer (the positive voltage on the gate pulls electrons in the Si to the surface and pushes the holes away). When the voltage is sufficiently high, and the oxide is sufficiently thin, electrons can tunnel from the semiconductor layer, through the oxide, to the gate. This is termed *gate leakage*. But, the type of tunneling depends upon the shape of the barrier. In panel (a), the electrons tunnel through

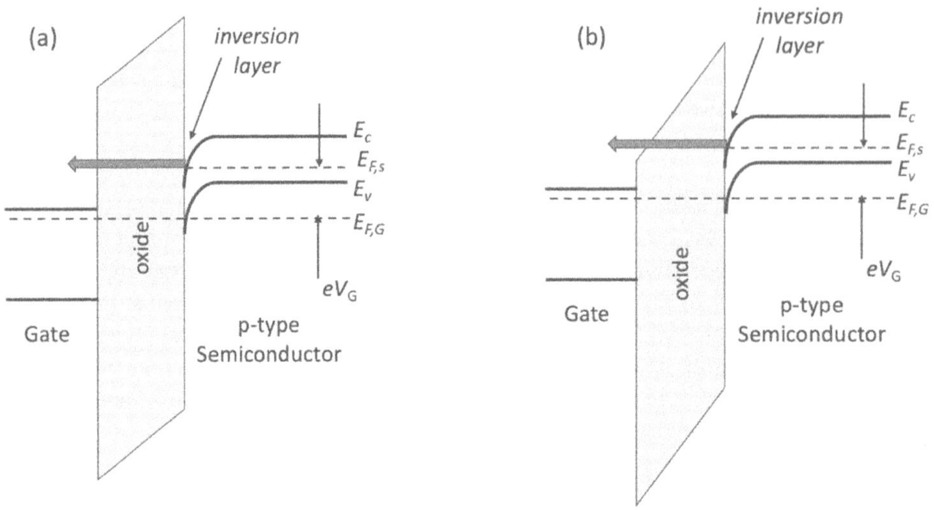

FIGURE 3.5 Gate tunneling in a metal–oxide-semiconductor field-effect-transistor. (a) Direct tunneling through the oxide from the semiconductor to the gate. (b) Fowler–Nordheim tunneling through only a portion of the gate. In the figure, E_c and E_v are the conduction band and valence band energies, respectively, of the semiconductor. $E_{F,G}$ and $E_{F,s}$ are the Fermi levels of the gate and the semiconductor, respectively. The difference in these latter two corresponds to the applied positive gate voltage V_G.

the entire thickness of the oxide, and this is just normal tunneling through a triangular-topped barrier. On the other hand, in panel (b), the electrons only tunnel through a part of the barrier, depending upon the strength of the electric field. This latter is Fowler–Nordheim tunneling, which was originally considered for field emission from a metal into vacuum (Fowler and Nordheim, 1928).

The current will depend upon a variety of factors, but can be estimated using the Duke tunneling formula (Duke, 1969). This formula is basically an extension of the Landauer formula, discussed in Section 2.7.3, to three-dimensional devices. This means that the transverse momentum becomes important as well as the actual longitudinal tunneling process. This results in an integration over the transverse momentum which replaces the summation over the modes of propagation.

In the "real" device, such as the MOSFET discussed here, the tunneling electrons are those within a narrow energy range near the Fermi energy, where the range is defined by the applied voltage as indicated in the figure. For this simple view, the device is treated in the linear-response regime, even though the resulting current is a nonlinear function of the applied voltage. The general treatment assumes that the total tunneling probability through the entire structure is *coherent*. By coherent here, we mean that the tunneling through the entire barrier is an energy- and momentum-conserving process, so no further complications are necessary. Hence, the properties of the barrier are completely described by the quantity $T(k)$. It is true that the presence of localized states in the oxide can lead to more complicated tunneling, but this is not the purpose of the present discussion. In equilibrium, where there is no applied bias, the left-going and right-going currents are equivalent and there is no net current. By requiring that the energy be conserved during the process, we can write the z-component of energy as (we take the z-direction as that of the tunneling current)

$$E = \frac{\hbar^2 k_z^2}{2m} = \frac{\hbar^2 k_{1z}^2}{2m} + constant, \tag{3.18}$$

where the constant accounts for the bias and is negative for a positive potential applied to the left of the barrier, as shown in Figure 3.5. Thus, the subscript "1" corresponds to the gate region of the device. The two wave vectors are easily related to one another by this equation, and we note that the derivative of this equation allows us to relate the velocities, using the standard formula

$$v_z = \frac{1}{\hbar}\frac{dE_z}{dk_z}, \tag{3.19}$$

on the two sides. In particular, we note that this gives

$$v_z(k_z)dk_z = v_z(k_{1z})dk_{1z}. \tag{3.20}$$

The current flow through the barrier is related to the tunneling probability and to the total number of electrons that are available for tunneling. Thus, the flow from the left to the right is given by

$$J_{LR} = 2e\int \frac{d^3k}{(2\pi)^3} v_z(k_z)T(k_z)f(E_L), \tag{3.21}$$

where the factor of 2 is for spin degeneracy of the electron states, the $(2\pi)^3$ is the normalization on the number of k states (related to the density of states in k space), and $f(E_L)$ is the electron distribution function *at the barrier*. Similarly, the current flow from the right to the left is given by

$$J_{RL} = 2e\int \frac{d^3k_1}{(2\pi)^3} v_z(k_{1z})T(k_{1z})f(E_R). \tag{3.22}$$

We know that the tunneling probability is reciprocal at the same energy, regardless of the direction of approach, and these two equations can be combined, using (3.20), as

$$J = 2e\int \frac{d^3k}{(2\pi)^3} v_z(k_z)T(k_z)\left[f(E_R + eV_a) - f(E_R)\right], \tag{3.23}$$

where we have related the energy on the left to that on the right through the bias, as shown in Figure 3.5, and expressed in (3.19). In the following, we will drop the subscript "R" on the energy, but care must be used to ensure that it is evaluated on the right of the barrier. This last result is basically the Duke tunneling equation.

At this point, we want to simplify some of the relationships in (3.23). First, we note that the energy is a scalar quantity and can therefore be decomposed into its z-component and its transverse component, as

$$E = E_z + E_\perp \tag{3.24}$$

and

$$d^3k = d^2k_\perp dk_z. \tag{3.25}$$

We would like to change the differential dk_z to one on dE_z. This is done through

$$dk_z = \left(\frac{dE}{dk_z}\right)^{-1}\frac{dE}{dE_z}dE_z. \tag{3.26}$$

The second term on the right-hand side is unity, so it drops out. The first term may be evaluated as

$$\frac{dE}{dk_z} = \frac{\hbar^2 k_z}{m} = \hbar v_z. \tag{3.27}$$

The velocity term here will cancel that in (3.23), and we can write the final representation of the current as

$$J = \frac{e}{\pi\hbar}\int\frac{d^2k_\perp}{(2\pi)^2}\int dE_z T(E_z,E_\perp)\left[f(E_R+eV_a)-f(E_R)\right]. \tag{3.28}$$

This equation contains two main parts. The first part is the supply function

$$\frac{e}{\pi\hbar}\int\frac{d^2k_\perp}{(2\pi)^2}\int dE_z\ldots\left[f(E_R+eV_a)-f(E_R)\right], \tag{3.29}$$

which essentially governs the number of carriers that are available to tunnel. This term will remain almost constant for most tunneling problems, although it has dependences upon the device structure and doping densities.

The second part of (3.28) is the actual tunneling probability, and it is in this term that we are primarily interested at this time. This is the term

$$T(E_z,E_\perp). \tag{3.30}$$

This is written as a function of both the longitudinal and transverse energies, because the tunneling depends upon the total energy. The tunneling probability depends upon the evanescent wave parameter γ in (3.1), and this depends upon the energy distance between the top of the barrier (V_0) and the actual energy of the tunneling carrier E. This energy is the total energy. As the particle obtains increased transverse momentum, its transverse energy rises, and this reduces the longitudinal energy of the particle (for a fixed total energy). Thus, the tunneling energy of the particle depends upon the transverse energy as well as the longitudinal energy. This is a result of having more than a single dimension in the problem. Nevertheless, it is this tunneling probability (3.30) that we want to examine in this section, particularly as it varies between the two situations of Figure 3.5.

Consider first the direct tunneling case of panel (a) of Figure 3.5, as shown in Figure 3.6. Note that the direction of tunneling and the structure have been reversed so that the gate is now on the right-hand side of the figure in order to ease the calculation (we recall that the tunneling probability is symmetrical in direction, so this does affect the result for T). Now, the evanescent tunneling parameter can be written as

$$\gamma(z) = \sqrt{\frac{2m}{\hbar^2}\left[(E_w - F_s z) - E\right]}, \tag{3.31}$$

where F_s is the electric field applied across the oxide and E_w is the energy at the top of the oxide layer. The energy is that of the tunneling carrier, which is assumed to be at the z-axis position, and the decomposition of this energy into longitudinal and transverse parts is done outside of the oxide, so that it does not affect the integration that is required in (3.17). The integral now becomes (with the change in variables)

$$\int_0^w \gamma(z)\,dz = \int_0^w \sqrt{\frac{2m}{\hbar^2}\left[(E_w - F_s z) - E\right]}\,dz. \tag{3.32}$$

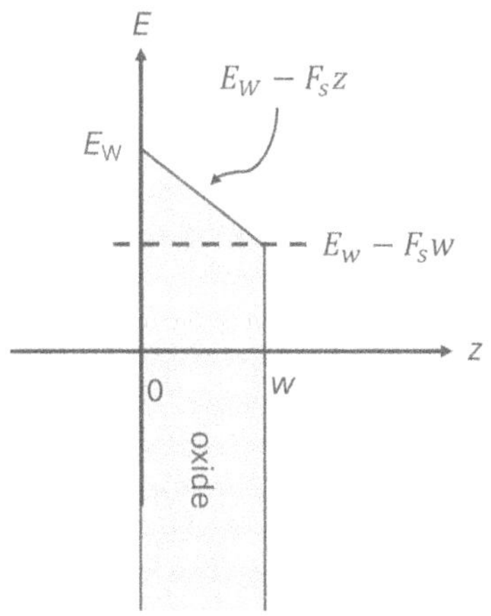

FIGURE 3.6 The pertinent energy vs. position diagram for direct tunneling through the triangular potential of Figure 3.5(a). The tunneling energy E is taken to be at the z-axis.

This is easily integrated and the tunneling coefficient (3.17) becomes

$$T = exp\left\{ -\frac{4}{3F_s}\sqrt{\frac{2m}{\hbar^2}}\left[\left(E_w - E\right)^{3/2} - \left(E_w - E - F_s w\right)^{3/2} \right] \right\}. \tag{3.33}$$

In many cases, this can be expanded and simplified, but we leave it here in its proper form.

The Fowler–Nordheim tunneling through the barrier shown in Figure 3.5(b) is more complicated because the tunneling length depends upon the energy and the electric field. This is depicted in more detail in Figure 3.7, which is also reversed from panel (b) of the earlier figure. Again, the energy of the tunneling particle is taken to be at the z-axis. Now, the tunneling distance does not extend completely through the oxide, but ends at a point z_0, where the tunneling particle exits the oxide. This point is the bottom of a triangle, and is given by

$$z_0 = \frac{\left(E_W - E\right)}{F_s}. \tag{3.34}$$

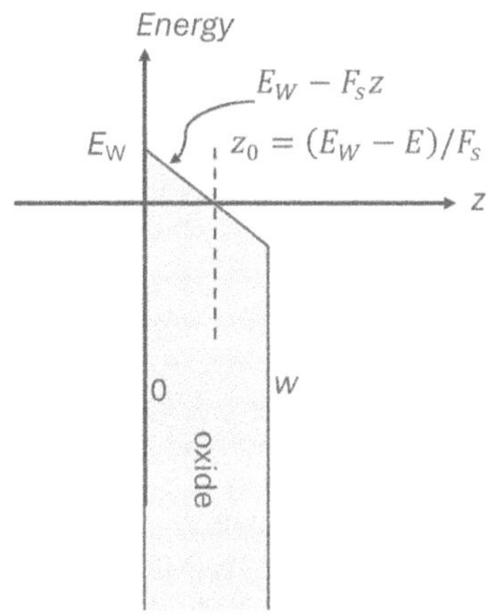

FIGURE 3.7 The pertinent energy vs. position diagram for direct tunneling through the triangular potential of Figure 3.5(b) for Fowler–Nordheim tunneling. The tunneling energy E is taken to be at the z-axis.

This changes the integral (3.32) only in the upper limit of integration, as this upper limit is now z_0 rather than W. What this does is to make the second term in (3.33) vanish, and the tunneling probability for Fowler–Nordheim tunneling becomes

$$T = exp\left\{-\frac{4}{3F_s}\sqrt{\frac{2m}{\hbar^2}}\left(E_w - E\right)^{3/2}\right\} = exp\left(-\frac{4}{3}\sqrt{\frac{2mF_s}{\hbar^2}}z_0^{3/2}\right). \tag{3.35}$$

The two types of tunneling can be identified by the different behavior of the tunneling coefficients.

3.1.5 The Scanning Tunneling Microscope

One of the most intriguing devices that uses quantum tunneling is the scanning tunneling microscope (STM). Electron microscopy was successfully developed in 1931 (Knoll and Ruska, 1932a, 1932b). Since this time, a variety of types of electron microscope have been developed. The STM was developed in 1981 by Gerd Binnig and Heinrich Rohrer at IBM's research laboratory in Zurich, Switzerland (Binnig *et al.*, 1982a, 1982b). Binnig and Rohrer subsequently shared the 1986 Nobel Prize in Physics with Ruska for these developments. The structure of a typical STM is illustrated in Figure 3.8. The tunneling tip (black cone) is mounted on a piezotube (the vertical cylindrical structure in the figure). When voltages are applied to this piezotube, it can be bent in the transverse plane and lengthened in the vertical plane. The purpose of the STM is to scan the surface of the sample (shown above the tip) to investigate the physics/chemistry of the sample's surface. The scan is accomplished by bending the piezotube in the transverse plane to enable a two-dimensional image. Tunneling is, of course, very sensitive to the distance between the tip and the sample. However, this distance is buried in the exponential portion of the tunneling coefficient as shown in (3.33) and (3.35). This is where the cleverness of the STM design appears. Usually, the piezotube is lengthened or shortened as the tip is scanned in order to keep the tunneling current constant. The current through the tip is measured, and as it changes, the signal is fed to a feedback control that changes the distance from tip to surface to maintain a constant current. The actual voltage of the feedback

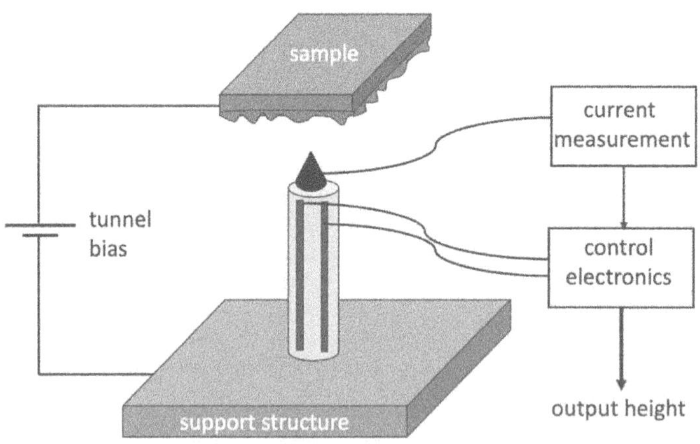

FIGURE 3.8 A schematic diagram of the scanning tunneling microscope. The tunnel tip is mounted on the top of a piezotube, which can be bent or lengthened by applied voltages. This allows the tip to scan the surface of the sample, and to move the tip to maintain a constant current. This current is measured and any variations are sent to the control electronics to move the tip up and down in order to maintain a constant current. This feedback voltage is a direct measurement of the tip-to-sample distance. Bias is applied as shown to provide the tunneling current.

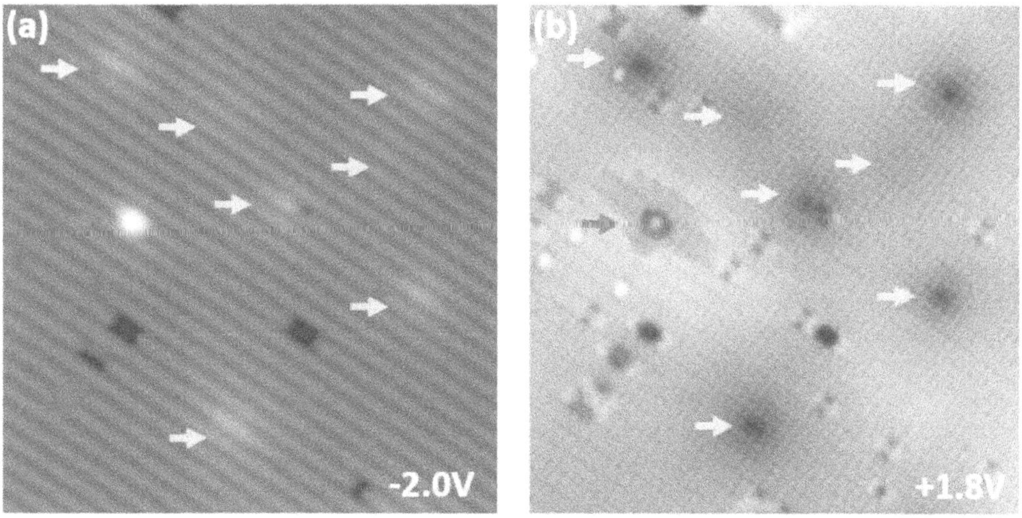

FIGURE 3.9 STM images of a Si (100) surface. The scan is a 20×20 nm^2 scan at constant current for the full states (a) and the empty states (b). The bias applied to the sample, relative to the tip, is indicated in the lower right corner of each image. The arrows in (a) indicate surface atoms (As), while the arrows in (b) is a dangling bond. The surface 2×1 reconstruction into dimer rows is most easily seen in (a) (the dimers run from upper left to lower right). (Reproduced from Labidi *et al.* (2015) under the creative commons attribution license 3.0.)

(to control the length of the piezotube) is then a linear measure of the distance between tip and sample, and this voltage is the output signal to be plotted.

It was pointed out in the previous sections that the tunneling coefficient is symmetrical in the current direction. Hence, if the sample is biased positive with respect to the tip, electrons tunnel from the tip to the sample if the sample has empty states available to receive the electrons. On the other hand, if the sample is biased negatively with respect to the tip, electrons will tunnel from the sample to the tip if there are full states (to provide the electron) in the sample at the tunneling energy. Thus, the STM can measure either the full state spectrum or the empty state spectrum at the material surface. One can see the difference in Figure 3.9, which shows STM images of a Si (1 0 0) surface (Labidi *et al.*, 2015). In panel (a), the image is that of the full states on the surface, while panel (b) shows an image of the empty states. The bias applied to the sample (relative to the STM tip) is indicated in the lower corner of each image. The yellow arrows indicate dopant (As) atoms on the surface layer, while the red arrow indicates a surface dangling bond. Atoms on semiconductor surfaces often relax to different arrangements than in the bulk. If this rearrangement maintains the bulk unit cell, it is called a *relaxation*. However, if the surface order gives a different unit cell, it is referred to as *reconstruction*. The latter is the case here, where the surface cell is twice the size of the bulk cell, due to the fact that the dangling bonds[2] for the surface atoms tend to bond with neighbors to cause this reconstruction. In the case of Si (100), the surface atoms form dimers with a 2×1 reconstruction (in terms of the bulk lattice constant), and the dimer rows can be seen as the gray lines stretching from upper left to lower right in the images. In panel (b), the dopants appear black as they tend to have electrons on the bonds rather than empty states.

3.2 THE DOUBLE BARRIER

Let us now put together two tunnel barriers separated by a quantum well. The quantum well (that is, the region between the two barriers) will have discrete energy levels because of the confinement quantization, just as in Section 2.4. When the incident wave energy corresponds to one of these resonant energy states of the quantum well, the transmission through the double barrier will rise to a value that is unity (for equal barriers). This resonant tunneling, in which the transmission is unity, is quite useful as an energy filter in many applications.

There are two approaches to solving for the composite tunneling transmission coefficient. In one, the entire problem is resolved from first principles, matching the wave function and its derivative at four different interfaces (two for each of the two barriers). The second approach, which will be pursued here, uses the results of the previous sections, and one merely seeks knowledge as to how to put together the transmission matrices that already have been found. The reason we can pursue this latter approach effectively is that the actual transmission matrices found in the previous sections depend only upon the wave vectors (the k's and γ's), and the thickness of the barrier, $2a$. They do *not* depend upon the position of the barrier, so the barrier may be placed at an arbitrary point in space without modifying the transmission properties. Thus, we consider the generic problem of Figure 3.10, where we have indicated the coefficients in the same manner as that in which they were defined in the earlier sections. To differentiate between the two barriers, we have used primes on the coefficients of the right-hand barrier. The task is to now relate the coefficients of the left-hand barrier to those of the right-hand barrier. We note that both E and A' describe a wave propagating to the right. Denoting the definition of the thickness of the well region as b, we can simply relate these two coefficients a simple propagator (see Sec. 2.71 and 2.96).

$$A' = Ee^{ikb}, \tag{3.36}$$

where k is the propagation constant *in the well region*. Similarly, F and B' relate the same wave propagating in the opposite direction. These two can thus be related by

$$B' = Fe^{-ikb}. \tag{3.37}$$

These definitions now allow us to write the connection as a matrix exactly as (2.96)

$$\begin{bmatrix} E \\ F \end{bmatrix} = \begin{bmatrix} e^{-ikb} & 0 \\ 0 & e^{ikb} \end{bmatrix} \begin{bmatrix} A' \\ B' \end{bmatrix}. \tag{3.38}$$

Equation (3.38) now defines a matrix $\boldsymbol{M}_W$, where the subscript indicates the well region. This means that we can now take the matrices defined in Section 3.1 for the left-hand and right-hand regions and write the overall tunneling matrix as

$$\begin{bmatrix} A \\ B \end{bmatrix} = [M_L][M_W][M_R] \begin{bmatrix} E' \\ F' \end{bmatrix}. \tag{3.39}$$

From this, it is easy to now write the composite M_{11}^T as

$$M_{11}^T = M_{11}^L M_{11}^R e^{ikb} + M_{12}^L M_{21}^R e^{-ikb}, \tag{3.40}$$

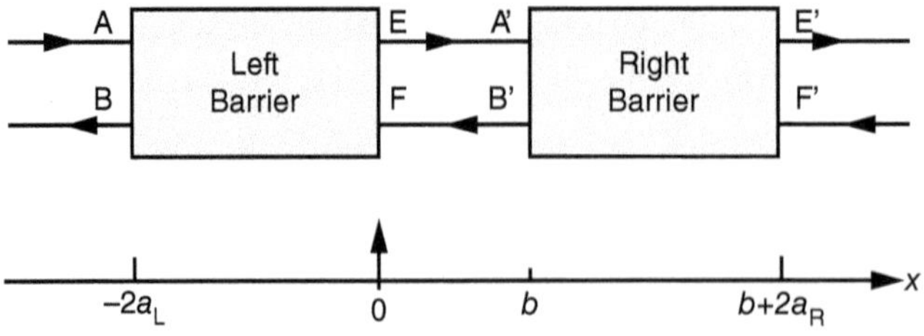

FIGURE 3.10 Two generic barriers are put together to form a double-barrier structure.

and it is apparent that the resonance behavior arises from the inclusion of the off-diagonal elements of each transmission matrix, weighted by the propagation factors. At this point, we need to be more specific about the individual matrix elements.

3.2.1 Equal Barriers

For the first case, the results of Section 3.1 can easily be used, where a simple rectangular barrier was considered. Here, it is assumed that the two barriers are exactly equal, so the same propagation wave vector k exists in the well and in the regions to the left and right of the composite structure. By the same token, each of the two barriers has the same potential height and therefore the same γ. We note that this leads to a magnitude-squared factor in the second term of (3.40), *but not in the first term* with one notable exception. The factor of e^{i2ka} does cancel since we are to the left of the right-hand barrier ($-a$-direction) but to the right of the left-hand barrier ($+a$-direction). Thus, the right-hand barrier contributes a factor of e^{-i2ka}, and the left-hand barrier contributes a factor of e^{i2ka}, so the two cancel each other. In order to simplify the mathematical details, we write (3.40) as

$$\begin{aligned} M_{11}^T = &\left[cosh(2\gamma a) - \frac{i}{2}\left(\frac{k^2-\gamma^2}{k\gamma}\right) sinh(2\gamma a)\right]^2 e^{ikb} \\ &- \frac{1}{4}\left[\left(\frac{k^2+\gamma^2}{k\gamma}\right) sinh(2\gamma a)\right]^2 e^{-ikb} \end{aligned} \tag{3.41}$$

While this latter form shows the explicit elements in the total transmission, it is easier both to understand and to manipulate if we leave the explicit terms out for the present.

To express this better, let us write the terms as a phasor; that is, let us write each of the various matrix elements as

$$M_{11} = m_{11} e^{i\vartheta}, \tag{3.42}$$

where

$$m_{11} = \sqrt{cosh^2(2\gamma a) + \left(\frac{k^2-\gamma^2}{k\gamma}\right)^2 sinh^2(2\gamma a)}, \tag{3.43}$$

and

$$\vartheta = atan\left[\left(\frac{k^2-\gamma^2}{k\gamma}\right) tanh(2\gamma a)\right]. \tag{3.44}$$

Then, the square magnitude of (3.41) can be rewritten as

$$\begin{aligned} \left|M_{11}^T\right|^2 &= \left|M_{11}\right|^4 + \left|M_{12}\right|^4 + 2\left|M_{11}\right|^2\left|M_{12}\right|^2 cos\left[2(kb+\vartheta)\right] \\ &= \left(\left|M_{11}\right|^2 - \left|M_{12}\right|^2\right)^2 + 4\left|M_{11}\right|^2\left|M_{12}\right|^2 cos^2(kb+\vartheta). \end{aligned} \tag{3.45}$$

The first term, the combination within the parentheses, is just the determinant of the individual barrier matrix and is unity for the simple rectangular barrier. This reduces to

$$\left|M_{11}^T\right|^2 = 1 + 4\left|M_{11}\right|^2\left|M_{12}\right|^2 cos^2(kb+\vartheta). \tag{3.46}$$

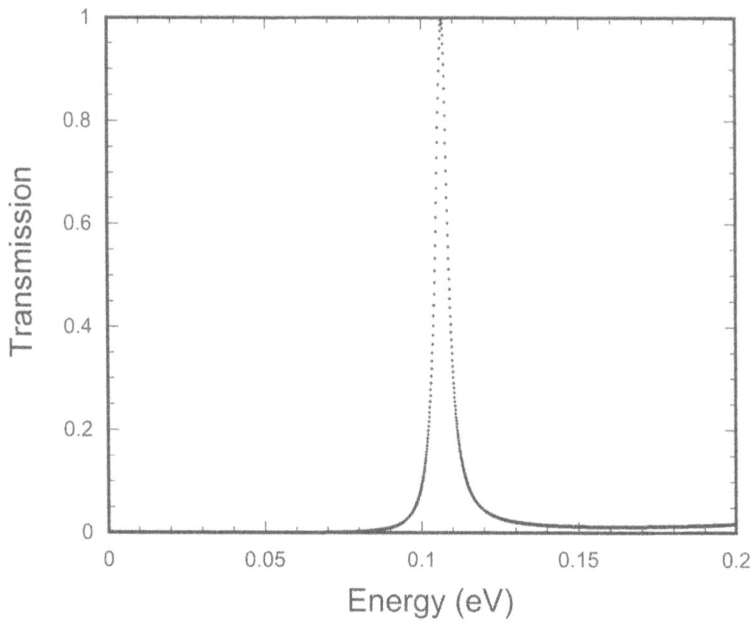

FIGURE 3.11 The transmission through a double-barrier system composed of AlGaAs barriers and a GaAs well. Here, the barrier height is 0.25 eV and the thicknesses are 4 nm. The well is taken to be 2 nm thick so that only a single resonant level is present in the well.

In general, the cosine function is nonzero, and the composite term of (3.45) is actually larger than that for the single barrier T_1, with

$$T \sim \frac{1}{4|M_{12}|^2}, \textit{ off resonance.} \tag{3.47}$$

However, for particular values of the wave vector, the cosine term vanishes, and

$$kb + \vartheta = (2n+1)\frac{\pi}{2} \rightarrow T_{total} = 1. \tag{3.48}$$

These values of the wave vector correspond to the resonant levels of a finite-depth quantum well (the finite-well values are shifted in phase from the infinite-well values by θ, which takes the value $-\pi/2$ in the latter case). Hence, as we supposed, the transmission rises to unity at values of the incident wave vector that correspond to resonant levels of the quantum well. In essence, the two barriers act like mirrors, and a resonant structure is created just as in electromagnetics. The incoming wave excites the occupancy of the resonance level until an equilibrium is reached in which the incoming wave is balanced by an outgoing wave and the overall transmission is unity. This perfect match is broken up if the two barriers differ, as we see below. In Figure 3.11, we plot the transmission for a double quantum well around the resonant level. The barriers are 0.25 eV high and 4 nm thick. The well is taken to be 2 nm thick. Here, it is assumed that the well is GaAs and the barriers are AlGaAs, so that there is only a single resonant level in the well. It is clear that the transmission rises to unity at the value of this level. We will return below to the shape of the resonant transmission peak, and how it may be probed in some detail experimentally.

3.2.2 Unequal Barriers

When the two barriers are different in any way, the results are more complicated, and greater care is necessary in the mathematics. One way is, of course, to actually compute the two matrices M_L and M_R

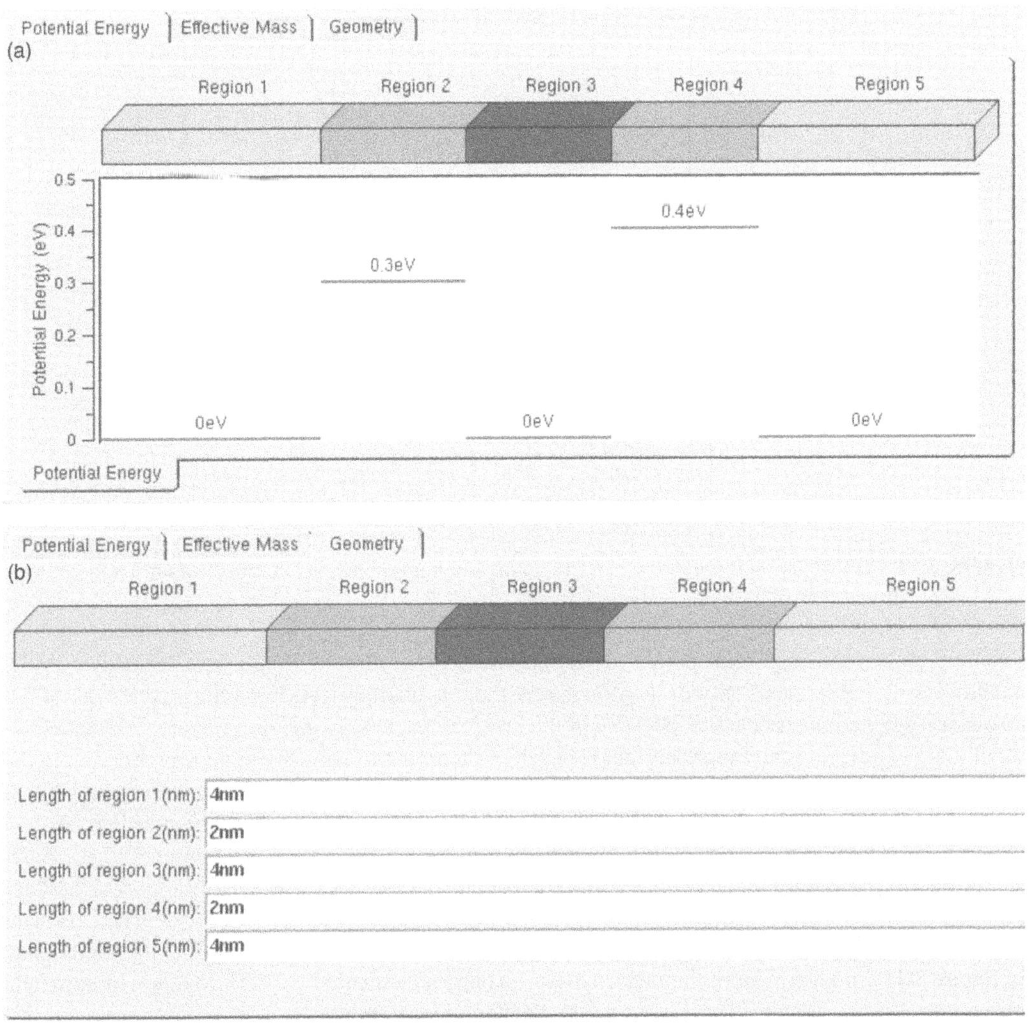

FIGURE 3.12 The use of a tool from NanoHub.org is illustrated here for the double barrier. (a) The frame for selecting the potentials for two barriers in the "piece-wise constant potential" tool. (b) The frame for selecting the geometry (thicknesses) of the individual regions.

given in (3.39) and work through the matrix mathematics to find the transmission coefficient. The result is an evaluation of (3.40). However, with the modern tools available, there is a second (perhaps more efficient) way and that is to go once more to Nanohub.org and use the "Piece-Wise Constant Potential Tool," as pictured in Figure 3.12. In panel (a), we have an input selection panel that allows one to set the potential values in each of five regions (left and right of the barriers, central well, and two barriers). Individual wave vectors for the regions to the left and right of the composite barrier, as well as in the well will be selected automatically as will the decay constants under the barrier. This panel shows the two potential barriers as 0.3 and 0.4 eV (which are set by clicking on the colored line and adjusting to the desired potential). The decay constants of the two barriers differ, according to their height. the thicknesses of each layer may be set in the "Geometry" pull-down tab and may also be different as shown in panel (b) of the figure. For example, we have set the five dimensions to 4, 2, 4, 2, and 4 nm, respectively. This tool will allow for a series of barriers to be created and studied, not just two barriers, which is the case of interest here. In fact, almost any complicated set of barriers and wells may be chosen, so long as the potential is flat in each region.

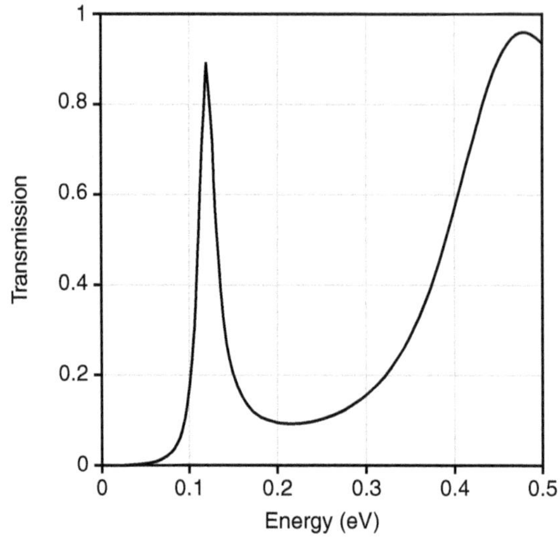

FIGURE 3.13 Transmission through the barriers defined in Figure 3.12. While the resonance still appears, it no longer has unity transmission.

There are several important points to be discussed. First, there will still be a resonance in the quantum well, but its position will be affected by the unequal barriers. In addition, the peak of the resonance will no longer be unity, because of the unequal barriers. In the optics analog, the reflectivity of the two mirrors is no longer perfect for each of the mirrors. Thus, we expect a reduction in the height of the resonance. But the resonance is not exponentially reduced, only linearly reduced; that is, it is attenuated, but not exponentially attenuated by the unequal barriers in the problem. We see this in Figure 3.13, where we plot the transmission for the barrier defined in Figure 3.12(a). Here, we have utilized the ability to download the data from the tool as a text file and replot it with another graphics package. There is a large peak at about 0.13 eV, which appears at the energy of the resonant level in the structure. However, the height is only about 0.88–0.9, which is less than the unity peak in Figure 3.11, and arises from the factors discussed above—unequal barriers and lack of complete phase matching at the reflection points of the barriers. One can give an approximate form for this reduction through the relationship

$$T = \frac{4T_L T_R}{\left(T_L + T_R\right)^2}, \tag{3.49}$$

where T_L and T_R are the transmissions of the two independent left and right barriers. It is clear that if the two are equal, then (3.49) gives a value of unity. But, if they are not equal, the peak transmission begins to be attenuated. If one is significantly larger than the other, it is thought that the result will be

$$T \sim 4\frac{T_{min}}{T_{max}}, \tag{3.50}$$

but this is different from what (3.49) would give (the latter should cancel the factor of four).

The opposite extreme is reached when we are away from one of the resonant levels. In this case, the maximum attenuation is achieved when the cosine function has the value of unity, and the resulting minimum in overall transmission is given approximately by the value

$$T = \frac{T_L T_R}{2}. \tag{3.51}$$

It is clear from these discussions that if we are to maximize the transmission on resonance in any device application, it is important to have the transmission of the two barriers equal under any bias conditions that cause the resonance to appear.

3.2.3 The Shape of the Resonance

It is apparent from the shape of the transmission curves in Figures 3.11 and 3.13 that there is a very sharp transmission peak that coincides with the resonant energy level in the quantum well. We can analyze this resonance more by assuming it to have a Lorentzian shape around the maximum, that is

$$\frac{1}{T(E)} = 1 + \left(\frac{E - E_0}{\Delta E}\right)^2, \tag{3.52}$$

where E_0 is the resonance energy and ΔE is the half-width of the resonance at half-height. In fact, this form holds quite accurately over orders of magnitude in transmission (Price, 1999). In essence, this means that we can write $|M_{11}|^2 = PQ$, with $P = Q^*$. From (3.42), we identify

$$M_{11} = Q = Q_0\left[1 - i\left(\frac{E - E_0}{\Delta E}\right)\right], \quad M_{11}^* = P = Q_0\left[1 + i\left(\frac{E - E_0}{\Delta E}\right)\right]. \tag{3.53}$$

Thus, when the energy E is swept through the resonance energy E_0, the phase angle arg(P) is swept through an amount approaching π. Price (1999) has pointed out that this has an analogy with formal wave propagation theory, so that one can assign a *transit time* for passage of an electron through the double-barrier structure as

$$t_{transit} \sim \hbar \frac{d\left[arg(P)\right]}{dE} \rightarrow \frac{\hbar / \Delta E}{1 + \left(\frac{E - E_0}{\Delta E}\right)^2}. \tag{3.54}$$

While it is quite difficult to measure the transit time of the electron through the resonant structure, creative methods have been found to measure the phase shift. In Chapter 1, we discussed the Aharonov–Bohm effect in a semiconductor nanostructure. Indeed, experiments have embedded a semiconductor quantum dot in one leg of the AB ring (Yacoby *et al.*, 1995). The quantum dot is a two-dimensional electron gas at the interface of a GaAlAs/GaAs heterostructure, which is further confined by lateral in-plane gates (the quantum dot will be treated in more detail in Chapter 9). Propagation through the quantum dot is facilitated by two *tunneling* quantum point contacts which, along with the dot itself, are located in one leg of the Aharonov–Bohm (A–B) loop. Measurements of the transmitted current through the A–B ring give the total phase through each branch in terms of the interference between the two legs. This phase interference is measured by varying the magnetic field through the ring. When the size of the dot is tuned by a *plunger* gate, the resonant energy is swept through the Fermi energy of the remaining parts of the system. By this gate voltage tuning, the resonance shape is swept through the transmission energy, and a change in phase by a factor of π is found, as expected.

3.3 SOME TUNNEL DEVICES

One of the attractions of tunneling devices is that it is possible to apply textbook quantum mechanics to gain an understanding of their operation, and still achieve a reasonable degree of success in actually getting quantitative agreement with experimental results. The concept of the "tunnel diode" goes back several decades and is usually implemented in heavily doped *p–n* junctions (Esaki, 1958). In this case, the tunneling is through the forbidden energy gap, as we will see below. Here, the tunneling electrons make a transition from the valence band, on one side of the junction, to the conduction band on the other side. More recently, effort has centered on resonant-tunneling devices which can occur in a material with a single carrier type. Each of these will be discussed below. In many cases, the approach will utilize the Landauer formula of Section 2.6.3. In other cases, the Duke tunneling formula from Section 3.1.4 is used. These of course are quite similar as mentioned in the referenced sections.

3.3.1 The Esaki Diode

Tunneling in a diode essentially arises in a very heavily-doped *p–n* junction, so the built-in potential of the junction is larger than the band gap. This is shown in Figure 3.14(a). When a small bias is applied, as shown in Figure 3.14(b), the filled states on one side of the junction overlap empty, allowed states on the other side, which allows current to flow (assuming the actual barrier is thin so tunneling

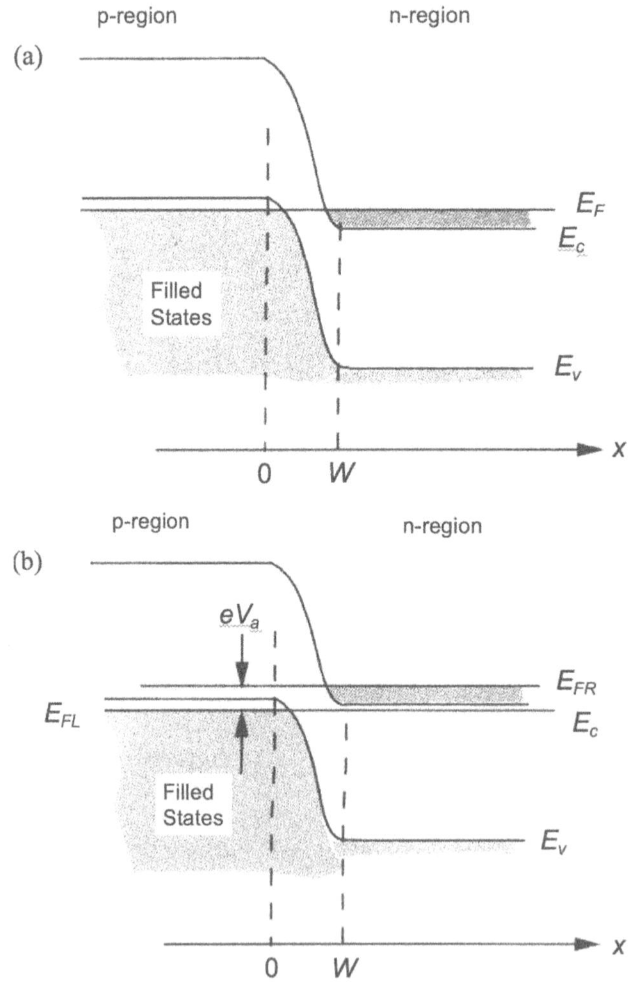

FIGURE 3.14 The band line-up for degenerately doped *p–n* junctions (a), and the possible tunneling transitions for small forward bias (b). These tunneling transitions come from the full bottom of the conduction band (right-hand side) to the empty states at the top of the valence band (left-hand side). These must conserve energy so must come from full states to empty states.

is allowed). So far, this is no different from a normal junction diode, other than the fact that the carriers tunnel across the forbidden gap at the junction rather than being injected. However, it may be noted from Figure 3.14(b)that continuing to increase the forward bias (the polarity shown) causes the filled states to begin to overlap states in the band gap, which are forbidden. Thus, the forward current returns to zero with increasing forward bias, and a negative differential conductance is observed. When combined with the normal *p–n* junction injection currents, an *N*-shaped conductance curve is obtained, which leads to the possibility of the use of the device for many novel electronic applications. In the reverse bias direction, the overlap of filled and empty (allowed) states continues to increase with all bias levels, so no negative conductance is observed in this direction of the current. When the electric field in the barrier region is sufficiently large, the probability of tunneling through the gap region is nonzero; for example, tunneling can occur when the depletion width W is sufficiently small.

One view of the tunneling barrier is that it is a triangular potential, whose height is approximately equal to the band gap, and whose width at the tunneling energy is the depletion width W. In Section 2.5, we found that a triangular-potential region gave rise to wave functions that were Airy functions. The complications of these functions provide a strong argument for the use of the WKB approximation, and we do so here. We can take the decay coefficient as

$$\gamma(x) = \begin{cases} \sqrt{\dfrac{2mE_G}{\hbar^2}\left(1 - \dfrac{x}{W} + \dfrac{E_\perp}{E_G}\right)}, & 0 < x < W, \\ 0, & elsewhere. \end{cases} \tag{3.55}$$

Here, the energy gap E_G has been factored out of the potential term and the electric field evaluated as E_G/eW. The last term in the square root accounts for the transverse energy, since the tunneling coefficient depends upon only the *z*-component of momentum (the *z*-component of energy must be reduced below the total energy by the transverse energy as discussed in Section 3.1.4). This expression must now be integrated over the tunneling region, which produces

$$T \approx exp\left[-\frac{4W}{3}\sqrt{\frac{2mE_G}{\hbar^2}}\left(1 + \frac{3E_\perp}{2E_G}\right)\right], \tag{3.56}$$

where we have expanded the radical to lowest order and retained only the leading term in the transverse energy since it is considerably smaller than the band gap. It turns out that the result (3.56) is not sensitive to the actual details of the potential, since it is measuring the area under the potential curve. Different shapes give the same result if the areas are equal. Recognizing this assures us that the approximation (3.56) is probably as good as any other. We can rewrite (3.56) as

$$T \approx T_0 exp\left(-\frac{E_\perp}{E_0}\right), \quad E_0 = \frac{E_G}{2W}\sqrt{\frac{\hbar^2}{2mE_G}}. \tag{3.57}$$

(The leading term in the square brackets is independent of the transverse energy and goes into T_0.) This can now be used to find the current.

The current is determined using the form (3.28). We first will tackle the transverse energy integral. To lowest order, we note that the term involving the Fermi–Dirac functions is mainly a function of the longitudinal *z*-component of the energy, which we will show below, so the transverse terms are given by

$$\int_0^{E_F - eV_a} \frac{d^2k_\perp}{(2\pi)^2} exp\left(-\frac{E_\perp}{E_0}\right) = \frac{mE_0}{2\pi\hbar^2}\left[1 - exp\left(-\frac{E_{FL} - eV_a}{E_0}\right)\right]. \tag{3.58}$$

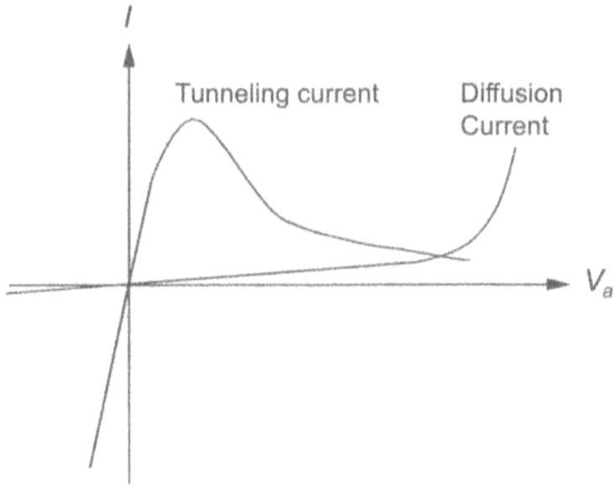

FIGURE 3.15 The tunneling current is an additional current to the normal diode current. The contribution of this current to the overall current of a tunnel diode is illustrated here.

The limits on the above integral are set by the fact that the transverse energy can only increase up to the value set by the Fermi energy on each side of the junction (measured from the band edges) reduced by the longitudinal energy.

The longitudinal contribution may be found by evaluating the energies in the Fermi–Dirac integrals, through shifting the energy on one side by the applied voltage eV_a. This leads to the result, in the linear-response limit, that

$$\begin{aligned}\left[f(E_z+E_\perp)-f(E_z+E_\perp+eV_a)\right] &= \frac{eV_a}{k_BT}f(1-f).\\ &\approx -eV_a\frac{\partial f}{\partial E_z}\approx eV_a\delta(E_z-E_{FL})\end{aligned} \tag{3.59}$$

The last approximation is for strongly degenerate material (or equivalently, very low temperature). Then the integration over E_z gives just eV_a times the tunneling probability T_0. We can now put (3.58) and (3.59) in the general equation to obtain the total current density

$$J_z = \frac{e^2T_0}{\pi\hbar}\left(\frac{mE_{FL}}{2\pi\hbar^2}\right)\left(1-\frac{eV_a}{E_{FL}}\right)V_a. \tag{3.60}$$

As we discussed at the beginning of this section, the current rises linearly with applied bias, but then decreases as the electron states on the right-hand side begin to overlap the forbidden states in the energy gap, which cuts off the current. We show the tunneling current in Figure 3.15, along with the normal *p–n* junction current due to injection and diffusion.

3.3.2 The Resonant-Tunneling Diode

In the above device, the Esaki diode, tunneling was of electrons to hole states, as the electron traveled from the bottom of the conduction band to the top of the valence band. With heterostructures, it is possible to make tunneling devices in which only electron states (or only hole states) are involved. One such device is the resonant-tunneling diode (RTD). This device is one in which a double barrier, such as that in Section 3.2, is inserted into, say, a conduction band, and the current through the structure is metered via the resonant level. The latter corresponds to the energy at which the transmission rises to a value near unity. The structure of such a system, in the GaAs/AlGaAs/GaAs/AlGaAs/GaAs system with the AlGaAs forming the barriers, is shown in Figure 3.16. Typically, the barriers are 3–5 nm thick and about 0.3 eV high, and the well is also 2–5 nm thick.

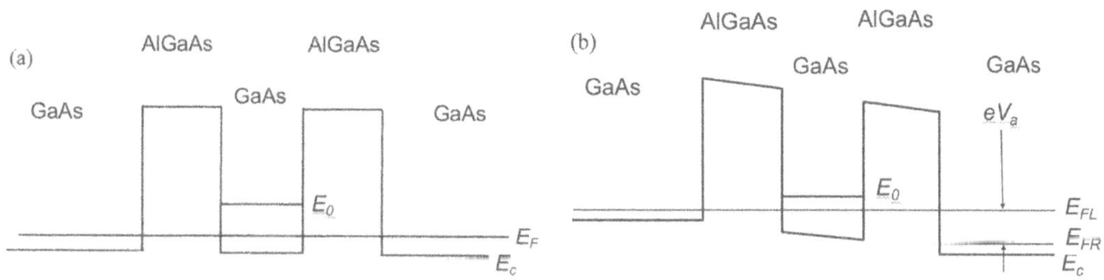

FIGURE 3.16 A typical double-barrier resonant-tunneling diode potential system, grown by heteroepitaxy in the GaAs–AlGaAs system. In (a), the basic structure is shown for an *n*-type GaAs well and cladding. In (b), the shape under bias is shown. Note that the bias has voltage dropped across both barriers as well as the central quantum well. The bias doesn't affect the two cladding layers as they are usually very heavily doped to reduce overall resistance.

To proceed, we will use the same approximations as used for the *p–n* junction diode, at least for the distribution function. The difference between the Fermi–Dirac distributions on the left-hand and right-hand sides, in the limit of very low temperature ($T \to 0$ K) gives

$$\left[f\left(E_z+E_\perp\right)-f\left(E_z+E_\perp+eV_a\right)\right]\sim eV_a\delta\left(E_z+E_\perp-E_{FL}\right). \tag{3.61}$$

We retain the transverse energy in this treatment, since we must be slightly more careful in the integrations in this model. The tunneling probability can also be taken as approximately a delta function, but with a finite width describing the nature of the actual line shape (an alternative is to use something like a Lorentzian line, but this does not change the physics). Thus, we write (we note that the transmission will be less than unity)

$$T\left(E\right)\cong E_W\delta\left(E_z+\frac{eV_a}{2}-E_0\right), \tag{3.62}$$

where we have assumed that the width of the transmission is E_W, and that the resonant level E_0 is shifted downward by an amount equal to half the bias voltage (everything is with reference to the Fermi energy on the left-hand side of the barrier, as indicated in the figure). Thus, the current can be written from (3.28) as

$$\begin{aligned}J&=\frac{e^2V_a}{\pi\hbar}\left(\frac{mE_W}{2\pi\hbar^2}\right)\int dE_z\int_0^{E_{FL}}\delta\left(E_z+E_\perp-E_{FL}\right)\delta\left(E_z+\frac{eV_a}{2}-E_0\right)dE_\perp\\&=\frac{e^2V_a}{\pi\hbar}\left(\frac{mE_W}{2\pi\hbar^2}\right)\int_0^{E_{FL}}\delta\left(E_z+E_\perp+\frac{eV_a}{2}-E_0\right)dE_\perp\\&=\frac{e^2V_a}{\pi\hbar}\left(\frac{mE_W}{2\pi\hbar^2}\right)\end{aligned}, \tag{3.63}$$

for $2(E_0-E_{FL})<eV_a<2E_0$. Outside the indicated range of applied bias, the current is basically zero, at least in the theoretical model. At finite temperature (or if a Lorentzian line shape for T is used), the current rises more smoothly and drops more smoothly. Essentially, the current begins to flow as soon as the resonant level E_0 is pulled down near to the Fermi energy on the left-hand side (positive bias is to the right), and current ceases to flow when the resonant level passes the bottom of the conduction band (widths of the order of k_BT should be seen on each side). This is shown in Figure 3.17.

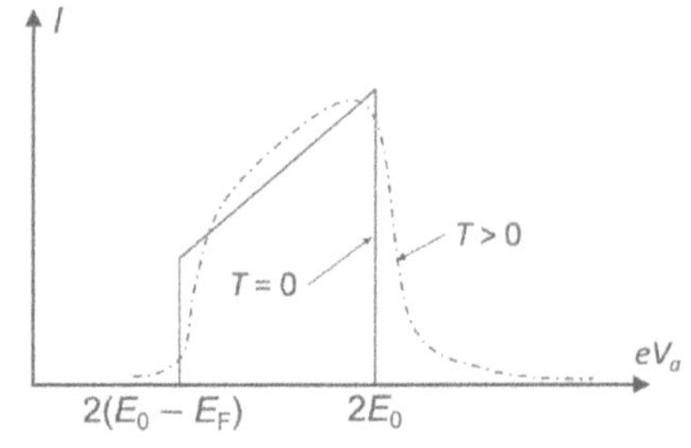

FIGURE 3.17 The theoretical curves for the simple model of (3.63) are shown for zero temperature and finite temperature.

In Figure 3.18, the results of experiments on RTDs grown on an InP substrate using InGaAs as the well and cladding layers (adjacent to the barriers) and AlAs for the barriers (Jacobs *et al.*, 2017). The quantum well was a 4.5 nm $In_{0.8}Ga_{0.2}As$ layer, and the barriers were 1.1 nm thick AlAs layers. The two cladding layers were undoped 20 nm layers of $In_{0.53}Ga_{0.47}As$, which is lattice matched to the InP substrate. Heavily doped (2×10^{19} cm^{-3}) $In_{0.53}Ga_{0.47}As$ layers were added to give low resistivity contacts. The results for several different devices, with varying areas, are shown in the figure, and the estimated cross-sectional areas are shown on the right of the curves. Clear negative differential conductance is seen in the figure. However, there is a plateau in the negative conductance regime in each curve. These plateaus are generally thought to be a result of the diodes, and their circuits, oscillating at high frequency (the oscillations are averaged out in a current–voltage measurement and give a plateau as an average value) (Young *et al.*, 1988). These oscillations involve both the negative conductance of the RTD and the circuit in which they are embedded.

3.3.3 Single-Electron Tunneling

In the above examples of tunneling devices, as well as in earlier sections, the result was always the current that arose from electrons (or holes) that tunnel through the barriers in the device. However, we have avoided asking the question, "Are there constraints on an electron (or hole) tunneling through a barrier?" That is, are there constraints other than the barrier itself which limits the transmission to an exponentially (often small) limiting function. The answer is, of course, yes. This is because we

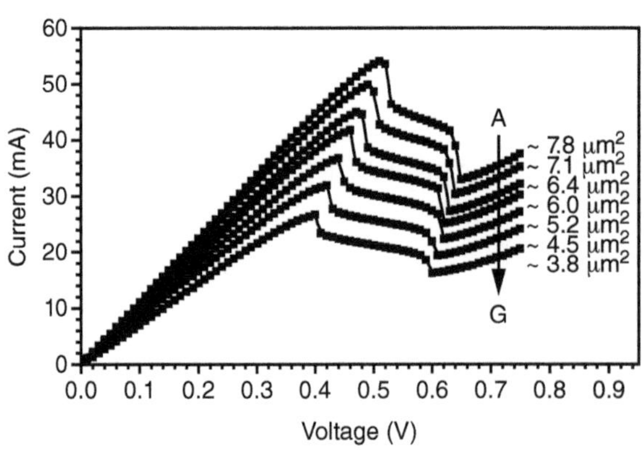

FIGURE 3.18 The I–V characteristics of single-etched RTDs measured in-line with the fabrication process. The area of the device for various etches A–G represents the total RTD area and is calculated from the peak current density of the material. (Reproduced from Jacobs *et al.* (2017) under the creative commons attribution license 4.0.)

have to answer the question about the amount of energy a particle needs in order to tunnel to a lower energy level (the other side of the barrier), such as illustrated in Figures 3.4 and 3.16(b).

As an example, consider tunneling through the insulator of a capacitor (in which we take the insulator to be an oxide such as SiO_2 that is found in MOS structures). The tunneling through the capacitor oxide is an example of a very simple physical system that can exhibit quite complicated behavior when it is made very small. The capacitor is formed by placing the insulator between two metals, or between a metal (gate) and a doped semiconductor in an MOS structure. What if the area of the capacitor is made very small, so that the capacitance is also quite small? It turns out that this can affect the operation of tunneling through the oxide significantly. When an electron tunnels through the oxide, it lowers the energy stored in the capacitor by the amount

$$\delta E = \frac{e^2}{2C}. \tag{3.64}$$

This energy change is coupled to a voltage change of

$$\delta V = \frac{e}{C}. \tag{3.65}$$

What this means is that the tunneling of the electron (or hole), at very low temperatures ($k_B T/e \ll \delta V$) cannot occur until a voltage equivalent to (3.65) is actually applied across the capacitor. If the voltage on the capacitor is less than this, no tunneling occurs because there is not sufficient energy stored in the capacitor to support the tunneling transition. When the capacitance is large, say $>10^{-12}$ F, this voltage is immeasurably small in comparison with the thermally induced voltages ($k_B T/e$). On the other hand, suppose that the capacitance is defined with a lateral dimension of only 50 nm. This area is 2.5×10^{-15} m^2, and the capacitor has a capacitance of 2.8×10^{-17} F. The required voltage of (3.65) is 5.7 mV. These capacitors are easily made, and the effects easily measured at low temperatures.

Some of the earliest experiments were made by Fulton and Dolan (1987) on such small-capacitor structures. These measurements showed the lack of tunneling until a sufficiently large voltage was applied. The blockage of current until a sufficiently large voltage is applied is known as the *Coulomb blockade*. As an example of this, in Figure 3.19, measurements of Pd-ZrO_2 granular nanoparticles deposited on an interdigitated array of gold lines (Bakkali and Dominguez, 2025). The nanoparticles are contacted by a conducting atomic force microscope operated in contact mode (the tip is in contact with a nanoparticle and measures the current to a nearby gold line). The figure shows the current for measurements on 100 nanoparticles, and the black line is the average current.

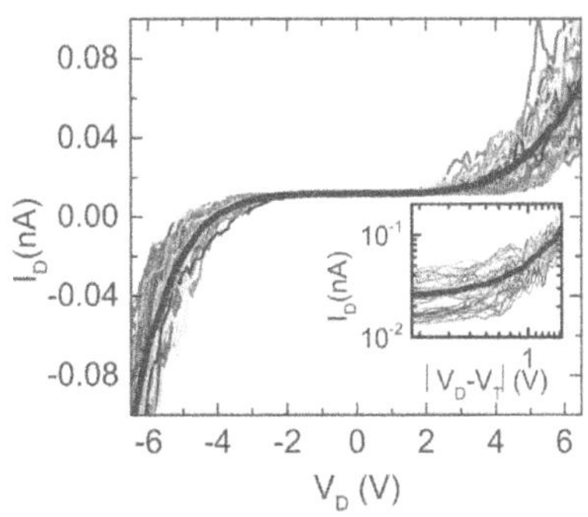

FIGURE 3.19 I–V characteristic at room temperature for group of samples with V_T values of 3.38 V. Data from 100 spectra taken in a 10 × 10 pixel area with a resolution of 27.34 nm per pixel. (Reproduced from Bakkali and Dominguez (2025) under the creative commons attribution license 4.0.)

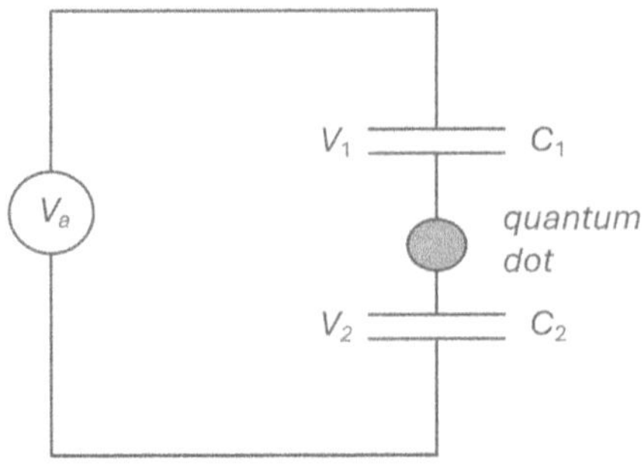

FIGURE 3.20 The two capacitors equivalent to a double-barrier tunneling structure. The tunneling barrier here is the insulator in each capacitor. The quantum dot shown is the equivalent of the quantum well in the double-barrier structure. For the single-electron case, we work with the circuit shown here rather than the quantum wave picture.

Generally, the current for Coulomb blockade follows a form $I \sim (V - V_T)^{\xi}$, where V_T is the turn-on voltage, often called the blockade voltage, and ξ is a parameter, expected to be unity in the simplest theory for one dimensional current flow. In Figure 3.19, $V_T \sim 3.38V$. The Coulomb blockade causes the offset of the current to the V_T in the small-capacitor case, as may be seen in the figure. This offset scales with area, and hence with C as expected in (3.65). Variations in granular size, and noise, give the differences in current above V_T.

To understand the Coulomb blockade somewhat more quantitatively, consider the double-barrier structure shown in Figure 3.20, as it may be created with extremely small lateral dimensions so that the capacitance of each barrier satisfies (3.64) as required for Coulomb blockade. This also prevents thermal fluctuations to induce electrons to cross the barriers. The circuit shown is the equivalent of the double-barrier structure of Figure 3.16, and in this section we will work with the circuit parameters rather than the wave parameters. The region between the two capacitors may be termed a *quantum dot*. In general, each of the two capacitors (barriers) will have some leakage. The leakage is not shown but can be considered by the addition of a shunt resistance in parallel with each capacitor. This leakage, or tunneling resistance, represents the current flow when the electrons tunnel through the barrier. In this sense, these resistances are different from ordinary resistances. In the latter, charge and current flow is essentially continuous. Here, however, the current flow only occurs when the electrons are tunneling, and this is for a very short time. The charge on the two capacitors is related to the two voltages, and is given by

$$\begin{aligned} Q_1 &= C_1 V_1, \\ Q_2 &= C_2 V_2. \end{aligned} \tag{3.66}$$

The net charge on the quantum dot island is given by

$$Q_{dot} = Q_2 - Q_1 = -ne, \tag{3.67}$$

where n is the net number of electrons on the island. The sum of the junction voltages is the applied voltage V_a, so that combining the two equations above, the voltage drops across the two junctions are just

$$\begin{aligned} V_1 &= \frac{1}{C_1 + C_2}\left(C_2 V_a + ne\right), \\ V_2 &= \frac{1}{C_1 + C_2}\left(C_1 V_a - ne\right). \end{aligned} \tag{3.68}$$

The electrostatic energy stored in the two capacitors, and the quantum dot, can be written as

$$E_s = \frac{Q_1^2}{2C_1} + \frac{Q_2^2}{2C_2} = \frac{1}{2\left(C_1 + C_2\right)}\left(C_1C_2V_a^2 + Q_{dot}^2\right). \tag{3.69}$$

In addition, when one wants to talk about switching voltage and/or charge on capacitors, the energy contributed by the power supply must also be considered, because any change in (3.69) must be provided by the response of this power supply. This is an integral over the power delivered to the tunnel junctions during the charging process, or

$$E_a = \int dtV_a\left(t\right) = V_a\Delta Q, \tag{3.70}$$

where ΔQ is the total charge transferred from the voltage source to the capacitors. This includes the integer number of electrons on the quantum dot as well as the continuous polarization charge that builds up in response to the voltages on the capacitors (resulting in an electric field in the island). A change in the charge on the island due to one electron tunneling through C_2 changes the charge on the island to $Q' = Q - e$, and $n' = n - 1$. From (3.69), the voltage change on C_1 results in $V_1' - V_1 - e/C_{eq}$, where $C_{eq} = C_1 + C_2$. Therefore, a polarization charge flows in from the voltage source $\Delta Q = -eC_1/C_{eq}$ to compensate. The total work done to pass n_2 charges through C_2 is then

$$E_a\left(n_2\right) = -n_2eV_a\left(\frac{C_1}{C_{eq}}\right). \tag{3.71}$$

A similar argument is used to generate the total work done to pass n_1 charges through C_1, which is

$$E_a\left(n_1\right) = -n_1eV_a\left(\frac{C_2}{C_{eq}}\right). \tag{3.72}$$

Combining the two results (3.71) and (3.72) with (3.69), the total energy of the complete circuit, including that provided by the voltage source, is given by

$$E\left(n_1,n_2\right) = E_s - E_a = \frac{1}{2C_{eq}}\left(C_1C_2V_a^2 + Q_{dot}^2\right) + \frac{eV_a}{C_{eq}}\left(C_1n_2 + C_2n_1\right). \tag{3.73}$$

With this description of the total energy in terms of the charge on each of the capacitors (given by the factors n_i), we can now look at the *change* in the energy when a particle tunnels through either capacitor. At low temperature, the tunneling transition must take the system from a state of higher energy to a state of lower energy. The change in energy for a particle tunneling through C_2 is found from (3.73) by changing n_2 as

$$\begin{aligned} \Delta E_2^{\pm} &= E\left(n_1,n_2\right) - E\left(n_1,n_2 \pm 1\right) \\ &= \frac{e}{C_{eq}}\left[-\frac{e}{2} \pm \left(Q_{dot} - V_aC_1\right)\right]. \end{aligned} \tag{3.74}$$

The value of the charge on the dot Q_{dot} is *prior* to the tunneling process. Similarly, the change in energy for a particle tunneling through C_1 is given by

$$\begin{aligned} \Delta E_1^{\pm} &= E\left(n_1,n_2\right) - E\left(n_1 \pm 1,n_2\right) \\ &= \frac{e}{C_{eq}}\left[-\frac{e}{2} \mp \left(Q_{dot} - V_aC_2\right)\right]. \end{aligned} \tag{3.75}$$

According to this discussion, only those transitions are allowed for which $\Delta E_i > 0$; e.g., the initial state is at a higher energy than the final state.

If we consider a system in which the dot is initially uncharged, $Q_{\text{dot}} = 0$, then (3.74) and (3.75) reduce to

$$\Delta E_{1,2}^{\pm} = -\frac{e^2}{2C_{\text{eq}}} \pm \frac{eV_a C_{2,1}}{C_{\text{eq}}} > 0. \tag{3.76}$$

Initially, at low voltage, the leading term on the right-hand side makes $\Delta E < 0$. Hence, no tunneling can occur until the voltage reaches a threshold that depends upon the lesser of the two capacitors. For the case in which $C_1 = C_2 = C$, the requirement becomes $|V_{\text{a}}| > e/C_{\text{eq}}$. Tunneling is prohibited and no current flows below this threshold voltage. This region of *Coulomb blockade* is a direct result of the additional Coulomb energy, which is required for an electron to tunnel through one of the capacitors. This leads to the offset shown in Figure 3.19 as $V_T \sim e/C_{\text{eq}}$.

Now, consider the situation when an additional contact is added to the quantum dot, through another capacitor, as shown in Figure 3.21. A new voltage source, V_G, is coupled to the quantum dot through an *ideal* capacitor C_G (which is large enough that no tunneling is possible through this capacitor). This additional voltage source modifies the charge balance on the quantum dot, so that

$$Q_G = C_G\left(V_G - V_2\right), \tag{3.77}$$

and the charge on the dot is modified to be

$$Q_{dot} = Q_2 - Q_1 - Q_G = -ne. \tag{3.78}$$

This now creates a three-terminal single-electron device, which is often termed a *single-electron transistor.*

To understand the operation of this new device, it is necessary to rewrite the two voltages to include this new capacitor and voltage source, as

$$\begin{aligned} V_1 &= \frac{1}{C_{\text{T}}}\left[\left(C_G + C_2\right)V_a - C_G V_2 + ne\right], \\ V_2 &= \frac{1}{C_{\text{T}}}\left(C_1 V_a + C_G V_2 - ne\right), \end{aligned} \tag{3.79}$$

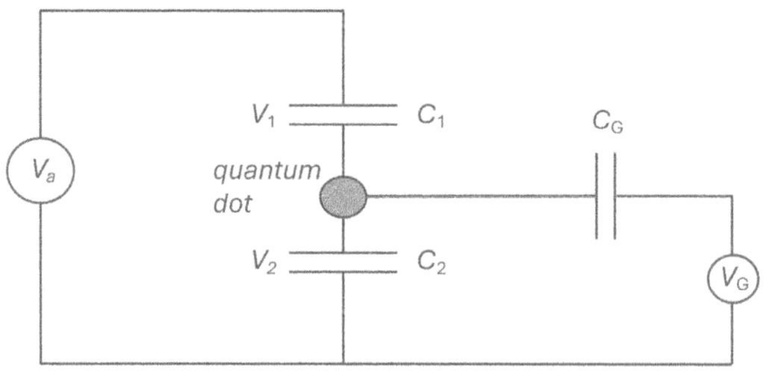

FIGURE 3.21 Adding an additional capacitor, C_G, and a gate voltage, V_G, creates a single-electron transistor. These additions create a circuit such that a gate bias is applied to directly vary the potential at the dot.

where now $C_T = C_{eq} + C_G$. The electrostatic energy (3.69) is also affected and now becomes

$$E_s = \frac{1}{2C_T}\left(C_1C_2V_a^2 + C_GC_2V_G^2 + C_GC_1\left(V_a - V_G\right)^2 + Q_{dot}^2\right). \tag{3.80}$$

The work done by the voltage sources during the tunneling includes the work done by the gate voltage and the additional charge flowing onto the gate capacitor. Equations (3.71) and (3.72) become

$$\begin{aligned} E_a\left(n_1\right) &= -n_1\left[\frac{eV_aC_2}{C_T} + e\left(V_a - V_G\right)\frac{C_G}{C_T}\right], \\ E_a\left(n_2\right) &= -n_2\left[\frac{eV_aC_1}{C_T} + \frac{eV_GC_G}{C_T}\right]. \end{aligned} \tag{3.81}$$

With these equations we can now rewrite the total energy (3.73) as

$$\begin{aligned} E\left(n_1, n_2\right) = \frac{1}{2C_T}\Big[&\left(C_1C_2V_a^2 + Q_{dot}^2\right) + C_GC_2V_G^2 + C_GC_1\left(V_a - V_G\right)^2 \\ &+ 2eV_a\left(C_1n_2 + \left(C_2 + C_G\right)n_1\right) - 2n_1C_GV_G\Big]. \end{aligned} \tag{3.82}$$

For the tunneling of an electron either C_1 or C_2, the energy changes are now given by

$$\begin{aligned} \Delta E_1^{\pm} &= E\left(n_1, n_2\right) - E\left(n_1 \pm 1, n_2\right). \\ &= \frac{e}{C_T}\left[-\frac{e}{2} \mp \left(Q_{dot} - V_a\left(C_2 + C_G\right) - C_GV_G\right)\right], \\ \Delta E_2^{\pm} &= E\left(n_1, n_2\right) - E\left(n_1, n_2 \pm 1\right) \\ &= \frac{e}{C_T}\left[-\frac{e}{2} \pm \left(Q_{dot} - V_aC_1 - C_GV_G\right)\right]. \end{aligned} \tag{3.83}$$

When we compare these results with those of (3.74) and (3.75), it is apparent that the gate voltage allows us to change the effective charge on the quantum dot, and therefore to shift the region of Coulomb blockade with V_G. As before, the condition for tunneling at low temperature is that the change in energy must be negative and the tunneling must take the system to a lower energy state. We now have two conditions that exist for forward and backward tunneling as

$$\begin{aligned} -\frac{e}{2} \mp \left(ne - V_a\left(C_2 + C_G\right) - C_GV_G\right) > 0, \\ -\frac{e}{2} \pm \left(ne - V_aC_1 - C_GV_G\right) > 0. \end{aligned} \tag{3.84}$$

The four equations (3.84) may be used to generate a stability plot in the (V_a, V_G) plane, which shows stable regions corresponding to each value of n, and for which no tunneling can occur. This diagram is shown in Figure 3.22 for the case in which $C_G = C_2 = C$ and $C_1 = 2C$, so that $C_T = 4C$. The lines represent the boundaries given by (3.84). The trapezoidal shaded areas correspond to regions where no solution satisfies (3.84) and hence where Coulomb blockade exists. Each of the regions corresponds to a different number of electrons on the quantum dot, which is stable in the sense that this charge state does not change easily. The gate voltage allows us to "tune" between different stable regimes, essentially adding or subtracting one electron at a time to the dot region. For a given

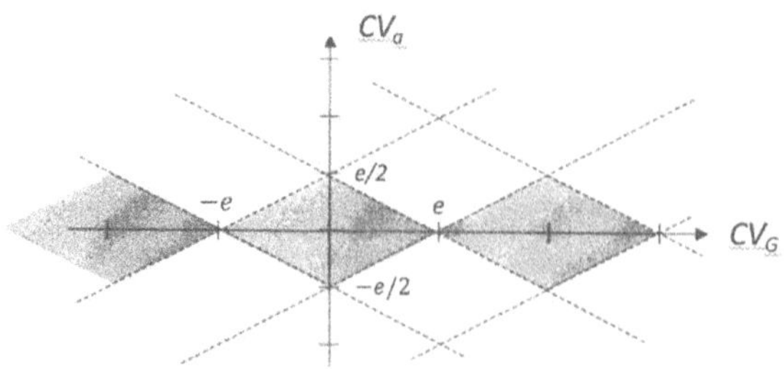

FIGURE 3.22 A stability diagram for the quantum dot in the single-electron transistor. The parameters are discussed in the text, and it is assumed that the temperature is $T = 0$. The shaded regions are where the dot electron number is stable and Coulomb blockade exists.

gate bias, the range of V_a over which Coulomb blockade occurs is given by the vertical extent of the shaded region. The width of this blockaded region approaches zero as the gate charge approaches half-integer values of a single-electron charge, and here tunneling occurs easily. It is at these gate voltages that tunneling current peaks can be observed as the gate bias is varied.

In Figure 3.22, the distance between the current peaks for tunneling through the structure is $\Delta V_G = e/C = e/C_G$. Between these peaks the number of electrons on the quantum dot remains constant. We will return to the quantum dot in Chapter 9, where we begin to worry about the quantum levels within the dot itself. To this point, we have ignored the fact that the quantum dot may be so small that the energy levels within the dot are quantized. Yet, this can occur, and the behavior of this section is modified. As remarked, we return to this in Chapter 9, where we deal with spectroscopy of just these quantum levels, and again in Chapter 11, where quantum dots appear in many forms as possible qubits for quantum computing.

3.3.4 Josephson Tunneling

The Josephson junction (1962) is a tunnel junction in which the materials on either side of the insulator are superconductors. Superconductors will be discussed in Chapter 4, but it is enough to know that these are generally low-conductivity metals whose resistivity goes to zero at low temperature, typically a few, or a few tens of, degrees Kelvin. This results from pairing of electrons with opposite spin (again to be discussed in Chapter 4) into a composite particle, the Cooper pair (Cooper, 1956), that can move through the lattice with no dissipation. The Josephson junction is a tunneling junction placed between two superconducting wires (that each have no resistivity) at low temperature. This tunneling produces new (and strange) results.

These superconducting leads produce zero loss in d.c. operation. The formation of the Cooper pairs lowers the energy of the entire electron gas below the Fermi energy, and this leads to a gap opening at the Fermi energy—the gap lies between the full states of Cooper pairs (at T = 0 K) and the empty single-electron states. This gap is 2Δ, where Δ is the condensation energy per electron for the pair, and is referred to as the superconducting gap.

Normally, with Josephson junctions, the tunneling is carried out by these Cooper pairs, but both the Cooper pair tunneling and single-electron tunneling can occur under the appropriate conditions. The Josephson junction itself operates as shown in Figure 3.23(a). As can be seen, no tunneling occurs until $eV_a > 2\Delta$ at low temperature. As the temperature increases, both unpaired single electrons and Cooper pairs exist below the gap, and the normal (unpaired) electron can tunnel giving a very small current.

However, these Cooper pairs are quantum mechanical objects, with extended wave functions. As with normal tunneling, if the barrier is relatively thin, say 1–2 nm, then the superconducting wave functions can extend through the barrier, so that they interact with those wave functions on the other side, in a coherent manner meaning that there is a connection between the phases of the wave functions on either side of the junction (just as in normal tunneling). An unusual effect in the Josephson junction is that this coherent mixing of the wave functions on either side of the junction can produce a current at zero bias! This is termed the d.c. Josephson current. This current has a fixed magnitude that can be modulated by a magnetic field as

$$I_J(B) = I_{J0} cos\left(\frac{eBA}{h}\right), \tag{3.85}$$

where $BA = \Phi$ is the flux flowing through the area of the junction. (One will note this phase relationship is basically the same as that of the Aharonov–Bohm effect in Chapter 1, and is a result of factors depending explicitly upon the flux enclosed.) As before, h/e is the quantum unit of flux, so that the flux can be quantized in a quantum system. The interesting aspect is that the current peak I_{J0} is proportional to the single-particle tunneling coefficient through the junction (Kuper, 1968).

We note that if a bias is applied, nothing much happens to the normal curve of Figure 3.23(a), except a few unpaired electrons may tunnel through. This may be seen in the other three panels of Figure 3.23. Panel (b) shows the band line-up at zero bias, while panel (c) has a small bias. No electrons can tunnel because of the energy gap. Once the bias is above 2Δ, as shown in panel (c), Cooper pairs and single electrons can tunnel into the empty upper band.

This voltage has a significant effect on the coherent flux of the Josephson current (3.85). This follows from the relationship

$$\hbar\frac{d\Phi}{dt} = 2eV_a, \tag{3.86}$$

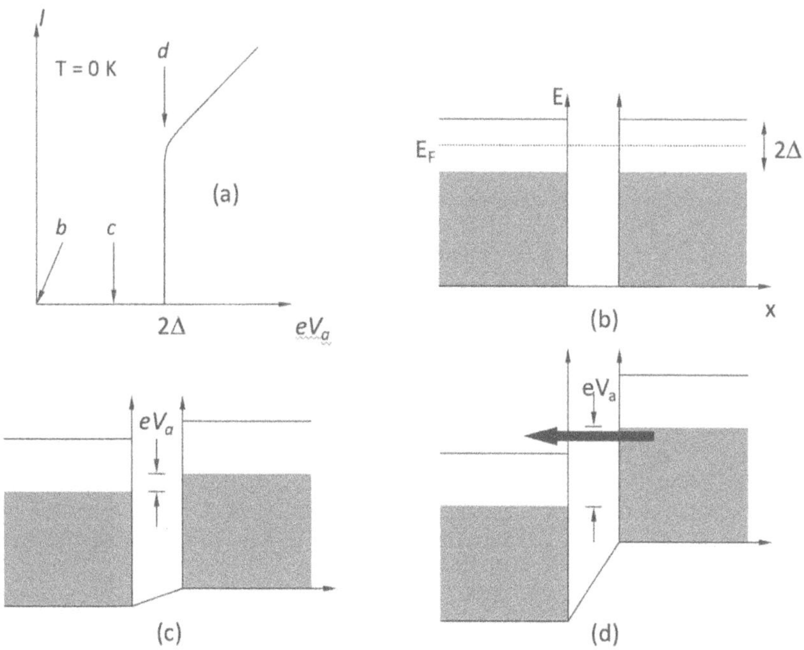

FIGURE 3.23 (a) The current–voltage curve of the Josephson junction at T = 0 K. The letters refer to the other pictures in the figure. (b) At zero bias, no tunneling can occur because the full Cooper pair states see only full Cooper pair states across the junction. (c) At small bias, the full states see only forbidden states in the energy on the other side of the junction, and no current can flow. (d) When $eV_a > 2\Delta$, Cooper pairs can tunnel, but the pair breaks up into two free electrons above the gap. Coherent Josephson pair tunneling at zero bias is not shown.

where the factor of 2 arises from the two electrons in the Cooper pair. Now, this means that the associated flux is changed by the voltage and this produces an a.c. signal, known as the a.c. Josephson effect, and leads to the important relation

$$\hbar\omega = 2eV_a. \tag{3.87}$$

Hence, the Josephson junction can be a microwave source. It has also been noticed that there is a peak in the spectrum at $\omega = 4\Delta/\hbar$, which is called a Riedel peak in the response of the junction (Riedel, 1964). We return to this device in Chapter 4 and again in Chapter 11, where the Josephson junction is a prime device in today's quantum computing.

NOTES

1 In fact, one can find several illustrative videos at YouTube.com which illustrate this classical optical tunneling effect.
2 Here, the surface atoms each have two dangling bonds and these tend to attach to neighboring Si atoms to form the dimers.

REFERENCES

Bakkali, H. and Dominguez, M. 2025. Unveiling single-electron tunneling in nanogranular ultra-thin films for sensor applications via hyperspectral conductive AFM. *Mat. Lett.* **382** 137932.

Binnig, K., Rohrer, H., Gerber, C. and Weibel, E. 1982a. Tunneling through a controllable vacuum gap. *Appl. Phys. Lett.* **40** 178–180.

Binnig, K., Rohrer, H., Gerber, C. and Weibel, E. 1982b. Surface studies by scanning tunneling microscopy. *Phys. Rev. Lett.* **49** 57–61.

Cooper, L. N. 1956, Bound electron pairs in a degenerate fermi gas. *Phys. Rev.* **104** 1189–1190.

Duke, C. B. 1969. *Tunneling in Solids.* (New York: Academic Press).

Esaki, L. 1958. New phenomena in narrow germanium p-n junctions. *Phys. Rev.* **109** 603–604.

Ferry, D. K. 2026. *The Physics and Technology of Semiconductor Devices'.* (London: IET Publishing).

Fowler, R. H. and Nordheim, L. 1928. Electron emission in intense electric fields. *Proc. Roy. Soc. London A.* **119** 173–181.

Fulton, T. A. and Dolan, G. J. 1987. Observation of single-electron charging effects in small tunnel junctions. *Phys. Rev. Lett.* **59** 109–112.

Jacobs, K. J. P., Stevens, B. J., Baba, R. *et al.* 2017. Valley current characterization of high current density resonant tunneling diodes for terahertz-wave applications. *AIP Adv.* **7** 105316.

Josephson, B. D. 1962. Possible new effect in superconductive tunneling. *Physics Lett.* **1** 251–253.

Knoll, M. and Ruska, E. 1932a. Beitrag zur Geometrischen Elektroneoptik. I. *Ann. Phys.* **404** 607–640.

Knoll, M. and Ruska, E. 1932b. Das Elektronenmikroskop. *Zeit. Phys.* **78** 318–339.

Kuper, C. G. 1968. *An Introduction to the Theory of Superconductivity* (Oxford: Clarendon Press) Sec. 8–2.

Labidi, H., Taucer, M., Rashidi, M., Koleini, M. *et al.* 2015. Scaning Tunneling Spectroscopy Reveals a Silicon Dangling Bond Charge State Transition. *New J. Phys.* **17** 073023.

Price, P. J. 1999. Quasi-bound states and resonances in heterostructures. *Microelectron. J.* **30** 925–934.

Riedel, E. 1964. Zum Tunneleffeckt bei Supraleitern im Mikrowellenfeld. *Z. Naturforsch.* **19a** 1634–1635.

Yacoby, A., Heiblum, M., Mahalu, D., and Shtrikman, H. 1995. Coherence and phase sensitive measurements in a quantum dot. *Phys. Rev. Lett.* **74** 4047–4050.

Young, J. F., Wood, B. M., Liu, H. C., *et al.* 1988. Effect of circuit oscillations on the DC current-voltage characteristics of double barrier resonant tunneling structures. *Appl. Phys. Lett.* **52** 1398–1400.

PROBLEMS

1. Consider a single rectangular barrier in which the barrier height is 2.5 eV. A free electron of energy 0.5 eV is incident from the left. If the barrier is 0.2 nm thick, what are the tunneling and reflection coefficients?

2. For a potential barrier with $V(x) = 0$ for $x > |a/2|$, and $V(x) = 0.3$ eV for $x < |a/2|$, plot the tunneling probability for E in the range 0–0.5 eV. Take the value $a = 5$ nm and use the effective mass of GaAs, $m^* = 6.0 \times 10^{-32}$ kg.
3. For a potential barrier with $V(x) = 0$ for $x > |a/2|$, and $V(x) = 0.4$ eV for $x < |a/2|$, plot the tunneling probability for E in the range 0–0.5 eV. Take the value $a = 5$ nm and use the effective mass of GaAs, $m^* = 6.0 \times 10^{-32}$ kg.
4. Consider the potential barrier discussed in problem 1. Suppose that there are two of these barriers forming a double-barrier structure. If they are separated by 4 nm, what are the resonant energy levels in the well? Compute the tunneling probability for transmission through the entire structure over the energy range 0–0.5 eV.
5. Suppose that we create a double-barrier resonant-tunneling structure by combining the barriers of problems 1 and 2. Let the barrier with $V_0 = 0.3$ eV be on the left, and the barrier with $V_0 = 0.4$ eV be on the right, with the two barriers separated by a well of 4 nm width. What are the resonant energies in the well? Compute the tunneling probability through the entire structure over the energy range 0–0.5 eV. At an energy of 0.25 eV, compare the tunneling coefficient with the ratio of the tunneling coefficients (at this energy) for the barrier of problem 2 over that of problem 1 (i.e., the ratio T_{min}/T_{max}).
6. Calculate the transmission coefficient of the double barrier structure, using a tool at NanoHub.org, for the following three cases:
   - (a) $V1 = V2 = 0.4$ eV, $L1 = L2 = 4$ nm, $W = 5$ nm.
   - (b) $V1 = 0.6$ eV, $V2 = 0.4$ eV, $L1 = L2 = 4$ nm, $W = 5$ nm.
   - (c) $V1 = 0.6$ eV, $V2 = 0.3$ eV, $L1 = 4$ nm, $L2 = 5$ nm, $W = 5$ nm.
   - Notation: Vi is the height of the barriers, Li describes the width of the barriers ($i = 1$ for barrier 1, and $i = 2$ for the second barrier) and W is the well width. Assume that the potential energy term $V(x)$ is zero in the source, well and detector regions. Assume that the effective electron mass equals to $m = 6 \times 10\text{-}32$ kg.
7. In the WKB approximation, show that the tunneling probability for a double barrier (well of width b, barriers of width $2a$, as shown in Figure 3.5, and a height of each barrier of V_0) is given by

$$T = \frac{4}{\left(4\theta^2 + 1/\left(4\theta^2\right)\right)\cos^2 L + 4\sin^2 L}$$

where

$$\theta = \exp\left(\int_b^{b+2a} \gamma(x)\mathrm{d}x\right)$$

and

$$L = \int_0^b k(x)\mathrm{d}x.$$

What value must b have so that only a single resonant level exists in the well?
8. Consider a potential barrier which is described by $V(x) = -Ax^2$. Using the WKB method, compute the tunneling coefficient as a function of energy for $E < 0$.

4 Periodic Potentials

The atoms in nearly all carbon compounds and nearly all semiconductors are covalently bonded. Covalent bonding involves the sharing of electrons among different atoms. In the hydrogen molecule (H_2), for example, the covalent bond contains two electrons. One electron is contributed from each atom of the molecule. Each of these two electrons moves back and forth between the two atoms (this is what the chemists call the resonating bond), and on average, is situated between the atoms. In this special case, the two electrons come from the $1s$ level of the atoms. When we move into solids, in particular crystalline solids, the nature of the sharing becomes a little bit more complicated. This is because the potential energy in which the atom sits becomes periodic throughout the lattice. That is, in crystalline solids, the atoms are arranged on a periodic lattice, and thus the atomic potentials are also periodic on this lattice. If we focus on an atom in say Si at point A and examine the potential landscape in which this atom sits, it will be exactly the same as for any other atom in the crystal. When we decide to solve the Schrödinger equation for this system, we now need to involve a very great number of atoms and electrons in the solution. One can easily understand that this is an intractable problem. But we make some approximations through which the problem can be attacked in a manner that is sufficient to gather some results. The most important one is the adiabatic approximation, in which we realize that the atoms are heavy while the electrons are light. Thus, we can conceive that the atom moves around rather slowly, while the electrons are zipping around with a greater speed. For all practical purposes, then, we can make the assumption that the atoms are frozen in their atomic positions when we solve for the electron quantum mechanics. Similarly, we may assume that the electrons instantaneously respond to the atomic motion (infinitely fast) when we solve the latter quantum mechanics. Here, we will not deal with the motion of the atoms themselves, but deal with the electrons and the periodic potential that they see. There is, of course, one other part, and that is the interaction between an atom and the electrons that are nearby. This interaction is typically handled with perturbation theory, which will be treated in Chapters 7 and 8.

When we say that a potential is periodic, we mean that for any two atoms, separated by a distance a in one dimension, the potential will have this periodicity so that the potential satisfies

$$V(x+a)=V(x), \tag{4.1}$$

where a is the spacing between atoms and is said to be the *lattice constant*. If we consider the two atoms mentioned, one at x and one at $x+a$, then the time-independent Schrödinger equations for these two atoms can be written as

$$\begin{aligned} -\frac{\hbar^2}{2m}\nabla^2\psi(x)+V(x)\psi(x)=E\psi(x), \\ -\frac{\hbar^2}{2m}\nabla^2\psi(x+a)+V(x+a)\psi(x+a)=E\psi(x+a). \end{aligned} \tag{4.2}$$

Now, let us put (4.1) into the second line of (4.2), so that it becomes

$$-\frac{\hbar^2}{2m}\nabla^2\psi(x+a)+V(x)\psi(x+a)=E\psi(x+a). \tag{4.3}$$

DOI: 10.1201/9781003717195-4

Since one cannot really tell which atom is which in such a crystalline solid, there is a requirement that $|\psi(x)|^2$ must be the same at each and every atom, since it represents the probability for an electron to sit at that atom. This means that

$$|\psi(x)|^2 = |\psi(x+a)|^2. \tag{4.4}$$

Since the squared magnitude of the two wave functions must be the same, the two functions themselves can differ only by a quantity whose squared magnitude is unity, that is, by only a phase factor. Thus, we may say that, at most,

$$\psi(x+a) = e^{i\varphi}\psi(x). \tag{4.5}$$

How do we determine the above phase factor? Are there other constraints that the periodic potential imposes upon the wave function? These two questions are at the heart of the matter, and the heart of this chapter is to address these questions. The answers are the basic tenets of what is called condensed matter physics, and this includes the properties of semiconductors that are important to the electronic world of semiconductor devices. We will begin with the phase factor by examining periodicity on at least two levels.

4.1 ATOMS ON A LATTICE

Assume a linear chain of atoms that are uniformly spaced by a lattice constant a, as shown in Figure 4.1. (This figure can be compared to Figure 2.15 and the discussion of Section 2.7.) Below this chain of atoms, the periodic potential is plotted as a guide to the reader to understand how each atom sits in a potential minimum. At this point, the atoms are stable, although of course they have thermal vibrations around each minimum. But those are not our concern here. The atoms are assumed to sit at their stable minimum. Each atom is assumed to have a single electron, which naturally orbits the atom (but interacts with the electrons on the neighboring atoms). As we move from one atom to the next in the chain, the wave function for the electron at the next atom is related to the wave function at the current atom by (4.5). Hence, the periodicity of the potential has introduced one level of periodicity into the problem. But we now have to face the problem of boundary conditions. We handle this by introducing the concept of periodic boundary conditions. With periodic boundary conditions, atom N in the figure folds back onto atom 0, so that these are the same atom, as shown in Figure 4.2. That is, we wrap the linear array of atoms into a ring structure, so that the last becomes the first (atom N is also atom 0, so that there are only N atoms in the chain). This means that the wave function must adopt a phase in such a way that

$$\psi(x+Na) = \psi(x+L) = \psi(x). \tag{4.6}$$

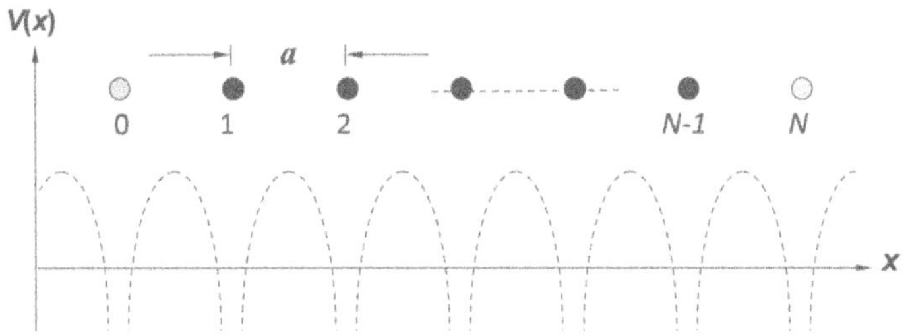

FIGURE 4.1 A one-dimensional chain of atoms and the corresponding periodic potential (dashed) for this system. For the arguments in the text, it has been assumed that there are N atoms in the chain, and that there is an atom 0. Atoms 0 and N have been shaded in order to highlight the periodicity.

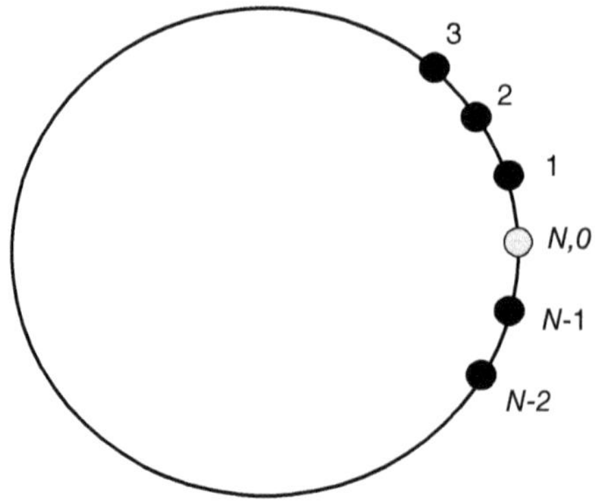

FIGURE 4.2 The linear chain of atoms, shown in Figure 4.1, is here circled upon itself so that atom N is now also atom 0. This illustrates the imposition of periodic boundary conditions. With the N atoms, the number of each atom is arbitrary.

If (4.5) and (4.6) are combined, the result leads to

$$N\varphi = 2\pi \rightarrow \varphi = \frac{2\pi}{N}. \tag{4.7}$$

The value given on the far right is the smallest value that the phase can have. If one moves two atoms, the total phase change is twice this minimum. Hence, the phase can take on the values

$$\varphi = \frac{2\pi n}{N}, \quad n = 1,2,\ldots,N. \tag{4.8}$$

We cannot allow n to go past N, because this begins to produce repetitions of each phase value. So, the smallest allowed value for φ is $2\pi/N$, and there are N independent values. (Note that we don't need a value of 0, because $n = N$ produces 2π, which is the equivalent of 0 phase.)

In (4.6), another notation was introduced, with $L = Na$. Let us replace N with the corresponding value given by this relation, so that (4.8) becomes

$$\varphi = \frac{2\pi n}{L}a, \; n = 1, 2, \ldots, N. \tag{4.9}$$

The units on the prefactor $2\pi n/L$ are inverse distance (1/cm). In electromagnetics, this inverse distance is called the wave number k. This is the same k that was encountered in the previous chapters when dealing with light and particle waves. This is the wave number of the electron waves, and it is the electron waves that are of interest in the present discussion. Just as in Chapter 2, the periodic potential allows just a fixed number of values for this wave number. If there are N atoms in the chain, there can be only N distinct values for k.

Now let us adopt an index j, which designates one of the atoms in the chain; which atom it refers to doesn't really matter, as this index just indicates an atom and its two neighbors. This is going to lead to a shorthand notation in which

$$|j\rangle = \psi(x_j), \tag{4.10}$$

so that the so-called ket term on the left, with the location subscript as its value, represents the wave function at that indicated position. This is often called "Dirac notation." The adjoint of this wave function is given the "bra" description

$$\langle j| = \psi^*(x_j). \tag{4.11}$$

The basis set for the future expansion of the wave function is one in which each electron's wave function is localized upon a single atom, so that orthonormality for electrons on different atoms appears as

$$\langle i|j\rangle = \int dx\psi^*(x_i)\psi(x_j) = \delta_{ij} = \delta(x_i - x_j). \tag{4.12}$$

(Here, the bra and ket together give a "bracket" which denotes the expectation value for the normalization of the wave function.) In the absence of the chain, the wave function for an individual atom is highly localized upon that atom. Since each atom and its electron are the same at each site in the chain, they have the same electronic energy. Hence, in the Hamiltonian for the system, one may write

$$H_0|j\rangle = H_0\psi(x_j) = E_0|j\rangle = E_0\psi(x_j), \tag{4.13}$$

where H_0 is the Hamiltonian operator

$$H_0 = -\frac{\hbar^2}{2m}\nabla^2 + V(x). \tag{4.14}$$

In the case of the isolated atom in free space, the atomic potential rises to a value of zero at infinite distance from the atom. In the lattice, this does not happen because at a distance $a/2$ from the atom, the potential from the neighboring atom appears. The interaction between the two potentials means that neither rises to the value of zero, and this modification produces a different potential in the region between the atoms. This difference is periodic in the lattice and will produce an interaction between the electrons on neighboring atoms. It is not at all clear just what this potential will look like, but it is represented by the interaction potential V_P. This now modifies the Schrödinger equation for the atom to

$$H_0|j\rangle + V_P|j+1\rangle + V_P|j-1\rangle = E|j\rangle. \tag{4.15}$$

We now operate with the adjoint operator $\langle j|$, and integrate over space, to get

$$E_0 + \langle j|V_P|j+1\rangle + \langle j|V_P|j-1> = E. \tag{4.16}$$

Equations (4.5) and (4.8) tell us that

$$|j\pm 1\rangle = e^{\pm ika}|j\rangle, \tag{4.17}$$

where the exponential is a "translation operator,"[1] and the integral over V_P now gives

$$\langle j|V_P|j\pm 1\rangle = e^{\pm ika}\langle j|V_P|j\rangle \equiv -Ae^{\pm ika}. \tag{4.18}$$

Hence, the energy of (4.15) can be written as

$$\begin{aligned} E &= E_0 - A\left(e^{ika} + e^{-ika}\right) \\ &= E_0 - 2Acos(ka) \end{aligned} . \tag{4.19}$$

This energy structure is plotted in Figure 4.3. The band is $2A$ wide (from lowest energy to highest energy), and is centered about the single site energy E_0, which says that the band forms by spreading around this single atom energy level. The band contains N values of k, as there are N atoms in

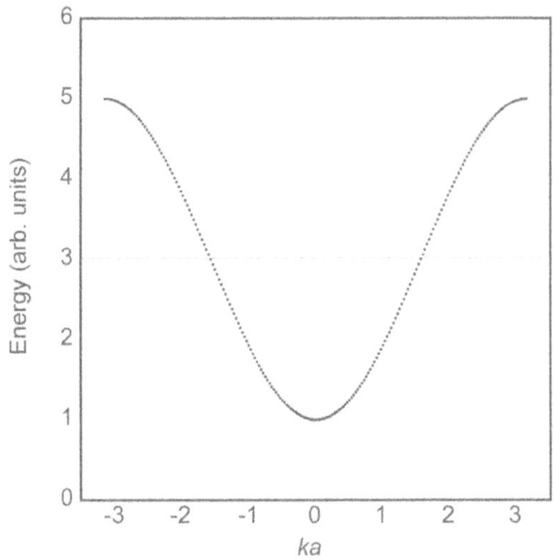

FIGURE 4.3 Plot of the energy for the linear atomic chain from (4.15). Here, arbitrary values of $E_0 = 3$ and $A = 1$ have been chosen for the plot.

the chain, and this is the level of quantization of the momentum variable, as was shown earlier. The minimum of the band appears at $ka = 0$, and maxima appear at $ka = \pm\pi$.

The fact that the electron has a spin was mentioned already in Chapter 2. It was actually discovered already in 1922 that the electron has an extra degree of motion that gave two possible states to each spatial set of quantum numbers (thus four quantum numbers in three dimensions). This extra quantum number was later shown to be due to the electron "spinning" on its own axis (Stern and Gerlach, 1922). This property of spin will be much more fully discussed in Chapter 10. Here, it is important to note that Pauli (1925) showed that only two particles with half-integer spin (referred to as fermions, such as electrons, as they satisfy Fermi–Dirac statistics) could occupy the same quantum state (note that this was even before the arrival of the proper quantum theories). While Pauli assumed only an additional quantum number, Uhlenbeck and Goudsmit identified it as the electron spin (1925). (It is known that Ralph Kronig suggested this to Pauli at one of the latter's seminars, but Pauli dismissed the idea as not being realistic.) The spin is important here for the simple reason of counting full states in the band that was calculated above. In the one-dimensional band of Figure 4.3, there are N discrete values of k. But, according to Pauli's exclusion principle, each value of k can hold two electrons, so the band can hold $2N$ electrons. But, each atom has been assumed to have a single electron in its outer shell (the bonding electrons are in the outer shell), so there are only N electrons. So, these electrons will fill only one-half of the available states in the band, and this becomes a metal. Some atoms with a single electron in their outer shell are Li, Cu, and Au; all are metals. While we think that this half-filled property is a result of the number of atoms, it is actually related to the number of unit cells (each cell has dimension a). This will become clearer in the next two sections.

4.2 THE DIATOMIC LATTICE

Now, a second atom will be added to the unit cell, so that we have a diatomic basis. This is illustrated in Figure 4.4. That is, the second atom is added to the right of each A atom at a distance b. The lattice is not periodic in this distance b, nor is it periodic in the distance $a - b$. It is only periodic in the distance a, so that the combination of atoms A and B forms the unit cell with two atoms in each unit cell. The lattice can be defined with either of the two atoms, but each lattice site contains a *basis* of two atoms. This will significantly change the electronic structure. Now there will be two wave functions, one for each of the two atoms in the basis, which will define the energy bands.

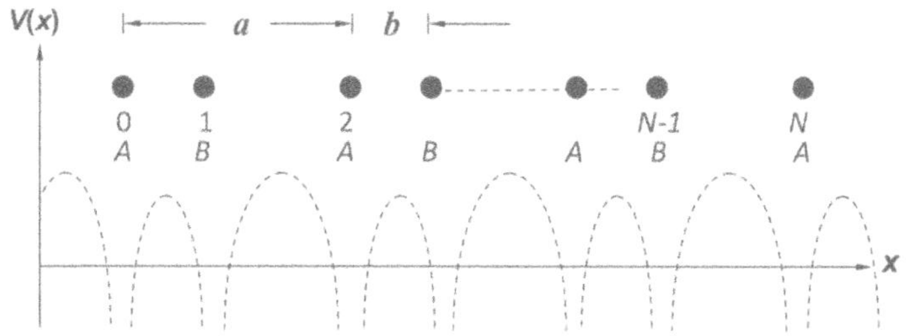

FIGURE 4.4 A linear chain of atoms, in which the spacing is not the same. Now, we require a basis of one A atom and one B atom, although the lattice constant remains a.

There are two distances that are now involved in the atom interactions. There is one interaction in which the two atoms are separated by b and one interaction in which the two atoms are separated by $a-b$. The lattice constant, however, remains a. To proceed, we adopt a slightly more complicated notation. The A atoms will be indexed by the site j, when j is an even number (including 0). Similarly, the B atoms will be indexed by the site j when j is an odd number. We have to write two equations, one for when the central site is an A atom and one for when the central site is a B atom. It really doesn't matter which sites we pick, but we take the adjacent sites so that these equations become

$$\begin{aligned} H_0|j\rangle + V_P|j+1\rangle + V_P'|j-1\rangle &= E|j\rangle, \\ H_0|j+1\rangle + V_P'|j+2\rangle + V_P|j\rangle &= E|j+1\rangle. \end{aligned} \tag{4.20}$$

Here, the first line is taken to have an even j and to represent the A atoms. Hence, the interaction V_P occurs over the distance b, while the interaction V_P' occurs over the distance $a-b$, so it will be different from V_P. The general case will appear for atoms A and B being different, so that we take

$$\begin{aligned} \langle j|H_0|j\rangle &= E_A, \qquad j\ even, \\ \langle j|H_0|j\rangle &= E_B, \qquad j\ odd. \end{aligned} \tag{4.21}$$

As in the previous section, the first of these equations is pre-multiplied by the complex value of the even site j, and the second equation is pre-multiplied by the complex value of the site (j + 1), and the equations are integrated. This leads to

$$\begin{aligned} E_A + \langle j|V_P|j+1\rangle + \langle j|V_P'|j-1\rangle &= E, \\ E_B + \langle j+1|V_P'|j+2\rangle + \langle j+1|V_P|j\rangle &= E. \end{aligned} \tag{4.22}$$

There are four interaction integrals to evaluate, and these are given by the parameters

$$\begin{aligned} \langle j|V_P|j+1\rangle &= e^{ikb}\langle j|V_P|j\rangle \equiv e^{ikb}A_1, \\ \langle j|V_P'|j-1\rangle &= e^{ik(b-a)}\langle j|V_P'|j\rangle \equiv -e^{ik(b-a)}A_2, \\ \langle j+1|V_P'|j+2\rangle &= e^{-ik(b-a)}\langle j|V_P'|j\rangle \equiv -e^{-ik(b-a)}A_2, \\ \langle j+1|V_P|j\rangle &= e^{-ikb}\langle j|V_P|j\rangle \equiv e^{-ikb}A_1. \end{aligned} \tag{4.23}$$

The above sets of interactions and equations lead to a secular determinant, which must be solved,

$$\begin{vmatrix} (E_A - E) & \left(A_1 e^{ikb} - A_2 e^{ik(b-a)}\right) \\ \left(A_1 e^{-ikb} - A_2 e^{-ik(b-a)}\right) & (E_B - E) \end{vmatrix} = 0. \tag{4.24}$$

It is clear that the two off-diagonal elements are complex conjugates of each other, and this assures that the Hamiltonian is Hermitian and the energy solutions are real. This basic requirement must be satisfied, no matter how many dimensions there are in the lattice, and is a basic property of quantum mechanics—in order to have real measurable eigenvalues, the Hamiltonian must be Hermitian. Two bands are formed from this diatomic lattice, and these are mirror images around an energy midway between the lowest energy of the upper band and the highest energy of the lower band. The energy is given by

$$E = \frac{E_A + E_B}{2} \pm \sqrt{\left(\frac{E_A - E_B}{2}\right)^2 + A_1^2 + A_2^2 - 2A_1A_2\cos(ka)}. \tag{4.25}$$

If the two atoms are the same, the first term in the square root vanishes. On the other hand, if the two atoms are different, then there is an additional term leading to the broadening of the bandwidth in energy. The two energy bands, separated by the ± sign, give a gap in which the center of the gap is the average value of the two atomic energies. In Figure 4.5, the energy bands are plotted for $E_A = E_B = 5$, and $A_1 = 2$, $A_2 = 1$ are used for the various values. These two bands are mirrored around 5, and each has a bandwidth of 2.

If there are N unit cells in this chain, as in the monatomic lattice, each band contains N states of momentum. Thus, the number of values of k is given by the number of unit cells, and not by the number of atoms, so that $k = N$. As before, each value of k can hold two electrons, which have opposite spins. Since each unit cell contains two atoms, there are two electrons in the cell, one from each of the two atoms (it is assumed that each has a single electron in the outer shell as before). Now, there are $2N$ electrons to occupy the $2N$ values of k. These will all reside in the lower of the two bands in Figure 4.5. Hence, the lower band is completely full, while the upper band is completely empty. This is a semiconductor if the band gap is not too large.

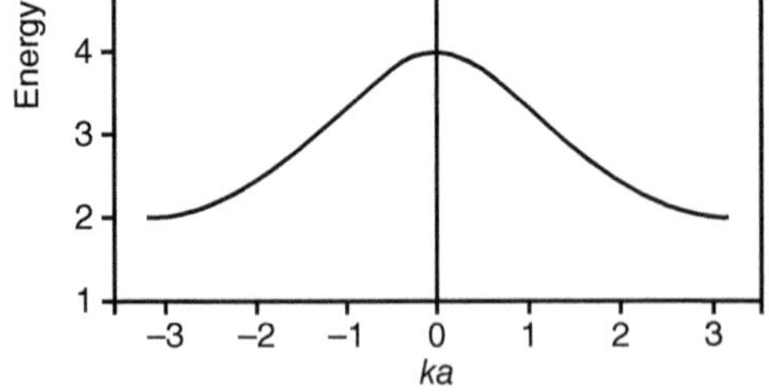

FIGURE 4.5 The resulting energy bands for the one-dimensional chain with two atoms per unit cell. Here, (4.25) is plotted for the values $E_A = E_B = 5$, $A_1 = 2$, $A_2 = 1$. The two bands are mirrored around the value 5, and each has a bandwidth of 2.

4.3 BONDS AND BANDS

Let us summarize a bit on the above developments and think about what they mean to us. First, when we had a single electron/atom in the unit cell, we found that a single band formed and there were N states in this band, as there were N unit cells in the linear array. Now, the founders of quantum mechanics always seemed to find that the electron possessed an additional quantum number. In fact, this knowledge preceded quantum mechanics. At the end of the nineteenth century, Pieter Zeeman was studying the optical absorption spectrum of various atomic gases and found that a magnetic field split the absorption line into a pair of lines, whose energy difference was linearly dependent upon the magnetic field (Zeeman, 1897a, 1897b, 1897c). Three-dimensional studies of the atomic structure of a simple atom like H could not produce this effect. Others repeated the experiments with the same results. For example, in 1921, Otto Stern and Walter Gerlach created an experiment with a very non-uniform magnetic field and sent a collimated beam of Ag atoms through it. In a normal (classical) system with a magnetic moment (as the electrons moving around the atom would have), a magnetic field will cause the magnetic moment to precess, much like a top spinning on its axis (this is discussed further in Chapter 10). In an inhomogeneous magnetic field, the force on the top and bottom of the "top" are different, and this should cause a deflection of the atom, which is detected by a sensitive screen. Because the magnetic moment is random in such an array of atoms, one expects to see a line on the screen, with the length being determined by the maximum range of this moment. What they found, however, was just two spots (Stern and Gerlach, 1921), just as Zeeman had found.

That is, in one dimension, the electron seemed to have two quantum numbers, whereas one-dimensional confinement only involved one of these (the value of k and its corresponding energy). Similarly, in three dimensions, the electron seemed to have four quantum numbers, whereas three-dimensional confinement only accounted for three of these. What was this fourth quantum number? As discussed above, this extra quantum number proved to be the electron spinning on its own axis, with the vector of the spin either up or down.

The importance of this extra quantum number seemed to be confined to fermions, that is, confined to particles with half-integer spin angular momentum ($\pm\hbar/2$), and which satisfied Fermi–Dirac statistics. It was Pauli (1926) who postulated that each spatial quantum state could have only two of the particles, provided that their extra quantum number was oppositely directed. Fermi–Dirac statistics have been mentioned several times already in discussions of the Landauer equation and tunneling current, but without specifically defining what they are. With these statistics, the probability of finding an electron in a quantum state with energy E is given by

$$P(E) = \left[1 + exp\left(\frac{E - E_F}{k_B T}\right)\right]^{-1}, \tag{4.26}$$

where E_F is termed the Fermi energy—the energy at which the probability is exactly 0.5.

In Section 4.2, the diatomic model was discussed and resulted in two bands, the lower of which was completely full and the upper of which was completely empty. In this configuration, the Fermi energy will lie exactly at the middle of the band gap, which in the model was at $(E_A + E_B)/2$, because this energy level lies just as far from the nearest empty state as it does from the nearest full state, so it has the probability of 0.5 of being full. The advantage of semiconductors as a canvas for electronic devices is that this position of the Fermi energy can be changed by the addition of an impurity with a different number of bonding electrons than silicon in the normal case.

The two-band model above is quite similar to the case of silicon. Silicon crystallizes in the diamond structure, so that the unit cell contains two atoms. But each atom has four bonding electrons. Since this crystal is three-dimensional, each k state in the lower band can hold a pair of electrons for each of the three directions, or $(2)^3 = 8$ electrons. Since each of the two atoms in the unit cell has four bonding electrons, they provide this 8 electrons so that the lower (valence) band is completely full and the upper (conduction) band is completely empty at $T = 0$. Thus, this material will be an insulator or a semiconductor, the difference being the size of the energy gap between the two bands.

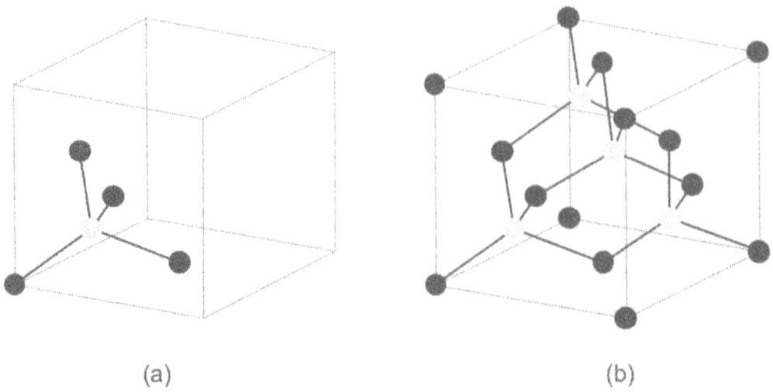

FIGURE 4.6 (a) The tetrahedral bond surrounding the central atom (light gray). (b) Groups of these atoms fit together for the diamond structure, which is a face-centered cubic lattice with a basis of two atoms (the lower left corner plus the light gray atom).

If electrons can be thermally excited over the gap, then it is a semiconductor. Otherwise, it is an insulator. In silicon, the energy gap is about 1.1 eV, so electrons can be excited into the conduction band at a reasonable temperature. But, as mentioned, it provides a blank canvas upon which desired properties can be 'painted' by the addition of donors (an extra electron) or acceptors (short one electron) at desired positions on the canvas.

We have referred to the four outer shell electrons of silicon as being *bonding* electrons. This is because the bonding in this case is covalent bonding, in which the four outer shell electrons bond with electrons from four nearest neighbors. These tend to form four hybrid wave functions which connect to these four nearest neighbors in the diamond lattice. These four neighbors form a tetrahedron, so we refer to this as tetrahedral bonding, which is common for nearly all semiconductors. The tetrahedral bond is depicted schematically in Figure 4.6.

4.4 QUASI-PARTICLES AND EFFECTIVE MASS

It is clear in Figures 4.3 and 4.5 that our electron no longer obeys the energy–momentum relationship of a free particle, and does not obey what is expected in classical physics. The presence of the periodic potential, to which the electrons must respond even though they are bonding to nearby atoms, changes the dynamics of the electrons. Since they do not respond as classical (or free) particles, they are often called quasi-particles. This term is applied to a great many types of particles that are found in quantum mechanics, but here these quasi-particles are quasi-electrons. They have exactly the same charge as a free electron, but their mass is different. We say the mass is different because we need to adjust the momentum so that it corresponds to the forces that are represented in the band structure of the above figures.

The band structure itself is a sort of filter response. It represents those frequencies (energies) for which a wave can propagate through the crystal, so it corresponds to a filter's "pass band." On the other hand, the band gaps correspond to those frequencies (energies) for which a wave cannot propagate through the crystal, so they correspond to a filter's "stop band." States can still exist within the band gaps, but these are evanescent states just as those that exist beneath a potential barrier. Since the primary interest is in the waves that propagate through the crystal, one tends to ignore the evanescent states.

It was remarked above that the quasi-electron, or quasi-particle that represents the electron, has a different mass, which varies so that the momentum of this particle satisfies the energy variation of the energy band. It has been emphasized through the previous chapters, and this chapter, which the electron has been treated as a wave. In the discussion of wave-particle duality in Section 1.1.2, it was emphasized that the wave has a momentum, due to de Broglie, that takes the value

$$p = \hbar k. \tag{4.27}$$

In strict duality between the wave and the particle, this means that the momentum of our quasi-particle needs to have exactly the value of (4.27). But, the particle momentum is typically expressed as mv. Hence, in the present situation, wave-particle duality now requires that the quasi-particle have an effective mass m^* that satisfies both of these momentum definitions, so that we require

$$m^* v = \hbar k. \tag{4.28}$$

The velocity that arises from the band structure was defined in (3.19) of Chapter 3. Using this in (4.28) means that the quasi-particle has to satisfy

$$m^* \frac{1}{\hbar}\frac{dE}{dk} = \hbar k, \tag{4.29}$$

or

$$\frac{1}{m^*} = \frac{1}{\hbar^2 k}\frac{dE}{dk}. \tag{4.30}$$

This is not what appears in most semiconductor device books, nor is it what is usually taught in courses dealing with these devices. But, this is the requirement of wave-particle duality.

To address the difference between (4.30) and what is commonly taught, consider again Figures 4.3 and 4.5. Electrons that are excited into the conduction band of Figure 4.5 will exist at the bottom of the upper band, which occurs at $k = 0$. But, at this point, we also have $\frac{dE}{dk} = 0$. Thus, (4.30) is indeterminate. It has zeroes in both the numerator and the denominator. Mathematics tells one to use L'Hopital's rule (1696)

$$\frac{1}{k}\frac{dE}{dk} \rightarrow \left(\frac{dk}{dk}\right)^{-1} \frac{d}{dk}\left(\frac{dE}{dk}\right) = \frac{d^2E}{dk^2}. \tag{4.31}$$

This gives the limiting form (for small k) of (4.30) to be

$$\frac{1}{m^*} = \frac{1}{\hbar^2}\frac{d^2E}{dk^2}, \tag{4.32}$$

and this is the form normally found in those books and classes mentioned above. But, it must be remembered that this is an approximation valid only at small k, and not throughout the band. Indeed, (4.32) gives rise to very strange results if applied where it is not valid!

4.5 ELECTRON PAIRING—SUPERCONDUCTIVITY

Superconductivity was mentioned at the end of Chapter 3, where the Josephson junction (1962) and the tunneling through this junction were discussed. Superconductors were said to be generally low-conductivity metals whose resistivity goes to zero at low temperature, typically a few, or a few tens of, degrees Kelvin. The presence of superconductivity in many materials was discovered by Kammerlingh Onnes (1910) early in the last century when studying mercury at low temperatures. He observed that the resistivity of Hg dropped abruptly to zero as the temperature was lowered to near 4K.

This results from the pairing of electrons with opposite spin (again to be discussed in Chapter 10) into a composite particle, the Cooper pair (Cooper, 1956), that can move through the lattice with no dissipation. The Josephson junction is a tunneling junction placed between two superconducting wires (each of which has no resistivity) at low temperature. This tunneling produces new (and strange) results. At 3K, Onnes observed that the resistance of Hg was less than 10^{-6} ohm, and he proposed to call the effect superconductivity. The onset, indicated by the abrupt drop in resistivity,

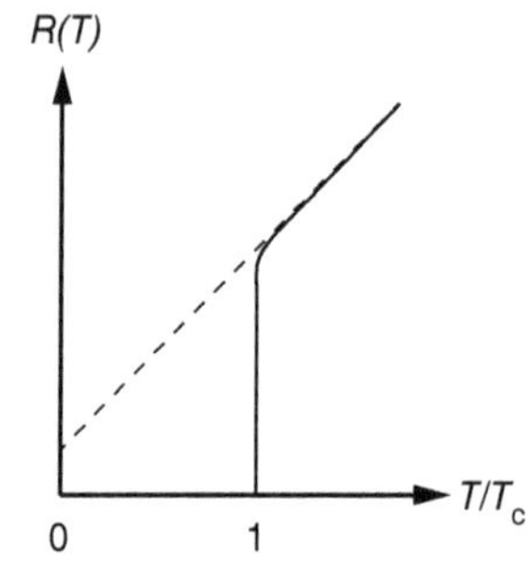

FIGURE 4.7 The resistivity of the material drops abruptly to zero as the temperature is lowered below the critical temperature T_c, as indicated in this figure. This signals the onset of superconductivity.

is considered to be a phase transition, and the critical temperature below which it occurs is called the *transition temperature* T_c. Our interest here lies in the concept of superconductivity, and the important quantum mechanical effect that modifies the simple energy bands discussed in the previous sections of this chapter. This important modification was already mentioned in Section 3.3.4, and that is the pairing of electrons into Cooper pairs and the opening of a gap in the simple energy bands discussed in the first sections of this chapter.

4.5.1 Observable Properties

The most easily observed property of superconductors is the transition temperature, T_c. If one plots the resistance as a function of temperature, there is a clear drop in the resistance at this transition temperature (see Figure 4.7). This drop is often a reduction of the resistance by orders of magnitude. Just above the transition temperature, there is a gradual drop in resistance, which gives a "rounding" to the curve, and this is caused by thermal fluctuations at the transition temperature. These fluctuations are believed to be due to the fact that not all of the electrons are paired except at absolute zero temperature. Above T_c, on average, none of the electrons are paired, so that the fraction that are paired is a function of temperature. In this region of enhanced conductivity above T_c, it is thought that small regions of the material are beginning to become superconducting, with the entire sample becoming so at further reductions of the temperature.

In general, metals that have very high conductivity do not become superconductors. On the other hand, metals that are poor conductors, such as Pb, Ta, Nb, and Hg, become quite good superconductors. Some of the transition temperatures are given in Table 4.1. Also shown are a few members of a new class of materials known as high-temperature superconductors, which have transition temperatures of tens of degrees Kelvin, up to ~200K at high pressure. These latter materials tend to be cuprates and ceramics.

A magnetic field can be used to destroy the superconductivity. That is, there is a critical magnetic field H_c above which the superconductivity vanishes. Moreover, it is found that this critical field varies with temperature as

$$H_c = H_{c0}\left[1-\left(\frac{T}{T_c}\right)^2\right], \tag{4.33}$$

where H_{c0} is the critical field at absolute zero of temperature. Comparing different materials, it is found that in general, a higher critical temperature will lead to a higher critical magnetic field. This critical magnetic field will also limit the amount of current that the superconductor can carry, since the current gives rise to a magnetic field by Ampere's law. However, there are other materials, such

TABLE 4.1
Superconducting Transition Temperatures

Element	T_c (K)	Compound	T_c (K)
Pb	7.2	Nb_3Sn	18.05
V	3.72	V_3Ga	16.5
Ti	0.39	V_3Si	17.01
Nb	9.1	NbN	16.0
Ta	4.48	InSb	1.9
Hg	4.15	Nb_3Al	17.5
Zn	0.85	$YBa_2Cu_3O_7$	92.0
Sn	3.72	$HgBa_2Ca_7Cu_3O_8$	134

as the compounds in Table 4.1, where the superconductivity does not end abruptly. In these materials, a lower critical magnetic field signals the beginning of the process, and an upper critical field signals when superconductivity has finally gone away.

4.5.2 Pairing and Gaps

When we first discussed Josephson effects in Section 3.3.4, it was remarked that the electrons would pair up as Cooper pairs (Cooper, 1956). Normally, electrons repel each other due to the Coulomb interaction between them. However, in metals, there are typically something like a few times 10^{22} electrons/cm^3. Such a high electron concentration usually heavily screens the Coulomb interaction, so that it is a relatively weak interaction, especially if the electrons are not fairly close to one another. Cooper hypothesized that because of this weakness, electrons could actually interact with each other if there was another positive interaction. He suggested that one electron could interact with the atoms of the crystal in a manner that distorted the local atomic potential, creating a potential well into which a second electron could be drawn. He thought the process would work best if the two electrons had opposite spins. Thus, for example, the two electrons in the same momentum state have opposite spins and would satisfy this requirement.

The combination of the two electrons, via the atomic interaction, creates the Cooper pair. Because the two electrons have opposite spins, the net spin of the Cooper pair is zero. This makes the Cooper pair into what is called a *composite boson*. Bosons are not required to satisfy the Pauli exclusion principle, so there is no limit to the number of bosons that can exist in any quantum state. These particles satisfy Bose–Einstein statistics, given by

$$P(E) = \left[1 - exp\left(\frac{E}{k_B T}\right)\right]^{-1}. \tag{4.34}$$

Moreover, when we form the Cooper pair, the interaction lowers the energy of the two electrons (this lowering of energy is why the pair is stable—it costs energy to break the pairing). This lowering of the energy is the pair formation energy, which was called Δ in Section 3.3.4. The result of this process is observable by a gap that opens in the energy spectrum. We noted in the previous section that the energy band in Figure 4.2 was half-filled. Thus, the Fermi energy would sit at 3 eV in that figure. With the formation of the Cooper pairs, a gap will open around the value of 3 eV, as shown in Figure 4.8. The value of the gap is 2Δ, so that a Cooper pair must gain this energy to transition above the gap and become a pair of free electrons. The interesting point is that the two electrons in the Cooper pair need not be close to one another. Instead, it is generally thought that they can be

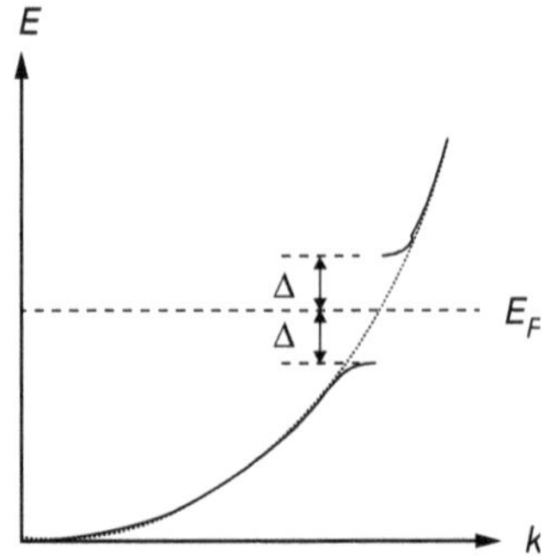

FIGURE 4.8 In superconductivity, the formation of Cooper pairs leads to the opening of a gap at the Fermi energy. This gap has a value of 2Δ, as shown.

some 100–400 nm apart, a length that is referred to as the *coherence length*. This means that there are a great number of other electrons near the Cooper pair, or at very low temperatures, the large number of Cooper pairs are all intertwined with one another. The energy gap that opens is also temperature dependent, varying as

$$E_G = E_{G0}\left(1 - \frac{T}{T_C}\right)^{1/2}, \tag{4.35}$$

where $E_{G0} = 2\Delta$.

It has been suggested that the Cooper pairs are the basic unit of superconductivity (Bardeen *et al.*, 1957a, 1957b). In the BCS theory, named for these three authors, the critical temperature is found to depend upon the density of states at the Fermi energy and the strength of the interaction between the lattice and the electron that leads to forming the Cooper pairs. Because all the states are full below the energy gap, one would not expect conduction, but the nature of the bosonic Cooper pairs seems to indicate they can move through the lattice without dissipation. That is, they cannot be scattered, so they move dissipation-free. And this seems to be the case. If one creates a coil of superconducting wire and induces a current through the wire, and then shorts the leads together, this current will continue to flow without dissipation. This current is known as a *persistent current* and is the heart of superconducting magnets.

4.5.3 An Incomplete Theory

The evidence for the theory of superconductivity is the BCS theory mentioned above. This theory depends upon the properties of the lattice motion. This motion depends upon the atoms in the lattice, and particularly their atomic weight. In fact, the variation in the critical temperature with the particular isotope[2] of the lattice atoms was an early indicator of the validity of the theory. But there are some superconducting materials that do not obey the expected isotope variation. One such is rhenium, and if this atom does not obey the isotope effect, there must be at least one other interaction that leads to superconductivity, although it has to be said that, of the number of suggested other theories, none has so far been proved entirely satisfactory.

A second objection to the BCS theory is the observation of superconductivity in the cuprates (the last two materials in the third column of Table 4.1). It is fairly recognized that the BCS theory does not work for these materials, nor do most of the alternatives that have been suggested over the past half-century. This suggests that an entirely new theory is required. Although there have been a few new suggestions, these have not met with acceptance by a skeptical community. It is fair to say at this point that the full understanding of the source of superconductivity in all materials just does not yet exist.

4.5.4 SQUIDS

A superconducting quantum interference device (SQUID) is a unique circuit in which one or more Josephson junctions are embedded in order to create coherence between the various tunneling currents. The Josephson junction was treated in Section 3.3.4 as a tunneling junction for Cooper pairs, which had some unique properties via the a.c. and d.c. Josephson currents. It was first developed as a very sensitive magnetometer to measure magnetic fields (such as in Magnetic Resonance Imaging, MRI, which remains commonly used today) (Jaklevic et al., 1964). The most common form of the SQUID is a wire loop with two Josephson junctions embedded in the loop, much as in Figure 4.9 (the basic SQUID is the "circle" with the two crosses, the latter of which are the Josephson junctions). When a bias current (I_b in the figure) is applied to the ring, this current divides between the two possible paths, so that $I_b/2$ flows through each arm of the circuit. Assuming the two junctions are equal, each has a critical current of I_{J0}, governed by (3.85). Now, if a small *external* magnetic field is applied, a screening current I_s (the dashed curve in Figure 4.9) will be induced (a by-product of Faraday's Law), which creates a change in the phase at each of the junctions. This screening current flows around the loop in the figure, so that one junction has a current $I_b/2+I_s$ and the other has a current $I_b/2-I_s$. Normally, there is no voltage at the output (indicated in green in the figure), since the loop is superconducting. But, as soon as one of the two currents exceeds the critical current I_J in that junction, a voltage will be developed since that junction is now resistive (the voltage is just above the critical voltage V_c shown in Figure 4.7).

If one now increases the flux so that it exceeds $\Phi_0/2$ (recall that Φ_0 is h/e and the flux is the product of the magnetic field and the enclosed area), a new behavior will arise. The magnetic flux contained within the loop is quantized and must be an integer multiple of Φ_0 (this behavior was seen in Chapter 1 in the Aharonov–Bohm effect as well). Now, the current through the junction must change so that it reflects the *difference* in the flux inside the loop (Φ_0) and the flux outside the loop $> \Phi_0/2$. This causes the junction current to decrease as the external flux is increased until it reaches Φ_0. When this occurs, the fluxes inside the loop and outside the loop are equal, and the effective magnetic field seen by the junction vanishes, with the result that the current has returned to zero. Thus, as the magnetic field is increased in amplitude, the junction current is going to oscillate with a period of Φ_0, with peak current at $(n+1/2)\Phi_0$ and a minimum current at $n\Phi_0$. Hence, it is possible to use these oscillations to measure the magnetic field outside the loop with the SQUID.

If the bias current I_b is raised so that it is larger than the critical current of the junctions, then the above behavior is not seen as the device works only in a resistive mode. Nevertheless, the output voltage will show oscillations that are periodic in the magnetic field, with a period of Φ_0. In this mode, the device acts as a flux-to-voltage converter, and sensitive measurements can be made on the actual magnetic field. Obviously, two SQUIDs can be connected in such a way that their voltages subtract and thus measure the derivative of the magnetic field. Four SQUIDS can measure the

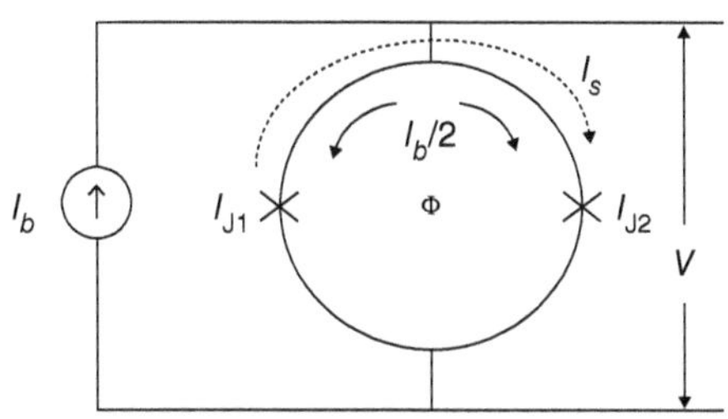

FIGURE 4.9 The basic circuit of a SQUID under current bias (I_b). The section that is the circle is the actual SQUID—a loop with two Josephson junctions (the X's) embedded in it. When a bias current I_b is passed through it, a voltage V is developed, which can have complicated behavior, discussed in the text. In the absence of a magnetic field, the two junction currents are each equal to $I_b/2$, as indicated by the two solid arrows.

second derivative and so on. Derivatives are often used for more sensitive measurements of magnetic field variations. If microwave fields are used, then the SQUID can work effectively with only a single Josephson junction.

As mentioned, SQUIDs are used to measure the variations in the magnetic field in MRI (other methods of measuring the magnetic field have arisen in recent years). They have also been used to measure the variations in the Earth's magnetic field in the oceans (obviously for detecting large magnetic objects under the sea surface). But, in recent years, the SQUID has been the basis for qubits in quantum computing, a topic we will return to in Chapter 11.

NOTES

1 The translation operator shifts the center of the wave packet by one lattice spacing in this case.
2 Isotopes of an element depend upon the number of neutrons in the nucleus. So different numbers of neutrons give different isotopes.

REFERENCES

Bardeen, J., Cooper, L. N., and Schrieffer, J. R. 1957a. Microscopic theory of superconductivity. *Phys. Rev.* **106** 162–163.

Bardeen, J., Cooper, L. N., and Schrieffer, J. R. 1957b. Theory of superconductivity. *Phys. Rev.* **108** 1175–1204.

Cooper, L. N. 1956. Bound electron pairs in a degenerate Fermi gas. *Phys. Rev.* **104** 1189–1190.

Gerlach, W. and Stern, O. 1922. Der Experimentelle Nachweis der Richtungsquantelung im Magnetfeld. *Zeit. Phys.* **9** 349–352.

Jaklevic, R. J., Lambe, J., Silver, A. H. and Mercereau, J. E. 1964. Quantum interference effects in Josephson tunneling. *Phys. Rev. Lett.* **12** 159–160.

Josephson, B. D. 1962. Possible new effect in superconductive tunneling. *Physics Lett.* **1** 251–253.

Kammerlingh Onnes, K. 1910. Further experiments with liquid helium. C. On the change of electric resistance of pure metals at very low temperatures etc. IV. The resistance of pure mercury at helium temperatures. *Konig. Nederland. Akad. Wetensch. Proc.* **13** 1274–1276.

de L'Hospitalopital, G. F. A. M. 1696. *Analyse des Infiniment Petits, Pour L'Intelligence des Lignes Courbes* (Paris: de L'Imprimerie Royale), pp. 145–146

Pauli, W. 1925. Über den Zusammenhang des Abschlusses der Elektronengruppen im Atom mit der Komplexstruktur der Spektren. *Zeit. Phys.* **31** 765–783.

Uhlenbeck, G. E. and Goudsmit, S. A. 1925. Ersetzung der Hypothese vom Unmechanischen Zwang durch eine Forderung bezüglich des inneren Verhaltens jedes einzelnen Elektrons. *Naturwiss.* **13** 953–954.

Zeeman, P. 1897a. On the influence of magnetism on the light emitted by a substance. *Phil. Mag., Ser. 5* **43** 226–239.

Zeeman, P. 1897b. Doublets and triplets in the spectrum produced by external magnetic forces. *Phil. Mag., Ser. 5* **44** 55–60.

Zeeman, P. 1897c. The effect of magnetization on the nature of light emitted by a substance. *Nature* **55** 347.

PROBLEMS

1. In (4.25), the values for which the right-hand side reaches −1 must be satisfied by the left-hand side having $\cos(ka) = -1$, which leads to the energies being those of an infinite potential well. Show that this is the case. Why? The importance of this result is that the top of every energy band lies at an energy defined by the infinite potential well, and the bands form by spreading *downward* from these energies as the coupling between wells is increased.
2. Using the "Periodic potential—Kronig-Penney Lab" at NanoHub.org, repeat the calculation leading to Figure 4.5, using the parameters in the text. Vary the mass from 0.1 to 1.0 (in units of the free electron mass) and determine how many bands appear over this range.

5 The Harmonic Oscillator

One of the most commonly used quantum mechanical systems is that of the simple *harmonic oscillator*. This is because it is very familiar to most people through their observations of the simple pendulum. The pendulum that people see is a very classical system of sinusoidal oscillations, at least when the amplitude of these oscillations is relatively small. Quantum mechanically, the motion described by the Schrödinger equation is more complex. Although quite a simple system in principle, the harmonic oscillator assumes almost overwhelming importance due to the fact that it is one of the few systems that can be solved exactly in quantum mechanics. Almost any interaction potential may be expanded in a Taylor series with the low-order terms (up to quadratic order) cast into a form described by the harmonic oscillator. As a result, the sinusoidal motion of classical mechanics is the simplest system that produces oscillatory behavior and, therefore, is found in almost an infinity of physical systems. Due to this single property, the harmonic oscillator has become important in describing quite diverse physical systems. These applications range from quantizing the lattice vibrations in a simple solid to quantizing the electromagnetic fields of Maxwell's equations.

In this chapter, we will develop the general mathematical solution for the wave functions of the harmonic oscillator and will introduce a simple operator algebra that allows us to obtain these wave functions and properties in a very usable and simple manner. Classically, the harmonic oscillator is described in terms of Hermite polynomials, as the differential equation is one of the variants of the Sturm–Liouville problem. We can obtain these functions without the normal complication of solving the detailed differential equation, as the operator approach makes life easier. We then turn to two classic examples of systems in which we use the results for the harmonic oscillator to explain the properties: the simple *LC*-circuit and vibrations of atoms in a crystalline lattice. We also discuss the motion of electrons in a quantizing magnetic field, which produces another system that reduces to a harmonic oscillator.

The simplest example of sinusoidal behavior in classical physics is that of a mass on a linear spring, in which the mass responds to a force arising from the extension or compression of the spring (Figure 5.1). The differential equation describing this is just

$$m\frac{d^2x}{dt^2} = F = -Cx, \tag{5.1}$$

where m is the mass and C is the spring constant. Instead of solving this equation classically, we instead write the total energy, using the square of the momentum $p = m(dx/dt)$ and the potential energy $V = Cx^2/2$ that arises from integrating the right-hand side, as

$$E = \frac{p^2}{2m} + \frac{m\omega^2 x^2}{2}, \tag{5.2}$$

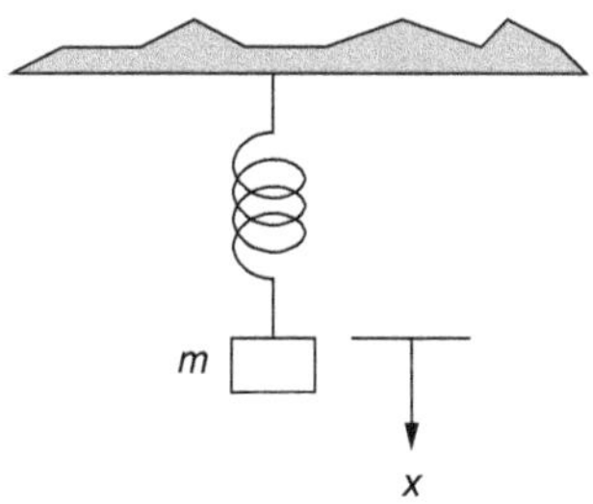

FIGURE 5.1 A simple mass on a spring, in which the mass will exhibit oscillatory motion in response to be displaced from its equilibrium position.

DOI: 10.1201/9781003717195-5

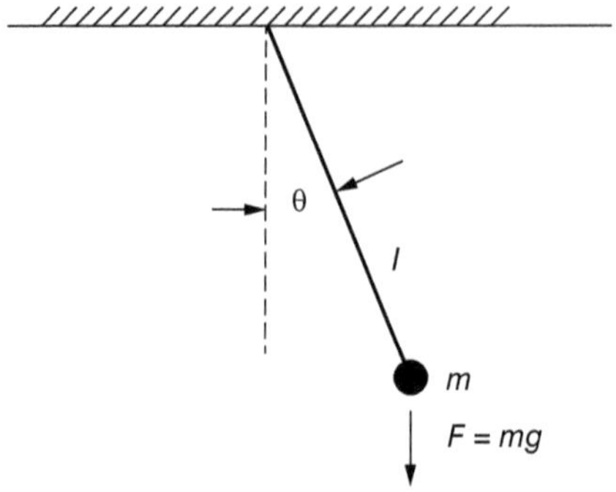

FIGURE 5.2 A simple pendulum, which also is a simple oscillatory motion (for small displacement) and obeys the harmonic oscillator equations.

where we have introduced the oscillator frequency ω through $C = m\omega^2$. This is the energy function that is used in the Schrödinger equation to find the quantum mechanical motion of the simple harmonic oscillator in the first section below.

To see the connection between this mass on a spring and the simple pendulum is relatively easy, as the pendulum is one of the first problems in introductory physics. Quite simply, a mass m is suspended by a rigid rod of length l from a fixed point, about which it can rotate. The angle that the rod makes with the vertical is given by θ. As the rod is pushed away from the vertical, gravity provides a restoring force of value mg, which is directed directly downward (Figure 5.2), and which leads to a force that tends to reduce the angular deflection of the rod, given by $-mg \sin \theta$.

5.1 THE WAVE FUNCTIONS

From the above discussion, the differential equation for the angular deflection θ of the pendulum may be written as

$$m\frac{d}{dt}\left(l\frac{d\theta}{dt}\right) = -mg\sin\theta. \tag{5.3}$$

The factor in the parentheses on the left-hand side is just the angular velocity of the pendulum, so the left-hand side is the time derivative of the angular momentum, and the right-hand side is the restoring force on the angle. We linearize this equation by expanding the sine function for small angles, so the resulting equation is just

$$m\frac{d}{dt}\left(l\frac{d\theta}{dt}\right) + mg\theta = 0. \tag{5.4}$$

We can easily develop the Hamiltonian for this, by remarking that this latter quantity is the sum of the kinetic and potential energies, just as for the mass on a spring discussed in the above paragraphs. (If we replace $l\theta$ with x, and mg/l with C, the last equation becomes (5.1).) The kinetic energy is obtained from the momentum as $p^2/2m$, and this is evaluated using the angular momentum in (5.3) and the mass of the pendulum (neglecting the mass of the rod itself). The potential energy is obtained by noting that the right-hand side of (5.3) is the force, which is obtained from the spatial derivative of the potential energy. In the linearized version of (5.3), this leads to

$$F = -mg\theta = -\frac{1}{l}\frac{\partial V}{\partial \theta}, \tag{5.5}$$

and

$$V(\theta)=\frac{1}{2}mgl\theta^2. \tag{5.6}$$

This leads to a Hamiltonian form

$$H=\frac{1}{2m}\left(ml\frac{d\theta}{dt}\right)^2+\frac{1}{2}mgl\theta^2 \tag{5.7}$$

To make (5.7) into a general Hamiltonian such as (5.2), which can be applied to many systems, we will make a few changes of variables. This begins with the change of variables discussed above to take the equation into the form for the mass on a spring. However, the gravity is replaced by a general frequency through the change $g/l \rightarrow \omega^2$. Finally, the momentum is introduced as a differential operator through

$$ml\frac{d\theta}{dt}\rightarrow -i\hbar\frac{d}{dx}. \tag{5.8}$$

This now leads us to the Hamiltonian of (5.2) and subsequently to the Schrödinger equation. The latter equation then becomes

$$-\frac{\hbar^2}{2m}\frac{d^2\psi}{dx^2}+\frac{m\omega^2x^2}{2}\psi=E\psi. \tag{5.9}$$

It is this formulation (5.9) of the Schrödinger equation, as well as the linearized-angle obtained around (5.7), that is found to occur in a great many applications. The recognition of this commonality is very important, since if we can recognize a known set of solutions to a new problem, it becomes quite easy to interpret the expected results for that problem. The harmonic oscillator is one of the most frequently studied problems for just this reason.

5.1.1 Introducing Generalized Operators

The Hamiltonian developed for the harmonic oscillator is doubly quadratic; that is, it contains one term quadratic in the momentum and one term quadratic in the position. We want to rearrange this Hamiltonian by introducing a set of operators that combine the position and momentum in a way that leads to a simpler approach to the problem. The rationale is based upon our familiarity with complex variables. So, let's examine this first so as to understand the methodology that is at play here.

Hopefully it was learned early in our careers that we could expand a doubly periodic set of variables using complex algebra, as

$$R^2+X^2=(R+iX)(R-iX), \tag{5.10}$$

where R might be, for example, resistance and X might be reactance. Now, each term in parentheses is a function in its own right and can be treated as such. One term might be an inductor and its series resistance and the other term a capacitor and its shunt resistance (or leakage or tunneling current). Nevertheless, each of the two terms in parentheses can be used as an impedance all on its own, although each has different functionality.

If we separate the double quadratic Schrödinger equation in this manner, we will change it from a single second-order equation into a pair of first-order equations. The important difference with the Schrödinger equation from the simple complex variable approach of (5.10) is the problem of operator order; that is, the commutation relations that have to be satisfied with quantum mechanics. Following the separation of (5.10), the right-hand side of (5.2) can be separated as

$$\frac{p^2}{2m}+\frac{m\omega^2x^2}{2}=\left(\sqrt{\frac{1}{2m}}p+i\sqrt{\frac{m}{2}}\omega x\right)\left(\sqrt{\frac{1}{2m}}p-i\sqrt{\frac{m}{2}}\omega x\right). \tag{5.11}$$

This is not the actual form we want the terms to be in (we need to rearrange some constants), so two operators will actually be defined in the desired form as

$$\begin{aligned} a &= \sqrt{\frac{m\omega}{2\hbar}}\left(x+i\frac{p}{m\omega}\right), \\ a^\dagger &= \sqrt{\frac{m\omega}{2\hbar}}\left(x-i\frac{p}{m\omega}\right). \end{aligned} \tag{5.12}$$

This rearrangement of the constants and signs is one of conformance to standard usage in the literature.

Now, we have to address the fact that x and p do not commute. For this purpose, we first write the product

$$\begin{aligned} a^\dagger a &= \frac{m\omega}{2\hbar}\left(x-i\frac{p}{m\omega}\right)\left(x+i\frac{p}{m\omega}\right) \\ &= \frac{1}{\hbar\omega}\left(\frac{p^2}{2m}+\frac{m\omega^2}{2}x^2\right)-\frac{1}{2} \end{aligned}. \tag{5.13}$$

Comparing this with (5.2), we can immediately rewrite the energy as

$$E=\frac{p^2}{2m}+\frac{m\omega^2}{2}x^2=\hbar\omega\left(a^\dagger a+\frac{1}{2}\right). \tag{5.14}$$

As will be shown later, the product of the two operators is the number of "particles" in a level, so this product will be called the *number operator:* $n = a^\dagger a$. It is important to realize that this product of operators (unusually) produces a non-operator (remember that the energy is *not* an operator, even though it contains operators), called a *c*-number (for constant number), that gives the quantum number of the energy level corresponding to a particular state when operating on the Hamiltonian eigenstates. We return to this in the next section.

The momentum and position operators satisfy a basic commutator relation since they are non-commuting operators. We expect that the above operators are also non-commuting since they arise from combinations of these non-commuting operators. The reverse product to (5.13) is found to be

$$\begin{aligned} aa^\dagger &= \frac{m\omega}{2\hbar}\left(x+i\frac{p}{m\omega}\right)\left(x-i\frac{p}{m\omega}\right) \\ &= \frac{1}{\hbar\omega}\left(\frac{p^2}{2m}+\frac{m\omega^2}{2}x^2\right)+\frac{1}{2} \end{aligned}. \tag{5.15}$$

If (5.13) is subtracted from (5.15), the result gives us the commutation relation

$$\left[a,a^\dagger\right]=aa^\dagger-a^\dagger a=1. \tag{5.16}$$

We note that both terms in (5.16) are *c*-numbers, from which the Hamiltonian can be defined. These operators can now be used to solve the Schrödinger equation in a greatly simplified manner.

5.1.2 Solving the Schrödinger Equation

Since the operator pair $a^\dagger a$ is a *c*-number, if we multiply a function by this quantity, we expect to obtain something proportional to the corresponding eigenvalue. That is, when the function is one of the eigenfunctions of the harmonic oscillator Hamiltonian, we may write

$$a^\dagger a\psi_n = \lambda_n\psi_n. \tag{5.17}$$

A similar result is obtained for the opposite choice of ordering, where λ_n is related to the eigenvalue still to be determined. The words "related to" is used because we have not yet assured ourselves that the wave function has been properly normalized. This will be addressed below. But, for now, we can certainly say that λ_n is positive, since from (1.28) we can write the expectation value as

$$\langle\psi_n|a^\dagger a|\psi_n\rangle = \langle a\psi_n|a\psi_n\rangle \geq 0. \tag{5.18}$$

At this point, some effort needs to be devoted to examine in detail the nature of these operators and the properties of the various eigenfunctions. To begin, let us examine the properties of the product $a^\dagger\psi_n$. We do this using (5.16) and (5.17), to demonstrate that

$$\begin{aligned} a^\dagger a\left(a^\dagger\psi_n\right) &= a^\dagger\left(aa^\dagger\psi_n\right) = a^\dagger\left(a^\dagger a+1\right)\psi_n \\ &= a^\dagger\left(\lambda_n+1\right)\psi_n = \left(\lambda_n+1\right)a^\dagger\psi_n\ . \end{aligned} \tag{5.19}$$

Thus, if the wave function (and state) ψ_n has the eigenvalue λ_n, then operating on this state with the operator $a^\dagger$ produces a new state with the eigenvalue $(\lambda_n + 1)$. In essence, we have "kicked" the energy of the initial state *upward* by one unit of $\hbar\omega$, which may be observed by reference to the Hamiltonian (5.14) and produced a new wave function corresponding to the level of higher energy. By the same token, if we operate with a, we find

$$a^\dagger a\left(a\psi_n\right) = \left(aa^\dagger - 1\right)a\psi_n = \left(\lambda_n - 1\right)a\psi_n. \tag{5.20}$$

Thus, operating with the operator a produces a new state with the eigenvalue $(\lambda_n - 1)$. In essence, we have "kicked" the energy of the initial state *downward* by one unit of $\hbar\omega$ and produced a new wave function corresponding to the level of lower energy. For these reasons, we normally refer to $a^\dagger$ as a *raising* operator (or *creation* operator since it creates one additional unit of energy in the system) and a as a *lowering* operator (or *annihilation* operator, since it destroys one unit of energy).

Now, there is always a lowest energy state, the *ground state*. In the harmonic oscillator, it corresponds to $n = 0$, the state in which the wave function corresponds to the zero-order Hermite polynomial. Thus, if we try to lower the energy, and kick the state down by one unit, by operating with the lowering operator, the result must be zero, as

$$a\psi_0 = 0, \tag{5.21}$$

as there can be no state below the lowest energy state. This means that $\lambda_0 = 0$. From (5.14), this means that the lowest energy is

$$E_0 = \frac{\hbar\omega}{2}. \tag{5.22}$$

Each step upward thus adds an energy $\hbar\omega$, which Planck (1900) denoted as the energy of the photon in his black-body radiation theory. As pointed out the above, the lowest energy state has $\lambda_0 = 0$. As each step adds a value of unity to this, repeated use of (5.19) leads to

$$\lambda_n = n. \tag{5.23}$$

At this point, a lot is known about the eigenvalues. Now, it is time to examine the eigenfunctions. To begin, consider the first operator of (5.12), which leads to the equation

$$\sqrt{\frac{m\omega}{2\hbar}}\left(x + i\frac{p}{m\omega}\right)\psi_0 = 0. \tag{5.24}$$

As remarked above, the zero results from the fact that there can be no state below the defined lowest state. This equation can be rearranged, and the operator form for p inserted, to give

$$-\frac{\hbar}{m\omega}\frac{d\psi_0}{dx} = x\psi_0. \tag{5.25}$$

This has an easy solution as

$$\psi_0(x) = C_0 exp\left(-\frac{m\omega}{2\hbar}x^2\right). \tag{5.26}$$

This is connected with the lowest (unnormalized, because we do not know the value of C_0) Hermite polynomial (only because solving the full differential equation is known to lead to such polynomials). The exponential term is a weighting function that multiplies the Hermite polynomials. By direct normalization with (1.28), we find that

$$C_0 = \left(\frac{m\omega}{\pi\hbar}\right)^{1/4}. \tag{5.27}$$

Now, by the use of (5.19), we can get the form of each eigenfunction by using the creation operator as

$$\psi_n \sim \left(a^\dagger\right)^n \psi_0 \sim \left(\frac{m\omega}{2\hbar}\right)^{n/2}\left(x - i\frac{p}{m\omega}\right)^n \psi_0. \tag{5.28}$$

We have not put an equality in this last equation because there is another factor in the normalization of the higher basis functions that differs from the lowest wave function due to the zero value of λ_0. We have to probe a little bit deeper to resolve this last factor in the normalization.

5.1.3 Normalizing the Wave Functions

To get the final parts of the normalization, we turn to the expectation values of the operators themselves. From the basic properties expressed in (5.19) and (5.20), we know that for unnormalized wave functions such that $\langle\psi_n|\psi_n\rangle = C_n^2$, we can write (we take the coefficients as real)

$$\begin{aligned}\langle\psi_n|a^\dagger\psi_k\rangle &= \langle\psi_n|\psi_{k+1}\rangle = \alpha_n\delta_{n,k+1}C_n^2 \\ \langle\psi_n|a\psi_k\rangle &= \langle\psi_n|\psi_{k-1}\rangle = \beta_n\delta_{n,k-1}C_n^2\end{aligned}. \tag{5.29}$$

The constants α_n and β_n still must be determined. These are related to the normalization of the eigenfunctions (the two eigenfunctions in the expectation value have different normalization constants). The delta functions assure that only wave functions which are nearest neighbors in the energy chain (up or down) appear in this process. Now, consider

$$\begin{aligned}\langle a\psi_n | a\psi_n \rangle &= \langle \psi_n | a^\dagger a\psi_n \rangle = nC_n^2 \\ &= \langle \psi_{n-1} | \psi_{n-1} \rangle = C_{n-1}^2\end{aligned}. \tag{5.30}$$

Of course, the two lines are two different evaluations of the normalization on the left-hand side. However, they allow us to connect the constants for two adjacent wave functions and lead to

$$C_{n-1} = \sqrt{n}C_n. \tag{5.31}$$

This process can be repeated with the creation operators as

$$\begin{aligned}\langle a^\dagger\psi_n | a^\dagger\psi_n \rangle &= \langle \psi_n | aa^\dagger \psi_n \rangle = (n+1)C_n^2 \\ &= \langle \psi_{n+1} | \psi_{k+-1} \rangle = C_{n+1}^2\end{aligned}, \tag{5.32}$$

and

$$C_{n+1} = \sqrt{n+1}C_n. \tag{5.33}$$

Now, one can rewrite Equation (5.29) as

$$\begin{aligned}\langle \psi_n | a^\dagger \psi_k \rangle &= \langle \psi_n | \psi_{k+1} \rangle = \sqrt{n+1}\delta_{n,k+1} \\ \langle \psi_n | a\psi_k \rangle &= \langle \psi_n | \psi_{k-1} \rangle = \sqrt{n}\delta_{n,k-1}\end{aligned}. \tag{5.34}$$

This now leads to the normalization of (5.28) which can be written, using the above results, as

$$\psi_n = \frac{1}{\sqrt{(n+1)!}}\left(a^\dagger\right)^n \psi_0 = \frac{1}{\sqrt{(n+1)!}}\left(\frac{m\omega}{2\hbar}\right)^{n/2}\left(x - i\frac{p}{m\omega}\right)^n \psi_0. \tag{5.35}$$

Since we already have ψ_0, n begins with 1 in (5.35) and goes upward in unit steps. In Figure 5.3, we plot the lowest four Hermite polynomials. Like the bound states in a quantum well, each wave function has an additional peak in order to insure orthogonality and normalization of the wave functions (termed orthonormality).

5.1.4 Hermite Polynomials

Several times in the above discussion, Hermite polynomials have been mentioned without really explaining the polynomial or their actual use in the wave functions. However, it is relatively clear from (5.35) that the creation operator contains both a term in x and a term in the derivative with respect to x. When these terms operate on the exponential function that exists in ψ_0, see (5.26), they will produce a power series in x, which has a finite length corresponding to the index n. This power series is the heart of the Hermite polynomial. Because of the constants that appear in the

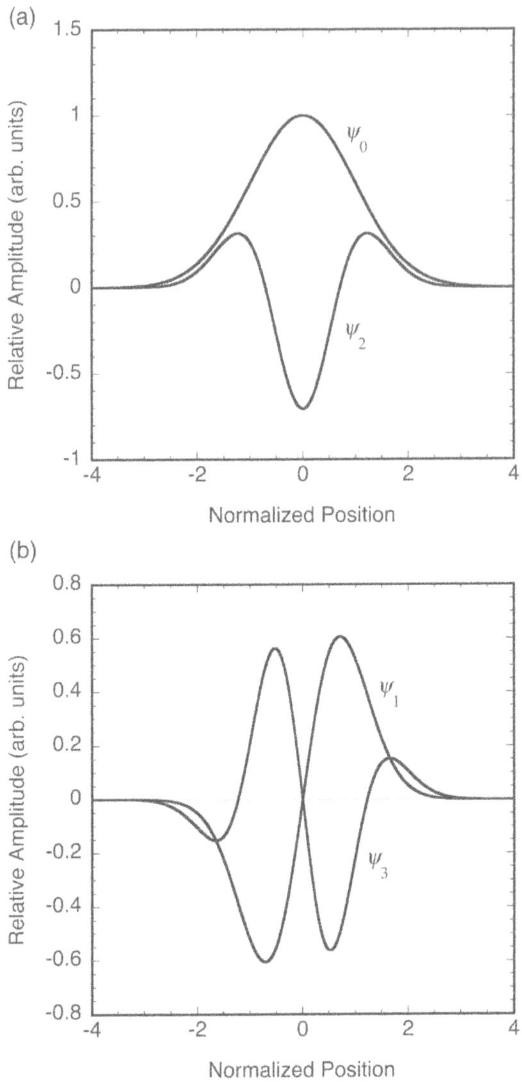

FIGURE 5.3 The lowest four wave functions for the harmonic oscillator. These are products of the appropriate Hermite polynomial and the exponential factor shown (5.26). The upper panel shows the even symmetry wave functions, while the lower panel illustrates the odd symmetry wave functions.

exponential, the terms in the Hermite polynomials are not strictly merely terms in x. Rather, they also involve the constants. If the reduced variable

$$\xi = \sqrt{\frac{m\omega}{\hbar}}x \tag{5.36}$$

is introduced, then the first few Hermite polynomials may be written as

$$\begin{aligned} &H_0(\xi) = 1, && H_1(\xi) = 2\xi, \\ &H_2(\xi) = 4\xi^2 - 2, && H_3(\xi) = 8\xi^3 - 12\xi, \\ &H_4(\xi) = 16\xi^4 - 48\xi^2 + 12, && H_5(\xi) = 32\xi^5 - 160\xi^3 + 120\xi. \end{aligned} \tag{5.37}$$

One may notice that the even indexed polynomials are functions of even powers while the odd indexed polynomials are functions of odd powers. These two sets are from the two basic solutions of the Hermite equation, just as sines and cosines result from Fourier series.

The wave functions are the product of a Hermite polynomial, the exponential function from (5.26), and a set of constants. They also produce orthonormal wave functions, although the proof of this lies deep in the properties of the Hermite polynomials and beyond the treatment here. Nevertheless, the orthogonality of these wave functions is established with the integral (Ferry, 1995)

$$\int_{-\infty}^{\infty} H_i(\xi) H_j(\xi) e^{-\xi^2} d\xi = \sqrt{\pi}\, 2^n n! \delta_{ij}. \tag{5.38}$$

The exponential term is just the square of that from (5.26) in the reduced coordinates. In terms of these Hermite polynomials, (5.35) may be rewritten as

$$\psi_n(x) = \frac{1}{\sqrt{2^n n!}} \left(\frac{m\omega}{\pi\hbar} \right)^{1/4} H_n\left(\sqrt{\frac{m\omega}{\hbar}} x \right) exp\left(\frac{m\omega}{2\hbar} x^2 \right). \tag{5.39}$$

5.2 THE LC CIRCUIT

A circuit that is far more familiar to electrical engineers is the resonant *LC*-circuit. However, this circuit produces precisely the behavior of the harmonic oscillator—oscillatory currents and voltage. The *LC*-circuit is one of the most pervasive systems found in electrical engineering. It represents not only a range of filters but also the resonant cavity of microwave circuits. In this section, we show that the quantum mechanics of the *LC*-circuit is really that of the linear harmonic oscillator. Consider the circuit of Figure 5.4. The familiar equations for current and voltage in the inductor and capacitor (directions are indicated in the figure, so this will change some signs) are

$$\begin{aligned} I &= -C\frac{dV}{dt}, \\ V &= L\frac{dI}{dt}. \end{aligned} \tag{5.40}$$

The energy stored in the circuit is given by

$$E = \frac{1}{2}LI^2 + \frac{1}{2}CV^2 = \frac{\Phi^2}{2L} + \frac{Q^2}{2C}, \tag{5.41}$$

in terms of the *magnetic flux* Φ and the charge Q. The first term on the right-hand side is the energy stored in the inductor, while the second term is that stored in the capacitor. In fact, this is the Hamiltonian that should be used in the Schrödinger equation. To these quantities, it is useful to add

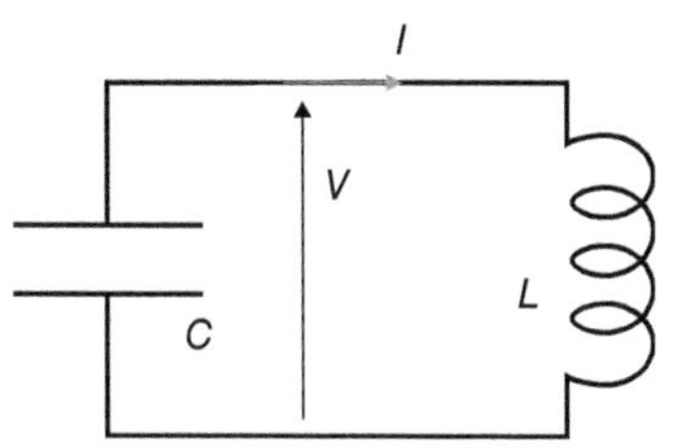

FIGURE 5.4 A resonant circuit composed of an inductor L and a capacitor C. The voltage V and current I are discussed in the text.

the resonant frequency $\omega = 1/\sqrt{LC}$. If we now relate L to the mass m, Q to the position x, and Φ to the momentum p, (5.41) is exactly (5.2). In this case, these relationships are *analogs* to one another. Indeed, (5.2) can be expressed here as

$$\frac{d^2\Phi}{dt^2} + \omega^2\Phi = 0. \tag{5.42}$$

Of course, this equation can be written for the charge, the voltage, and the current with various constants included. Nevertheless, this last equation is the harmonic oscillator equation. However, it is still necessary to identify the conjugate operators and their relationship to each other.

One obvious possibility is to take the charge Q (= CV) as one operator and then the current I = dQ/dt as the other (in analogy between the position and the velocity). However, this doesn't quite work out. The analogy would have the current as the velocity (of the charge) and not the momentum. We can be guided by our understanding that the two fundamental quantities in circuits (and, indeed, in electromagnetic theory in general) are the *charge* and the *flux*. It is these two quantities that are invariant in special relativity theory and it is these two that are subject to important conservation laws in circuit theory—the conservation of charge and the conservation of flux linkages. Moreover, the flux linkage in the present context is just the flux in the inductor, given by $\Phi = LI$. Here it is better to take the charge as the "coordinate" variable and take the flux as the "momentum" variable. If this is done, the energy is the last form of (5.41). By introducing the resonant frequency, this form can be written as

$$E = \frac{\Phi^2}{2L} + L\omega^2\frac{Q^2}{2}, \tag{5.43}$$

so that the commutator relationship becomes

$$\left[Q, \Phi\right] = i\hbar. \tag{5.44}$$

In this interpretation, the charge and flux are non-commuting variables, with the flux interpreted as the variation with respect to the charge. On the other hand, this relationship vanishes in the limit as $\hbar \to 0$. In this latter limit, charge and flux are independent quantities, but in the quantum case they are not independent and cannot be simultaneously measured. However, we should be very careful, as this limit cannot properly be interpreted as going to the classical limit, as there are cases in which this limit introduces singularities.

The interpretation of the LC-circuit as a linear harmonic oscillator means that the energy in the resonator is governed by Planck's relationship—energy can be increased or decreased only in units of the energy quantum $\hbar\omega = \hbar/\sqrt{LC}$. Thus, the photons that enter or leave an electromagnetic cavity (or a normal circuit) have well-defined energies in terms of the resonant frequency of the cavity. This is obvious in the measurement of the frequency of the oscillating cavity, where we measure the energy of the photons emanating from the cavity, just as was done for the study of black-body radiation discussed in Chapter 1. In many cases, such as was done by Planck (1900), it is often useful to consider the resonant circuit quantum mechanically rather than classically. The full understanding of this lies in the harmonic oscillator results of the previous sections.

With this discussion, it is now clear that electromagnetic waves satisfy the resonant LC-circuit, which itself is a linear harmonic oscillator. We previously established that the particles which satisfied the harmonic oscillator were *bosons*. Thus, it is clear that photons, the particles which are associated with electromagnetic waves, are bosons. The photons actually have spin, but it is quantized into values of ± 1, corresponding to the circular polarization of the plane wave of electromagnetics. Thus, we may say that bosons have *integer* values of spin, while fermions have half-integer values of spin.

This approach also opens the door to a treatment of almost any oscillatory behavior to be treated by a quantum approach. As here, one only needs to identify the proper *generalized* coordinates, which are then subjected to an uncertainty principle. It is this step that introduces the quantum nature of the subsequent solutions in terms of harmonic oscillator coordinates. Generalized coordinates have long been treated in classical mechanics, and their use naturally flows over to quantum mechanics.

5.3 QUANTIZING THE ELECTROMAGNETIC FIELD

Quantizing any classical field, whether it is the atomic motion in a crystal, or the electromagnetic fields with which nearly all physicists and electrical engineers are familiar, follows a similar path. Generally, the first point is to take a volume to consider as a method to introduce discretization in the wave number $\boldsymbol{k}$, which is going to be three-dimensional here, exactly as in the beginning of Chapter 4. The purpose of this is give rise to modes in the Fourier transform of the fields; these modes are precisely those studied in Sections 2.7.2 and 2.7.3. So, these first two steps are just the same as those used in previous chapters. Then, it is relatively simple to show that these modes satisfy a Hamiltonian of the same form as the harmonic oscillator, so that each mode may treated in this manner and quantized as in the preceding sections of this chapter. Thus, while quantizing the electromagnetic fields sounds daunting, it turns out to be no more than using the methods that have already been developed in the previous parts of this book. The daunting part is merely understanding the fields themselves, and their relationships with each other.

5.3.1 Electromagnetic Fields

The electromagnetic fields generally describe the generation and propagation of waves from frequencies that are extremely low to optical waves well above the visible spectrum. These are of course the fields accompanying photons and describe their propagation. As is known, the mass of these photons is zero. The waves generally are described by Maxwell's equations (Maxwell, 1865). The key fields are the electric field intensity $\boldsymbol{E}(\boldsymbol{r},t)$ and the magnetic field intensity $\boldsymbol{H}(\boldsymbol{r},t)$, which are determined by the vector potential $\boldsymbol{A}(\boldsymbol{r},t)$ and the scalar potential $\phi(\boldsymbol{r},t)$ through the equations (Balanis, 2024)

$$\begin{aligned} \boldsymbol{B}(\boldsymbol{r},t) &= \nabla \times \boldsymbol{A}(\boldsymbol{r},t), \\ \boldsymbol{E}(\boldsymbol{r},t) &= -\nabla \phi(\boldsymbol{r},t) - \frac{\partial \boldsymbol{A}(\boldsymbol{r},t)}{\partial t}. \end{aligned} \tag{5.45}$$

Here, the magnetic flux density is $\boldsymbol{B}(\boldsymbol{r},t) = \mu_0 \boldsymbol{H}(\boldsymbol{r},t)$ and is commonly used more than the magnetic field intensity $\boldsymbol{H}(\boldsymbol{r},t)$. The vector potential and the scalar potential are usually connected through what is called the Lorentz gauge equation

$$\nabla \cdot \boldsymbol{A}(\boldsymbol{r},t) - \mu_0 \epsilon_0 \frac{\partial \phi(\boldsymbol{r},t)}{\partial t} = 0, \tag{5.46}$$

where μ_0 and ϵ_0 are the permeability and permittivity, respectively, in free space, and satisfy

$$\mu_0 \epsilon_0 = \frac{1}{c^2}, \tag{5.47}$$

where c is the speed of light in free space. Normally, both the permeability and the permittivity can be different from their free space values when one considers various material systems, particularly semiconductors. This will, of course, change the speed of light in these systems.

In considering the quantization of electromagnetic fields, there is an assumption that the vector potential also satisfies the Coulomb gauge, with

$$\nabla \cdot \boldsymbol{A}(\boldsymbol{r},t) = 0, \tag{5.48}$$

which leads to $\boldsymbol{A}(\boldsymbol{r},t)$ being transverse to the direction of propagation, which itself is determined by the direction of $\boldsymbol{k}$. This also means that the scalar potential is time independent, and the electric field is normally determined from the vector potential. Normally, $\boldsymbol{A}(\boldsymbol{r},t)$ is parallel to the electric field, so that both $\boldsymbol{E}(\boldsymbol{r},t)$ and $\boldsymbol{H}(\boldsymbol{r},t)$ are transverse as well, and the wave is a plane wave. This plane wave is a rotating wave, in that the direction of the two fields rotates with the circular polarization of the plane wave, this polarization of right-hand, or left-hand, rotation defines the unit spin of the wave ($s = \pm 1$), with the plus sign for a right-hand circularly polarized wave. This allows us to define a unit polarization vector $\boldsymbol{e}_s$ according to

$$\boldsymbol{e}_s = \frac{1}{\sqrt{2}}\left(\boldsymbol{a}_x + si\boldsymbol{a}_y\right), \tag{5.49}$$

where $\boldsymbol{a}_x$ and $\boldsymbol{a}_y$ are the standard unit vectors in Cartesian coordinates.

There are some other relationships that are going to be useful in the following discussion. Many of these have already been used earlier in this text but are repeated here for convenience. That is, the wave momentum is given by

$$\boldsymbol{p} = \hbar \boldsymbol{k}, \tag{5.50}$$

and the wave number is related to the frequency through

$$\omega = 2\pi f = ck. \tag{5.51}$$

5.3.2 Quantization

Discretizing the wave momentum follows the discussion at the beginning of Chapter 4, in which periodic boundary conditions are imposed upon the fields. In three directions, this leads to the wave number taking on the values

$$k_x = \frac{2\pi n_x}{L}, k_y = \frac{2\pi n_y}{L}, k_z = \frac{2\pi n_z}{L}. \tag{5.52}$$

Thus, the normalizing volume is $\Omega = L^3$. Because the momentum is discretized, the energy will also be discretized and will deviate from the classical parabolic shape varying as p^2 of classical physics. To avoid this, it is usually necessary to make L sufficiently large that the energy remains in the lowest region of this artificial band structure so as to replicate the classical energy–momentum relationship.

With this form of discretization of the momentum, the Fourier integral of the spatially varying vector potential becomes a Fourier series which is summed over the values of the momentum. Since, the positive-going waves and the negative-going waves are usually treated differently, as discussed in Chapters 2 and 3, this means that the positive and negative magnitudes of the momentum should be treated as different Fourier modes here. Thus, the vector potential can be written as

$$\boldsymbol{A}(\boldsymbol{r},t) = \sum_{\boldsymbol{k}}\sum_{s}\left[\boldsymbol{e}_s \mathcal{A}_s(\boldsymbol{k},t)e^{i\boldsymbol{k}\cdot\boldsymbol{r}} + \boldsymbol{e}_s^* \mathcal{A}_s^*(\boldsymbol{k},t)e^{-i\boldsymbol{k}\cdot\boldsymbol{r}}\right]. \tag{5.53}$$

In quantizing the fields, the Fourier amplitudes ($\mathcal{A}$) are replaced with the creation and annihilation operators introduced in (5.12) but with different overall amplitudes. This is done through the quantization

$$\begin{aligned}\mathcal{A}_s(\boldsymbol{k},t) &\rightarrow \sqrt{\frac{\hbar}{2\omega\Omega\epsilon_0}}a_s(\boldsymbol{k}),\\ \mathcal{A}_s^*(\boldsymbol{k},t) &\rightarrow \sqrt{\frac{\hbar}{2\omega\Omega\epsilon_0}}a_s^\dagger(\boldsymbol{k}).\end{aligned} \tag{5.54}$$

These new creation and annihilation operators obey

$$\left[a_s(\boldsymbol{k}),a_{s'}^\dagger(\boldsymbol{k}')\right]=\delta_{\boldsymbol{k}\boldsymbol{k}'}\delta_{ss'}. \tag{5.55}$$

Of course, two different creation operators or two different annihilation operators commute.

Now, as mentioned above, each momentum state corresponds to a different propagating mode of the electromagnetic plane wave. In the space-time version, the total energy density is given by

$$\begin{aligned}E(\boldsymbol{r},t) &= \frac{\epsilon_0}{2}\left|\boldsymbol{E}(\boldsymbol{r},t)\right|^2+\frac{\mu_0}{2}\left|\boldsymbol{H}(\boldsymbol{r},t)\right|^2\\ &= \frac{\epsilon_0}{2}\left\{\left|\boldsymbol{E}(\boldsymbol{r},t)\right|^2+c^2\left|\boldsymbol{B}(\boldsymbol{r},t)\right|^2\right\}\end{aligned}. \tag{5.56}$$

This becomes the Hamiltonian when integrated over the volume, but if we introduce the quantization operators into the vector potential and compute the fields from (5.45), one arrives at

$$\begin{aligned}H(t) &= \frac{\epsilon_0}{2}\int_\Omega d^3\boldsymbol{r}\left\{\left|\boldsymbol{E}(\boldsymbol{r},t)\right|^2+c^2\left|\boldsymbol{B}(\boldsymbol{r},t)\right|^2\right\}\\ &= \frac{\epsilon_0\Omega\omega^2}{2}\sum_{\boldsymbol{k}}\sum_s\left[a_s^\dagger(\boldsymbol{k})a_s(\boldsymbol{k})+a_s(\boldsymbol{k})a_s^\dagger(\boldsymbol{k})\right]\\ &= \sum_{\boldsymbol{k}}\sum_s\hbar\omega\left[a_s^\dagger(\boldsymbol{k})a_s(\boldsymbol{k})+\frac{1}{2}\right].\end{aligned} \tag{5.57}$$

Thus, each polarization of each Fourier mode, defined by the momentum vector, is represented by a harmonic oscillator, which is quantized by the number of photons in that mode. As the photons are bosons (integer spin or polarization), they satisfy Bose–Einstein statistics, and any number of photons can occupy each mode. These are given as

$$P_{BE}=\left[exp\left(\frac{\hbar\omega}{k_BT}\right)-1\right]^{-1}. \tag{5.58}$$

5.4 LATTICE VIBRATIONS

In a solid, the motion of the atoms is much like the motion of free particles, with the important exception that the atoms are forced *on average* to retain positions within the solid that specify a particular crystal structure (we can also treat non-crystalline solids by similar methods, but our approach here will be limited to the crystalline solids, where the atoms are equally spaced on a *lattice*, such as in Figures 4.1 and 4.4). In any real solid, the lattice is a three-dimensional structure. However, when the motion is along one of the principal axes of the structure, it is usually possible to treat the atomic motion as a one-dimensional system. Although this is a simple model, its applicability can be extended to the real crystal if each atom represents the typical motion of an entire plane of atoms normal to the wave motion. In fact, the one-dimensional diatomic lattice begins exactly with Figure 4.4.

As in Chapter 4, the momentum wave number is already discretized due to the periodic boundary conditions and takes on the values given in (5.52). The normal model for the lattice differs from Figure 4.4 in that the atoms are assumed to be connected by springs, which directly, of course, connects to the harmonic oscillator. However, this differs from (4.19) in that the force of the spring depends upon the relative *difference* in the position of the atoms, and the zero motion has no energy associated with it. The use of the difference changes the sign in the second term of (4.19) and leads to a sine function rather than a cosine function, and the zero-energy set E_0 to zero (the factor of i is included in the constant A). As a result of these differences, the frequency of the vibrations for a Fourier space mode of momentum $\boldsymbol{k}$ is

$$\omega_k^2 = \frac{4C}{M} sin^2\left(\frac{ka}{2}\right). \tag{5.59}$$

where C is the spring constant and M is the atomic mass. (Traditionally, the momentum wave vector for the atomic motion is $\boldsymbol{q}$ rather than $\boldsymbol{k}$, and this will be used in the following.) At low wave number, this leads to

$$\omega_q \approx \sqrt{\frac{Ca^2}{M}}q = v_s q, \tag{5.60}$$

where v_s is the velocity of sound in the material.

There are actually three acoustic modes because of the three directions. These are normally aligned along the Cartesian axes. If the direction of propagation is in the z-direction, a wave whose atomic vibrational motion is also along the z-direction is referred to as a longitudinal acoustic mode. If the vibrational motion is normal to this z-direction, then it is referred to as a transverse acoustic mode, and there will be two of these along the y- and x-directions. Any arbitrary lattice motion is normally a combination of these three modes.

In the diatomic system of Figure 4.4, there will be two vibrational bands, just as the case for the electrons. One will be the sinusoidal band of (5.59) in which the two atoms are more or less in phase with one another. The other will arise from motion of the atoms with respect to each other, and correspondingly will have a much higher frequency. These are termed the optical modes. However, the fact that the forces depend upon the difference in the positions of the two atoms flips the upper band of Figure 4.5 so that the peak frequency (energy) is at $k = 0$. Again, there will be a longitudinal mode and two transverse modes. Normally, all optical modes are degenerate at $k = 0$, as are all acoustic modes.

There is one case where the degeneracy of the optical modes is raised at $k = 0$. This is when the atomic number of the two atoms is different. In this case, some charge transfer occurs (a small degree of ionicity enters the bond), and the longitudinal motion of the two against each other creates a polarization charge that adds to the normal electric fields (Ferry, 2020). The result of this is an additional term in the permittivity of the material that gives a static permittivity that is larger than the optical value of the permittivity (the lattice polarization frequency lies in the far infrared). In this case, the longitudinal optical phonon is termed the polar optical phonon, and its frequency is above that of the transverse modes. This is shown in Figure 5.5.[1]

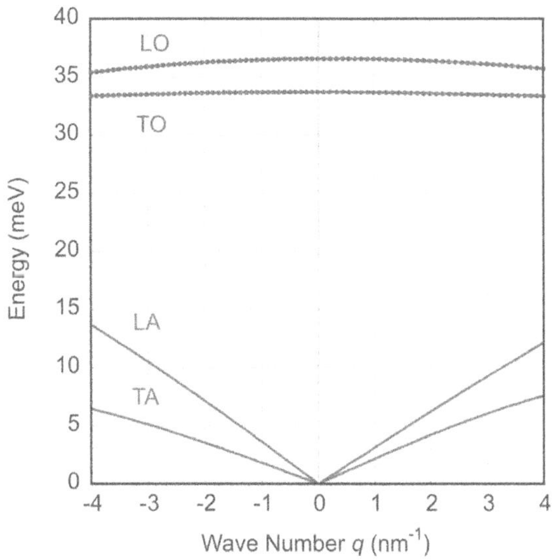

FIGURE 5.5 The various phonon energies around the $k = 0$ point (denoted normally by the symbol Γ). The parameters for GaAs were used to create the spectrum. The various phonon modes are indicated and the separation between the LO (polar in this case) and TO modes arises from the lattice polarization that leads to a larger static permittivity than the optical permittivity, as discussed in the text.

5.5 MOTION IN A QUANTIZING MAGNETIC FIELD

There is a very important example of a harmonic oscillator that should be considered, and this is the motion of an electron orbiting around magnetic field lines. Classically, the electron is pulled into a circular orbit such that the magnetic field is directed normal to the plane of the orbit, and the motion arises from the Lorentz force acting on the electron, as

$$\boldsymbol{F} = -e\left(\boldsymbol{E} + \boldsymbol{v} \times \boldsymbol{B}\right), \tag{5.61}$$

where $\boldsymbol{E}$ is the electric field (again, not to be confused with the energy), $\boldsymbol{v}$ is the electron velocity, and $\boldsymbol{B}$ is the magnetic flux density ($= \mu_0 \boldsymbol{H}$ in free space, as used above). The charge on the electron has been taken to be its proper negative value. Clearly, for motion in the $\boldsymbol{a}_\phi$ direction (in cylindrical coordinates) and a magnetic field in the $\boldsymbol{a}_z$ direction, the force is an inward centripetal force which causes the motion to be a closed circular orbit (ignoring the role of the electric field). Here, we want to examine the quantization of this orbit and will deal only with the electric field that is induced by the magnetic field B, ignoring any explicitly applied field. We will find that this leads to an interesting result that the density of electrons breaks up into a set of almost delta functions at the harmonic oscillator energies (which are called Landau levels for the magnetic quantization) and leads to a set of one-dimensional modes propagating at the edge of the sample. The Hall resistance (to be described below) is then quantized into the ratio of two fundamental constants, e and $\hbar$, and this result has caused a revolution and complete redesign of the international system of units (BIPM, 2019). Indeed, it may fairly be said that the new SI unit system is fundamentally based upon the quantum Hall effect that is discussed in this section.

The approach we follow is exactly the one we used in Section 1.3. The accelerative force on the wave momentum of the electron may be expressed as in (1.8) as

$$\hbar \frac{dk}{dt} = -e\left(\boldsymbol{v} \times \boldsymbol{B}\right). \tag{5.62}$$

In fact, we want to introduce the magnetic field through the vector potential. In Maxwell's equations, the magnetic field produces the electric field through Faraday's law, and manipulation of Maxwell's equations give the electric field from (5.45) as

$$\boldsymbol{E} = -\frac{\partial \boldsymbol{A}}{\partial t}. \tag{5.63}$$

However, any constant vector whose curl is zero can be added to the right-hand side. This would normally be the gradient of the scalar potential, but this term is not important at present. The electric field is determined from the magnetic field alone and hence from the vector potential that gives rise to the magnetic field. This suggests that if we now use the first line of (5.45) in place of the magnetic field term on the right-hand side of (5.62), the proper total momentum to use in the Hamiltonian is just the combined quantity

$$\boldsymbol{p} = \hbar \boldsymbol{k} + e\boldsymbol{A}, \tag{5.64}$$

which is usually termed the "Peierls' substitution" in solid-state physics, although the form has certainly been derived and developed by a great many workers.

It is now useful to use (5.64) for the momentum in the Hamiltonian (5.11), except that we do not include any external scalar potential. Nevertheless, the magnetic field will provide such a term. To proceed, one must actually take a functional value for the vector potential **A**. Our interest is in motion perpendicular to the magnetic field, which may be taken to be constant and oriented in the positive *z*-direction. Then, a vector potential may be defined as $\boldsymbol{A} = \boldsymbol{a}_y Bx$ which provides this magnetic field oriented in the correct direction. We note that this definition is only *sufficient*, as many other definitions of the vector potential could as easily describe a uniform field in the *z*-direction. However, this particular definition, called the *Landau gauge*, will suit our purposes sufficiently well. Furthermore, we will only treat the motion in the plane normal to the magnetic field.

In the absence of any potential, other than the vector potential giving rise to the magnetic field, Schrödinger's equation can be written as

$$-\frac{\hbar^2}{2m}\left(\nabla - \frac{eA}{i\hbar}\right)^2 \psi = E\psi, \tag{5.65}$$

where the wave momentum has been replaced by the normal gradient operator for the momentum. Expanding the momentum operator bracket, and using our form for the vector potential in the Landau gauge, leads to the differential equation

$$-\frac{\hbar^2}{2m}\left[\frac{\partial^2}{\partial x^2} - \frac{2eBx}{i\hbar}\frac{\partial}{\partial y} + \frac{\partial^2}{\partial y^2} - \left(\frac{eBx}{\hbar}\right)^2\right]\psi = E\psi. \tag{5.66}$$

(Note that here, E is again the energy.) Here, it may be observed that the magnetic field has indeed generated the quadratic term in position that is important to the harmonic oscillator. However, there are terms in the y-direction that have to be handled.

Traditional approaches (Landau, 1930) note that the quantization terms for the harmonic oscillator are in the x-direction, and there seems to be no confinement in the y-direction. This result is from the choice of the vector potential that has been used. Thus, it seems appropriate to define the wave function as

$$\psi(x, y) = e^{-iky}\phi(x). \tag{5.67}$$

With this substitution for the wave function, (5.66) becomes

$$-\frac{\hbar^2}{2m}\left[\frac{\partial^2}{\partial x^2} - \frac{2eBxk}{\hbar} + k^2 - \left(\frac{eBx}{\hbar}\right)^2\right]\phi = E\phi. \tag{5.68}$$

Our goal is to put this into a form reminiscent of the harmonic oscillator. To this end, we introduce the cyclotron frequency $\omega_c = eB/m$ and rewrite (5.68) as

$$-\frac{\hbar^2}{2m}\frac{\partial^2 \phi}{\partial x^2} + \frac{m}{2}\omega_c^2 \left(x - x_0\right)^2 \phi = E\phi, \tag{5.69}$$

where

$$x_0 = \frac{\hbar k}{eB} \tag{5.70}$$

is a displacement of the oscillator center from the normal $x = 0$.

From (5.67), k is the y-component of the wave motion. Comparison of (5.69) with (5.9) shows that the x-motion satisfies a harmonic oscillator equation, which yields Hermite polynomials as solutions. Thus, the energy levels are given as

$$E = \hbar\omega_c\left(n + \frac{1}{2}\right), \tag{5.71}$$

and the y-motion just shifts the center of the orbit. This circular motion is called cyclotron motion and the energy levels (5.71) are called Landau levels, as mentioned above. The corresponding wave function is just

$$\phi_n\left(x\right) = \frac{1}{\sqrt{2^n n!}}\left(\frac{m\omega_c}{\pi\hbar}\right)^{1/4} e^{m\omega_c\left(x - x_0\right)^2/2\hbar} H_n\left[\sqrt{\frac{m\omega_c}{\hbar}}\left(x - x_0\right)\right]. \tag{5.72}$$

The above solutions are, of course, just the normal harmonic oscillator wave functions shifted by the offset (5.70). Here, both the scaling factor and the offset depend upon the magnetic field, just as the cyclotron frequency does. Noting the argument of the Hermite polynomial in (5.72) gives a natural scaling length of the wave number (the product of the prefactor and x_0) described by $\sqrt{\hbar/eB}$, which is termed the *magnetic length.* To lowest order, the Hermite polynomial has its peak at the position $\sqrt{2n+1}$ times this basic length, so this is the natural cyclotron radius of the harmonic motion for a particular energy level.

5.5.1 The Hall Effect in Classical Terms

The form of the wave function (5.72) and the energy (5.71) do not give us a clear view of the orbiting nature of the electron in the magnetic field. True, it is a harmonic oscillator in the x-direction, but what about its other motion? As remarked earlier, the electron orbits around the magnetic field, and this motion is a solution of (5.62), which we can rewrite as

$$m\frac{d\boldsymbol{v}}{dt} = -e\left(\boldsymbol{v} \times \boldsymbol{B}\right). \tag{5.73}$$

The magnetic field is in the z-direction, according to the Landau gauge previously adopted, so that the motion in the (x, y)-plane is given by the equations

$$\begin{aligned} m\frac{dv_x}{dt} &= -ev_y B, \\ m\frac{dv_y}{dt} &= ev_x B. \end{aligned} \tag{5.74}$$

These two equations may then be solved for an arbitrary initial condition to give

$$v_x = v_0 cos(\omega_c t),$$
$$v_y = v_0 sin(\omega_c t). \tag{5.75}$$

where the velocity amplitude may be found from

$$E = \frac{1}{2}mv_0^2 = \hbar\omega_c\left(n + \frac{1}{2}\right) \tag{5.76}$$

or

$$v_0 = \sqrt{(2n+1)\frac{\hbar eB}{m^2}}. \tag{5.77}$$

Equation (5.75) can be integrated to fine the spatial positions of the particles as

$$x = \frac{v_0}{\omega_c} sin(\omega_c t),$$
$$y = -\frac{v_0}{\omega_c} cos(\omega_c t). \tag{5.78}$$

The radius of the orbit may now be found from these last equations to be

$$r^2 = x^2 + y^2 = \left(\frac{v_0}{\omega_c}\right)^2 = \sqrt{(2n+)\frac{\hbar}{eB}}. \tag{5.79}$$

The last line gives us the result hypothesized at the end of the previous section, $\sqrt{\hbar/eB}$ is a magnetic length that gives us the radius of the lowest energy level and provides a natural scaling of the size of the cyclotron orbit around the magnetic field. The quantization of the orbit radius arises directly from the quantization of the energy levels of the harmonic oscillator.

One of the important aspects of the quantization in a magnetic field is the fact that the continuum of allowed momentum states is broken up by this quantization. The allowed states are coalesced into the Landau levels, and each spin-degenerate Landau level can hold

$$n_s = \frac{1}{\pi r_0^2} = \frac{eB}{\pi\hbar} \tag{5.80}$$

electrons. In some magneto-transport experiments, this quantization can be seen as oscillatory magnetoresistance, in which the periodicity in $1/B$ arises from (5.80) as

$$\Delta\left(\frac{1}{B}\right) = \frac{e}{\pi\hbar n_s}. \tag{5.81}$$

This oscillatory magnetoresistance is termed the Shubnikov–de Haas effect.

5.5.2 Adding Lateral Confinement

While the above treatments still consider infinite semiconductors, the role of a finite size has important considerations, not the least of which is the Hall effect, and its quantum cousin the quantum Hall effect. To approach this, consider adding a simple harmonic oscillator potential in the x-direction, the opposite direction in which the magnetic field has already produced a harmonic oscillator behavior. When this additional potential is added to the system, (5.68) becomes

$$-\frac{\hbar^2}{2m}\left[\frac{\partial^2}{\partial x^2}-\frac{2eBxk}{\hbar}-k^2-\left(\frac{eBx}{\hbar}\right)^2+\frac{1}{2}m\omega^2x^2\right]\phi=E\phi. \tag{5.82}$$

It is clear that this can be rearranged, much along the lines of (5.68), to give

$$-\frac{\hbar^2}{2m}\frac{\partial^2\phi}{\partial x^2}+\frac{1}{2}m\Omega^2\left(x-x_0\right)^2\phi=\left[E-\frac{\hbar^2k^2}{2m}\left(1-\frac{4\omega_c^2}{\Omega^2}\right)\right]\phi, \tag{5.83}$$

with

$$\Omega^2=\omega^2+\omega_c^2,\qquad x_0=\frac{\hbar k\omega_c}{m\Omega^2}. \tag{5.84}$$

The motion remains that of a harmonic oscillator, but now the shift of the wave function and the energy levels are hybridizations of the confinement harmonic oscillator and the magnetic harmonic oscillator. Each energy level of the magnetic field, the Landau levels, is raised by the confinement to a higher value. In essence, this coupling of the two harmonic oscillators leads to enhanced confinement, and stronger confinement always costs energy—the result is that the energy levels lie at higher values. This becomes more important if the harmonic oscillator is nonlinear. That is, if the value of the parameter ω varies with distance from the center of the harmonic oscillator, then the energy levels are further increased as they are found farther from the center. This reaches the extreme in hard-wall boundary conditions at certain points on the x-axis. This is shown in Figure 5.6(a). Here, the Landau levels rapidly increase in energy as they approach the hard walls due to the extra confinement, just as in (5.84). The importance of this is the fact that an electron whose orbit is within the orbit radius of the edge will strike the wall and create a bouncing orbit that moves along the wall, as shown in Figure 5.6(b). Instead of being trapped in a Landau orbit, these bouncing orbits can carry current along the walls of the confining region and are called *edge states*.

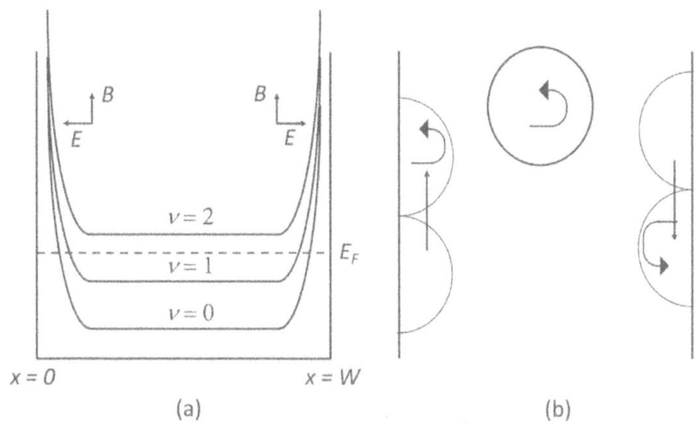

FIGURE 5.6 (a) The bending of the Landau levels at the edges of a confined sample. The electric field is shown for comparison at the two edges. (b) The confined and bouncing orbits for the situation of (a). The magnetic field is out of the page.

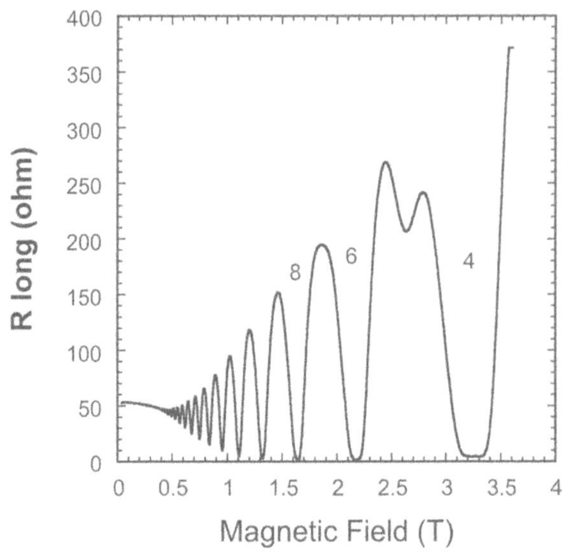

FIGURE 5.7 The longitudinal resistance for a quasi-two-dimensional electron gas in a high magnetic field. The oscillations are the Shubnikov–de Haas oscillations and correspond to the sequential emptying of Landau levels. Material parameters are discussed in the text.

When the Fermi level lies in a Landau level, there are many states available for the electron to gain small amounts of energy from the applied field and therefore contribute to the conduction process. On the other hand, when the Fermi level is in the energy gap between two Landau levels, the upper Landau levels are empty and the lower Landau levels are full. Thus, there are no available states for the electron to be accelerated into, and the conductivity drops to zero in two dimensions. In three dimensions it can be scattered into the direction parallel to the field (the z-direction), and this conductivity provides a positive background on which the oscillations ride. Figure 5.7 shows a typical measurement of the longitudinal resistance. The "zeros" of the longitudinal resistance correspond to the magnetic field for which there are full Landau levels. While this may seem confusing, the conductance in the plane normal to the magnetic field is a tensor that may be written as

$$\sigma = \begin{bmatrix} \sigma_{xx} & \sigma_{xy} \\ \sigma_{yx} & \sigma_{yy} \end{bmatrix}. \tag{5.85}$$

Typically, $\sigma_{xx} = \sigma_{yy}$ and $\sigma_{xy} = -\sigma_{yx}$. When this matrix is inverted to the resistivity (and also resistance when the dimensions are included) one obtains

$$\rho_{xx} = \frac{\sigma_{yy}}{\sigma_{xx}^2 + \sigma_{xy}^2}, \tag{5.86}$$

so that ρ_{xx} has zeros when $\sigma_{xx} = 0$ because the magnetic field makes $\sigma_{xy} \neq 0$. Although the longitudinal conductance vanishes, the transverse conductance does not, and this means that the longitudinal resistance also vanishes. The index that is shown (4, 6, 8, etc.) is that for spin-resolved levels, rather than spin-degenerate Landau levels. Hence, for the case of the zero at index 4 ($B = 3.25$ T, termed the $\nu = 4$ level), the $n = 0$ and $n = 1$ Landau levels (both doubly spin-degenerate) are full. The zero marked as 4 corresponds to the transition of the Fermi energy between the 5/2 and 3/2 (in units of $\hbar\omega_c$) levels in Figure 5.7. These measurements are for a quasi-two-dimensional electron gas at the interface of a GaAlAs/GaAs heterostructure. From (5.81), we can determine the areal density to be approximately 3.3×10^{11} cm^{-2}.

5.5.3 The Quantum Hall Effect

The zeros of the conductivity that occur in Figure 5.7 when the Fermi energy passes from one Landau level to the next-lowest level are quite enigmatic. They carry some interesting by-products. A full derivation of the quantum Hall effect is well beyond the level at which we are discussing the topic here. However, we can use a consistency argument to illustrate the quantization exactly, as well as to describe the effect that is observed. When the Fermi level lies between the Landau levels, the lower Landau levels are completely full, so that

$$E_F \sim \nu \hbar \omega_c, \tag{5.87}$$

where ν is an integer giving the number of *spin-resolved* Landau levels (as marked in Figure 5.7), and factor of 1/2 has been ignored. The presence of the edge states means that some carriers have moved to the edge of the sample, and this arises from including the electric field in the second of Equation (5.75) as

$$m\frac{dv_y}{dt} = -qE_y + ev_x B. \tag{5.88}$$

In a stable situation, the lateral acceleration must vanish at the edge, so that a transverse field must exist, given by

$$E_y = v_x B = -\frac{J_x}{ne}B. \tag{5.89}$$

Using (5.80) for the density in each Landau level, we have (inserting a factor of two to raise the spin degeneracy)

$$n_s = \nu\frac{eB}{2\pi\hbar} = \nu\left(\frac{e}{h}\right)B. \tag{5.90}$$

The density is constant in the material, so using (5.88) to define the *Hall resistance* gives

$$R_{xy} = -\frac{E_y}{J_x} = \frac{h}{\nu e^2}. \tag{5.91}$$

The quantity $h/e^2 = 25.81$ kΩ is a ratio of fundamental constants. Thus, the conductance (reciprocal of the resistance) increases stepwise as the Fermi level passes from one Landau level to the next-lower level. Between the Landau levels, when the Fermi energy is in the localized state region, the Hall resistance is constant (to better than 1 part in 10^7) at the quantized value given by (5.91) since the lower Landau levels are completely full. In Figure 5.8, the variation of the Hall resistance as a function of magnetic field for a typical sample is shown. These measurements were made in the same sample and geometry of Figure 5.7, so that they can be easily compared.

The magnetic field could, of course, be swept to higher values in both Figures 5.7 and 5.8. When this is done, new features appear, and these are not explained by the above theory. Klaus von Klitzing (1980) received the Nobel Prize for the discovery of the quantum Hall effect. In fact, in high-quality samples, once the Fermi energy is in the lowest Landau level, one begins to see fractional filling and plateaus, in which the resistance is a fraction of h/e^2 (Tsui *et al.*, 1982). This *fractional quantum Hall effect* is theorized to arise from the condensation of the interacting electron

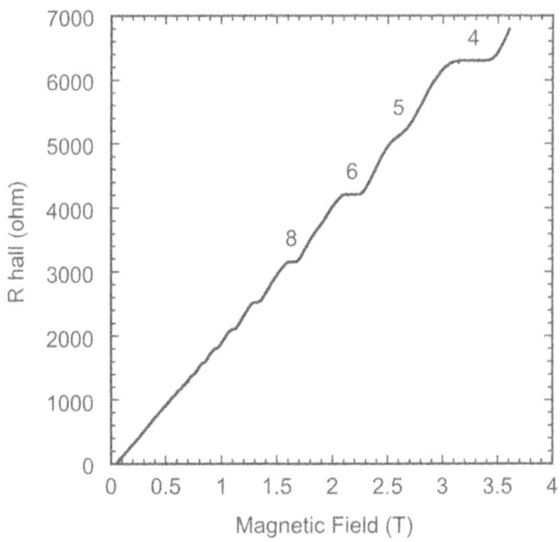

FIGURE 5.8 The Hall resistance of quantized Landau levels. The normalization is chosen so that the index value is of spin-resolved Landau levels. The first spin splitting is just being resolved at v = 5 (~2.6 T).

system into a new many-body state characteristic of an incompressible fluid (Laughlin, 1983). Tsui, Stormer, and Laughlin shared the Nobel Prize for this discovery. However, the properties of this many-body ground state are clearly beyond the present level.

5.6 THE MODERN STANDARD UNIT SYSTEM

The MKS system of units, what we commonly call SI units, was standardized in 1889 in order to provide some international agreement on what constituted a kilogram or a meter. Originally, the kilogram was a standardized hunk of precious metal that the French called *Le Grande K*. Several similar pieces of this material are kept in the US at the National Institute of Standards and Technology. Similarly, the meter was defined as one ten-millionth of the distance from the North Pole to the equator. Over the years, changes were gradually made in many of the various units that are part of the MKS hierarchy. What most of us knew and accepted in this system dates from 1960 when an updated international set of units was published.

Recently, however, this system has been changed once again, with new standards being based upon revelations from quantum mechanics, such as Planck's constant and the quantum Hall effect. In 2019, the world standardized the value of Planck's constant to $6.62607015 \times 10^{-34}$ J-s (BIPM, 2019[2]). Similarly, the meter has been standardized as the distance light travels in $1/c$ seconds, where c is the speed of light, standardized to 2.99792458×10^8 m/s. But this latter definition relies upon the second itself, which arises from the frequency separation of the Cesium-133 atom's outer electron spin resonance (transition from spin up to spin down), which is 9.1926317790×10^9 Hz. Now, the quantum Hall effect gives us the resistance of the $\nu = 1$ plateau to be 25812.807 ohms, and we know from (5.89) that this is h/e^2. Hence, standardizing this value and Planck's constant means that we have standardized a value for the electronic charge e. Since Planck's constant can also be given its value in terms of kg-m^2/s, standardization of this quantity plus the standard second and meter now defines the new value for the kilogram as well. And this can be extended to other standards for current, voltage, and so on.

The adopted standards in the new 2019 version of the SI uses seven *defined* constants. These include the frequency of the unperturbed ground state energy of the Cesium-133 atom, mentioned above; the speed of light in a vacuum (299792458 m/s), Planck's constant, the elementary charge (1.602176634 C), Boltzmann's constant ($1.380649 \times 10^{-23}\, J/K$), Avogadro's constant ($6.02214076 \times 10^{23}\, /mol$), and the luminous efficiency of monochromatic radiation of frequency 540×10^{12} Hz of 683 lm/W. Every other parameter that we consider a constant is derived from these seven quantities.

For example, the second is defined as $9192631770/f_{Cs-133}$, while the meter is defined to be $30.663319c/f_{Cs-133}$. In addition, the standard kilogram is now defined by h, since the units on Planck's constant are J-s, or kg-m²/s, a kilogram is now defined in terms of h by $1.4755214\times10^{40}(hf_{Cs-133}/c^2)$. Similarly, the ampere is given by the measured charge divided by its defined value, or 1 Ampere is $\approx 6.789687\times10^{8} f_{Cs-133}e$, energy is defined from h and the second, and the volt is defined from h, the second, and the ampere. There are no more stored hunks of mass and lengths of sticks; all comes from the definition of some fundamental constants.

NOTES

1 The plot in the figure was created using data determined from the band structure tool at Nanohub.org, with GaAs parameters.
2 An English-only version can be obtained from NIST—https://nvlpubs.nist.gov/nistpubs/Special Publications/NIST.SP.330-2019.pdf

REFERENCES

Balanis, C. 2024. *Advanced Engineering Electromagnetics, 3rd Ed.* (New York: Wiley).
Bureau International des Poids et Mesures. 2019. *The International System of Units (SI), 9th ed.*.
Ferry, D. K. 1995. *Quantum Mechanics: An Introduction for Device Physicists and Electrical Engineer*, 1st Ed. (Bristol: IOP Publishing).
Ferry, D. K. 2020. *Semiconductors: Bonds and Bands, 2nd Ed.* (Bristol: IOP Publishing).
Landau, L. 1930. Diamagnetismus der Metalle. *Zeit. Phys.* **64** 629–637.
Laughlin, R. B. 1983. Anomalous quantum Hall effect: An incompressible quantum fluid with fractionally charged excitations. *Phys. Rev. Lett.* **50** 1395–1398.
Maxwell, J. C. 1865. A dynamical theory of the electromagnetic field. *Phil. Trans. Royal Soc. (London).* **155** 459–512.
Planck, M. 1900. Zur Theorie des gesetzes der Energievertheilung im Normalspektrum. *Verhand. Deut. Phy. Gesell.* **2** 237–245.
Tsui, D., Stormer, H. L. and Gossard, A. C. 1982. Two-dimensional magnetoresistance in the extreme quantum limit. *Phys. Rev. Lett.* **48** 1559–1562.
Von Klitzing, K., Dorda, G. and Pepper, M. 1980. New method for high accuracy determination of the fine-structure constant based on quantized Hall Resistance. *Phys. Rev. Lett.* **45** 494–497.

PROBLEMS

1. Using the creation and annihilation operators, compute p^2 and x^2 for an arbitrary harmonic oscillator state wave function.
2. Determine the expectation values for the kinetic and potential energies in a harmonic oscillator in a state n. Make use of the creation and annihilation operators discussed in class. What can you say about T and V?
3. By methods similar to those we used for the evaluation of $\langle x^2 \rangle$, evaluate $\langle x^4 \rangle$ and $\langle p^4 \rangle$ for the case when $\psi(x) = \psi_n(x)$, a harmonic oscillator wave function.
4. Consider an electromagnetic resonator with a resonant frequency of 10^{10} Hz. What is the energy separation of the oscillator levels? How does this compare with the thermal energy fluctuation?
5. If the velocity of sound $d\omega/dq$ of a set of lattice vibrations is 10^5 cm/s, and the lattice constant is $a = 0.25$ nm, what is the phonon frequency at the zone edge $q = \pi/a$?
6. Small-amplitude vibrations of a diatomic molecule can be studied using as a model of the true molecular vibrational potential the simple harmonic oscillator potential:

$$V(x) = \frac{1}{2}k_f x^2,$$

where k_f is the force constant. For a typical diatomic molecule, the force constant is roughly $k_f = 1.0 J/m^2$, and the mass is $9.57m_0$.

a. Using the above values, calculate the value of the zero-point vibrational energy of a diatomic molecule in eV.
b. Calculate the energy spacing (in eV) between the ground (zero-point) vibrational state and the first excited vibrational state of the molecule.
c. Suppose the molecule undergoes a transition from the first excited vibrational state to the ground state, emitting a photon in the process. Calculate the energy, frequency, and wavelength of the emitted photon. In what range of the electromagnetic spectrum is this photon found?

7. Consider an electron near the surface of a solid. The surface presents an infinite potential barrier to the electron. Suppose the forces inside a metal are most accurately modeled by the simple harmonic oscillator (SHO) potential energy, so that the potential of the electron is:

$$V(x) = \begin{cases} \infty, & x < 0, \\ \frac{1}{2} m\omega_0^2 x^2, & x \geq 0. \end{cases}$$

a. What are the energies of the lowest three stationary states of this system? Explain how you arrived at your answer.
b. Do the eigenfunctions of this system have definite parity? Why or why not?
c. Sketch the eigenfunctions for the ground state and the first excited state of this system.

8. If one begins with the so-called symmetric gauge, where $\mathbf{A} = (eB/2)(-y\mathbf{a}_x + x\mathbf{a}_y)$, show that a more complicated harmonic oscillator solution results. Show that this still gives energy levels according to (5.70) and find the appropriate form of the wave functions that are the solutions.
9. Consider an electron moving in a z-directed magnetic field and constrained in a quadratic potential $\frac{1}{2} m\omega_0^2 x^2$. In the simplest case, in which the system is homogeneous in the y-direction, determine the energy levels of the electron.
10. A semiconductor sample measures 1 cm by 0.5 cm and is 0.1 cm thick. For an applied longitudinal (along the current direction) electric field of 1 V/cm, 5 mA of current flows. If a 0.5 T magnetic field is applied normal to the broad surface, a Hall voltage of 5 mV is developed. Determine the Hall mobility and the carrier density.
11. Consider a free electron 'gas' with an areal density of 2×10^{12} cm^{-2}. What is the periodicity (in units of $1/B$) expected for the Shubnikov–de Haas oscillation?

6 Operators and Bases

In the past few chapters, the Schrödinger equation has been developed and applied to the solution of a number of quantum mechanical problems, mainly to develop experience with the results of simple and common systems. The results from these examples can be summarized relatively simply. In general, quantization enters the problem through the non-commuting nature of conjugate operators such as position and momentum as well as from classical boundary quantization. This has led to the Schrödinger equation itself being the primary equation of motion for the wave function solution to the problem. In essence, duality is incredibly important and is the basic method by which the electron is treated as a wave, rather than as a particle, and the wave equation of interest is the Schrödinger equation. When boundary conditions are applied, either through potential barriers or through the form of the potential (as in the harmonic oscillator), the time-independent Schrödinger equation yields solutions that are often special functions. Examples of this are the sines and cosines in the rectangular-barrier case, the Airy functions in the triangular-well case, and the Hermite polynomials in the harmonic oscillator case.

It is generally true that any time we examine a bounded system (even a classical system), the allowed energy levels take on discrete values, each of which corresponds to a single one of the family of possible members of the special functions. Thus, in the rectangular-well case, each energy level corresponds to one of the sinusoidal Fourier harmonics; in the harmonic oscillator, each energy level corresponds to one of the Hermite polynomials. In the set of levels, there is always a lowest energy level, called the ground state. The higher energy levels are referred to as the excited states, even when some are occupied in thermal equilibrium.

It should not be surprising that the set of all possible solutions to a given problem formulation, for example, the set of all sines and cosines for the rectangular barriers, can be shown to form a complete set and thus can serve as a *basis set* for a many-dimensional *linear-vector-space* representation of the problem. In the preceding chapters, we have not employed this terminology but rather treated the expansion in terms of these functions in a manner that usually ignored the time variation. This is the normal time-independent Schrödinger *picture*, in which the so-called basis functions arise from solving the Schrödinger equation subject to the boundary conditions and these basis functions. In this picture, the operators are not time-varying, but their projection onto any of the basis functions is time-varying. Here, the basis functions can be thought of as the unit vectors of a coordinate system, and the amplitudes, such as those of can be thought of as amplitudes along each axis of this coordinate system. In the *Schrödinger picture*, the coordinate system rotates around the direction of the vector operator as a function of time. But this is a relative view—it could as easily be that the coordinate system is fixed and the vector operator rotates in the coordinate frame in such a way that the projection on any axis still varies with time in the prescribed manner. This picture, or view, is referred to as the *Heisenberg picture* of quantum mechanics (Heisenberg, 1925). However, there is a time variation, and we begin this chapter with a discussion of this.

With this chapter, we want to achieve a number of things. First, in the next two sections, we will examine how introducing the time variation leads us to be able to determine the time dependence of operators. For example, we introduced pseudo-position and pseudo-momentum operators in order to quantize some systems. In essence, this was the introduction of generalized conjugate operators, and we want to examine here the rules that allow us to do this. In fact, the *linear vector–space* description will also be introduced in these next two sections, although it will be formally be treated in a following section, where we will examine how operators function in this space. Then, we turn to using generalized operators, such as those encountered in Chapter 5 to actually solve cases with more complex potentials and study the symmetries that are introduced by these complex potentials

DOI: 10.1201/9781003717195-6

(where this can be done). In a following section, it is worthwhile to see how the Schrödinger equation can be formally connected to the Fokker–Planck equation of statistical mechanics to formalize the probability interpretation of quantum mechanics. Then, we look at a more complex way of actually studying the probabilities rather than just the wave function and how this can lead to a phase-space formulation of quantum mechanics.

6.1 TIME DEPENDENCE OF WAVE FUNCTIONS

The connection between the Heisenberg picture and the Schrödinger picture may be arrived at in a relatively simple manner. The wave function $\psi(x,t)$ evolves under the action of the Hamiltonian operator via the time-dependent Schrödinger Equation (1.18)

$$i\hbar\frac{\partial\psi}{\partial t} = H\psi = \left[-\frac{\hbar^2}{2m}\nabla^2 + V(r)\right]\psi \tag{6.1}$$

In a sense, this wave function evolves (in one dimension) as

$$\psi(x,t) = e^{-iHt/\hbar}\psi(x,0). \tag{6.2}$$

If we consider any arbitrary complete set of functions $\{\phi_n(x)\}$, for example, the set of basis states for an infinite potential well lying in the range $0 < x < a$, for which the solutions were found in Chapter 2, a basis function may be written as

$$\varphi_n(x,t) = \sqrt{\frac{2}{a}}sin\left(\frac{n\pi x}{a}\right)e^{-iE_nt/\hbar}, \tag{6.3}$$

in which the explicit time variation has been included, and in general, any wave function may be written in this situation as

$$\psi(x,t) = \sum_n c_n\varphi_n(x,t) = \sqrt{\frac{2}{a}}\sum_n c_n sin\left(\frac{n\pi x}{a}\right)e^{-iE_nt/\hbar}. \tag{6.4}$$

The basis functions ϕ_n create an infinite set of coordinates, and E_n is the energy of the nth state, while the coefficients $c_n(t)$ are the projections of the wave function onto each of these coordinates. Hence, the basis states create an infinite-dimensional coordinate system, which is known as the Hilbert space corresponding to these functions. However, as pointed out above, in this notation, the exponential operators lead to a rotation of the coordinate system in time (because they change the length of the various unit vectors, the sines, and/or the coefficients c_n). Nevertheless, it can be repeated that the basis states create an infinite-dimensional coordinate system, which is the Hilbert space corresponding to these functions.

As we will see, the difference between the Heisenberg picture and the Schrödinger picture is one of just where to assign the temporal exponential factor. That is, should it be connected with the basis states as has been done here for the Schrödinger picture, or should it be connected with any operator that is present in the system, as will be described below, for the Heisenberg picture. As mentioned in the last paragraph, it is perhaps more meaningful to assign the exponential factor to the coefficient of the basis function as

$$c_n(t) = e^{-i\omega_n t}c_n(0). \tag{6.5}$$

This is already a step toward the Heisenberg picture.

6.2 THE TIME DEPENDENCE OF OPERATORS

Already in Chapter 1, the expectation value of the position operator was introduced as an integration over the magnitude squared value of the function. This was not an abstract, single occurrence, or process. This can be done quite generally for any operator, and the expectation value of such a general operator A is given by the generalization of (1.28) as

$$\langle A \rangle = \int \psi^\dagger(x,t) A \psi(x,t) dx, \tag{6.6}$$

where the wave function here has a time dependence included. The general operator A is not written with a dependence on position and momentum, although it is quite likely to be a function of these generalized conjugate variables. Thus, $A = A(p,x)$ may well be some function of the position and of the momentum. An example of this is the quadratic harmonic oscillator potential energy, in which we can write $A = m\omega^2 x^2/2$, so the expectation value (6.6) would be the average of the potential energy. Nevertheless, the appearance of these operators might well confuse the points to be made here.

Now, the question that arises is just how does the average defined in (6.6) change with time in a system in which the wave function $\psi(x,t)$ evolves in time? In order to examine this question, a general approach for an operator needs to be developed. To begin, it will be assumed that the operator A is *not a function of time itself* (time will not be one of the explicit variables upon which it depends). Thus, the time variation of (6.6) can be expanded as

$$\begin{aligned} \frac{d\langle A \rangle}{dt} &= \int \frac{\partial \psi^\dagger(x,t)}{\partial t} A \psi(x,t) dx + \int \psi^\dagger(x,t) A \frac{\partial \psi(x,t)}{\partial t} dx. \\ &= \frac{i}{\hbar} \int H^\dagger \psi^\dagger(x,t) A \psi(x,t) dx - \frac{i}{\hbar} \int \psi^\dagger(x,t) A H \psi(x,t) dx \\ &= \frac{i}{\hbar} \int \psi^\dagger(x,t)(HA - AH)\psi(x,t) dx. \\ &= \frac{i}{\hbar} \langle [H,A] \rangle. \end{aligned} \tag{6.7}$$

The result (6.7) is very important and is the basis of the Heisenberg approach to quantum mechanics, but it is applicable to all quantum mechanics. Previously, one needed to solve directly for the wave function and then develop the expectation value of specific operators. Here, however, *we may choose to construct the wave junction with any convenient basis set.* The time variation of the expectation values of the operators is entirely determined by the commutator relation between the specific operator and the total Hamiltonian. *Thus, the expectation value of any operator that commutes with the Hamiltonian remains constant in time.* This is because in conservative (non-dissipative systems), the Hamiltonian is not a function of time and the total energy is conserved. Therefore, any operator that commutes with the Hamiltonian is similarly conserved and can also be simultaneously measured.

In treating conjugate operators, such as position and momentum, it was not generally possible (with a potential term present) to write the Hamiltonian as a function of only position or of only momentum. In general, the Hamiltonian contains both position (in the potential energy term) and momentum (in the kinetic energy term). Thus, we cannot choose a basis in which both commute with the Hamiltonian, so we cannot measure both simultaneously. Hence, non-commuting operators generally appear in a manner such that they do not commute with the Hamiltonian.

Equation (6.7) leads us to the general view that we can choose a basis set of functions, in terms of which to expand the total wave function, that is *convenient.* Thus, one could choose a set that is the natural set of expansion functions for the coordinate system and boundary conditions present. The choice is only important in that one must be able to evaluate the integrals inherent in the averages expressed in (6.7), and as expansions of the wave functions in, for example (6.4), to which we return in the next section. So far, the time variation in the wave functions has been ignored, although (6.7) begins to transfer this time variation to the operator itself. Let us continue this process.

The time-varying Schrödinger equation allows one to write directly the time variation through a simple exponential term, as was done in (6.4). We know that the energy arises in the argument of the exponential as the eigenvalue for the Hamiltonian (which must be summed over in most cases), so when we introduce this into the averages in (6.7), the operator form will be retained; for example, the formal solution of the time-varying Schrödinger equation is introduced into (6.7). This leads to

$$\begin{aligned}\frac{d\langle A\rangle}{dt} &= \frac{i}{\hbar}\int \psi^{\dagger}(x,t)(HA - AH)\psi(x,t)\,dx. \\ &= \frac{i}{\hbar}\int \psi^{\dagger}(x)e^{iHt/\hbar}(HA - AH)e^{-iHt/\hbar}\psi(x)\,dx. \\ &= \frac{i}{\hbar}\int \psi^{\dagger}(x)\left[H, e^{iHt/\hbar}Ae^{-iHt/\hbar}\right]\psi(x)\,dx.\end{aligned} \tag{6.8}$$

This has created a time-varying operator in the Heisenberg representation (where the wave functions rotate in the coordinate system defined by the basis functions), which is given as

$$A_H(t) = e^{iHt/\hbar}Ae^{-iHt/\hbar}, \tag{6.9}$$

when the operator does not have a specific time variation on its own. The subscript 'H' is often used to denote the Heisenberg representation of an operator, which varies with time only through the effects that arise from the non-commuting of this operator with the Hamiltonian. However, this subscript is just as often not explicitly given. From (6.9), one can work backward toward the starting Equation (6.6) by dropping the time derivatives, so the expectation of an operator can be written in the Heisenberg picture as

$$\langle A\rangle = \int \psi^{\dagger}(x)A_H(t)\psi(x)\,dx. \tag{6.10}$$

This last equation is the basis for the Heisenberg picture (or representation) of quantum mechanics. Here, the wave function is expanded in a convenient basis set of functions, which corresponds to a fixed coordinate system. Then, instead of the wave function evolving or varying with time, the operators themselves evolve in time (if they do not commute with the Hamiltonian). The averages are now computed using the arbitrary, fixed basis functions and the "time-varying" Heisenberg representation of the operators. In essence, if one expands the wave function in terms of the basis set, the above average is a matrix multiplication.

This approach can be applied in almost any situation. In fact, as a tutorial, let us apply this to determine the time variation of the creation and annihilation operators of the harmonic oscillator itself. Here, the Hamiltonian is given as

$$E = \sum_n \hbar\omega_n\left(a_n^{\dagger}a_n + \frac{1}{2}\right). \tag{6.11}$$

The operators with subscript n operate only on the wave function for that energy level. Hence, one can write the creation operator for level n

$$\begin{aligned}\frac{da_n}{dt} &= \frac{i}{\hbar}\hbar\omega_n\left[a_n^{\dagger}a_n, a_n\right] = i\omega_n\left(a_n^{\dagger}a_n a_n - a_n a_n^{\dagger}a_n\right) \\ &= i\omega_n\left(a_n^{\dagger}a_n a_n - 1 - a_n^{\dagger}a_n a_n\right) = i\omega_n\end{aligned}. \tag{6.12}$$

In the last line, the commutator relation (5.16) has been used. This leads easily to

$$a_n(t) = a_n(0)e^{i\omega_n t}. \tag{6.13}$$

In a similar manner, the annihilation operator becomes

$$a_n^\dagger(t) = a_n^\dagger(0)e^{-i\omega_n t}. \tag{6.14}$$

6.3 LINEAR VECTOR SPACES

When one talks about linear vector spaces, he/she is combining the concepts of state variables, used in linear-systems theory, and vector calculus, used in electromagnetic fields. In vector calculus, one chooses a set of basis vectors to define a coordinate system; these can be, for example, the unit vectors along the three Cartesian coordinate directions in a rectangular coordinate system. Then, any general vector can be written as the sum of components in each of these directions, as

$$\boldsymbol{B} = b_1\boldsymbol{a}_x + b_2\boldsymbol{a}_y + b_3\boldsymbol{a}_z, \tag{6.15}$$

although this approach is readily applied to any coordinate system, and especially so for those for which the differential equation is separable. For example, there are far more coordinate systems for which Laplace's equation is separable than there are for Poisson's equation (Morse and Feshbach, 1953).

In (6.15), the $\boldsymbol{a}_i$ are the unit vectors, and the b_i are the components of the vector B in each of the various directions defined by the unit vectors. The same ideas carry over to linear vector spaces of operators. We choose a basis set of functions that define the coordinate "directions," which are usually infinite in number. Then, one can expand any general function in terms of the components as in (6.4), with any time dependence incorporated. Again, the $\varphi_i(x)$ are the unit vectors of this space, and the c_i are the components of the "vector" in each of the various directions defined by the unit vectors. The difference between (6.4) and (6.15) is that the Cartesian coordinate system has only three unit vectors (is only three-dimensional) while the linear vector space generally has an infinite number of dimensions (Kaschner and Russell, 2023; von Neumann, 1932).

In the rectangular coordinate system for vectors, and in any orthogonal coordinate system (such as spherical or cylindrical coordinates), the unit vectors satisfy the relationship $\boldsymbol{a}_i \cdot \boldsymbol{a}_j = \delta_{ij}$, that is, they satisfy an orthonormality condition. Similarly, the basis vectors in the linear vector space also are required to be orthonormal (normally, but there actually are cases where this is not required, such as "fuzzy" sets (Zadeh, 1965) which are not orthonormal[1]). This is expressed as

$$\int \varphi_i^\dagger(x)\varphi_j(x)dx = \delta_{ij}. \tag{6.16}$$

Just as one finds the property of closure of a complete set in Fourier series, the basis set also satisfies a *closure* relation as

$$\sum_i \varphi_i^\dagger(x)\varphi_i(x') = \delta(x - x'). \tag{6.17}$$

The complete set of basis functions has many standard properties associated with it being a *linear* vector space. Some of these are the following: (1) when two basis functions are added, the order is unimportant; (2) when a basis function is multiplied by two c-numbers (non-operators), the order of these two numbers is unimportant; (3) when the sum of a number of basis functions is multiplied by a c-number, or an operator, each basis function is multiplied by this c-number or operator. Finally, the basis functions are linearly independent, so that a statement such as

$$\sum_i c_i\varphi_i(x) = 0, \tag{6.18}$$

implies that all $c_i = 0$. This leads us to a very important point. If the wave function can be written in terms of a single basis function as

$$\psi(x) = c_i \varphi_i(x), \tag{6.19}$$

the wave function is in a *pure* state. Otherwise, it may be said to be in a *mixed* state. If a given wave function cannot be expanded into a sum over a set of basis functions, it is said to be in an *entangled* state, which will be addressed below.

6.3.1 Hermitian Operators

It may be recalled from earlier discussions that $A^\dagger$ is the Hermitian adjoint of the operator A. If A were represented by a matrix, then the Hermitian adjoint would correspond to the transpose, complex conjugate of that matrix. Here, it has to be emphasized once again that all *measurable* quantities correspond to operators in which the Hermitian conjugate is equal to the operator itself; for example, the operator is said to be a Hermitian operator. This was mentioned in earlier chapters and again noted in the average value of the operator A given by (6.12). If the expectation value is a measurable quantity, then this average value must be a real quantity. With this in mind, we rewrite (6.10) to set the expectation value to its complex conjugate as

$$\begin{aligned} \langle A \rangle = \langle A \rangle^* &= \left[\int \psi^\dagger(x,t) A\psi(x,t)\,dx \right]^* \\ &= \int \left[A\psi(x,t) \right]^\dagger \psi(x,t)\,dx. \\ &= \int \psi^\dagger(x,t) A^\dagger \psi(x,t)\,dx. \end{aligned} \tag{6.20}$$

By comparison with (6.10), it is now obvious that for the expectation value to be a real quantity one must have $A = A^\dagger$. As mentioned, operators for which this relation holds are said to be Hermitian operators.

Now, consider the situation in which the wave function is described in such a manner that the operator A produces the eigenvalue a, where a is a c-number (not an operator but a scalar constant). Thus, the operator relation is expressed as

$$A\psi(x,t) = a\psi(x,t), \tag{6.21}$$

for which

$$\langle A \rangle = \int \psi^\dagger(x,t) A\psi(x,t)\,dx = \int \psi^\dagger(x,t) a\psi(x,t)\,dx = a. \tag{6.22}$$

By the same token, the expectation value of the square of the operator becomes

$$\langle A^2 \rangle = \int \psi^\dagger(x,t) A^2\psi(x,t)\,dx = \int \psi^\dagger(x,t) a^2\psi(x,t)\,dx = a^2. \tag{6.23}$$

The uncertainty of an operator is given by $\sqrt{A^2 - A^2}$. The uncertainty of the above operator is zero, so that an operator with an eigenvalue, such as in (6.21), has no uncertainty! Its expectation value is the eigenvalue.

Suppose that there are two such wave functions for which the operator A produces eigenvalues, so that

$$\begin{aligned} A\psi_1(x,t) &= a_1\psi_1(x,t), \\ A\psi_2(x,t) &= a_2\psi_2(x,t). \end{aligned} \tag{6.24}$$

This is, in fact, not allowed. To show this, the first equation of (6.24) is pre-multiplied by $\psi_2^\dagger$, while the adjoint of the second equation is post-multiplied by ψ_1. Then, the two equations are subtracted, the second from the first, as

$$\begin{aligned}(a_1 - a_2^*)\int \psi_2^\dagger(x,t)\psi_1(x,t)dx \\ = \int \psi_2^\dagger(x,t)A\psi_1(x,t)dx - \int\left[A\psi_2(x,t)\right]^\dagger \psi_1(x,t)dx \\ - \int \psi_2^\dagger(x,t)A\psi_1(x,t)dx - \int \psi_2^\dagger(x,t)A\psi_1(x,t)dx = 0\end{aligned} \tag{6.25}$$

There are two options for the integral on the left. The first option is that $\psi_2 = \psi_1$, which leads to $a_1 = a_2^* = a_1^*$. That is, if the two wave functions are the same, then the eigenvalue must be the same, and this eigenvalue must be real! The second option is that the eigenvalues actually are different, for which the integral itself must vanish, which means that the two wave functions are orthogonal to one another. Hence, they must be two different basis functions in the expansion of the wave function.

It is the above point that allows us to set up a "coordinate" system with basis functions that are orthonormal (orthogonal and normalized). This is the purpose of the linear-vector-space approach. One can choose any set of orthogonalized (and normalized) wave functions $\varphi_i(x)$ (only the spatial variation is kept here, as we will pursue the Heisenberg picture for clarity). These can correspond to the infinite set of wave functions found in the potential well problems of Chapter 2, or the set of Hermite polynomials of Chapter 5, as an example. Whichever is chosen, it may be expected that any operator A will give a set of eigenvalues a_i corresponding to each member of the set of wave functions. To illustrate this, let us expand the total wave function in a set of basis functions, in which the function $\varphi_i(x)$ could be either of these mentioned choices. This expansion is written

$$\psi(x) = \sum_n c_n \varphi_n(x), \tag{6.26}$$

which, of course, has appeared already many times. However, now, this expansion will be used to find the expectation of the operator A as

$$\begin{aligned}A &= \sum_{n,m} c_n^* c_m \int \varphi_n^\dagger(x) A \varphi_m(x) dx \\ &= \sum_{n,m} c_n^* c_m \delta_{nm} a_m \\ &= \sum_n |c_n|^2 a_n\end{aligned} \tag{6.27}$$

In going from the first line to the second, the fact that the integral is zero (orthogonality) unless $n = m$ has been used. This provides the Kronecker delta function in the second line, which is then used in the summation over the index m. Thus, the coefficients provide the weight, or probability amplitude, for each of the individual eigenvalues of the basis set. The set of eigenvalues a_i is termed the *spectrum* of the operator A. The connection with a normal spectrum is quite straightforward: in the normal case, the strength of a component at frequency ω is given by the corresponding Fourier coefficient. Here, the eigenvalue a_i is the generalized Fourier coefficient giving the strength of the operator A in the particular basis coordinate φ_i. The coefficients c_i give a normalization to the excitation of the particular mode φ_i by the total wave function.

One important aspect of the set of basis functions is their use. For example, if the eigenvalues are known for an operator, one can construct a set of basis functions that yield these eigenvalues. Thus, it is possible to construct the series (6.26) from the known eigenvalues. The time-independent Schrödinger equation is known to be a linear differential equation. Consequently, *any* basis set may be used to solve a problem, so long as this linearity exists. It is a fundamental property of the linear second-order differential equations that any solution found that satisfies both the equation and the boundary conditions is a correct solution. Thus, any set that is convenient to use will provide a correct answer, although the form may not be as pleasing as a different choice for the set.

6.3.2 Some Matrix Properties

Once a basis set of functions has been selected, then it is no longer necessary to continue to restate these functions. Rather, we need only to keep track of the expansion coefficients for the wave function. This leads to matrix operations. Consider, for example, a wave function expanded in the series given in (6.26). If one operates on this wave function with the operator A, it will produce the new wave function given by $A\psi_a = \psi_b$. Then, the process may be written as an equation as

$$\begin{aligned} A\psi_a(x) &= \sum_n a_n A\varphi_n(x) \\ &= \sum_m b_m \varphi_m(x) = \psi_b(x) \end{aligned}. \tag{6.28}$$

If the last two expressions (on the second line) are pre-multiplied by $\varphi_j^*(x)$, and integrated over space, one achieves the result

$$b_j = \sum_i a_i A_{ji}, \tag{6.29}$$

where the *matrix element* is

$$A_{ji} = \int \varphi_j^\dagger A\varphi_i dx. \tag{6.30}$$

As a result, the connection between the two wave functions (more particularly, between their coefficients) may be expressed via the matrix equation

$$[b] = [A][a], \tag{6.31}$$

where $[b]$ and $[a]$ are column matrices and $[A]$ is a square matrix, all of which could be of infinite size.

Extending the treatment above, one can also show that the expectation value of a given operator is found from a matrix equation involving the matrix of the operator, as

$$\begin{aligned} A &= [a]^\dagger [A][a] = [a^*]^T [A][a] \\ &= [b]^\dagger [A][b] \end{aligned}. \tag{6.32}$$

Here, the superscript T implies the transpose operation (which here produces the row matrix). Since the operator is Hermitian, the matrix representation of the operator is a Hermitian matrix, which implies that $A_{ij}^* = A_{ji}$.

Suppose that after carefully choosing a set of basis functions, it is discovered that another choice would have been preferred. Just as one can do coordinate transformations in vector algebra, one can do coordinate transformations in the linear vector spaces. Initially, the wave function is defined in terms of the basis set $\varphi_{1i}(x)$ according to

$$\psi(x) = \sum_i a_i \varphi_{1i}(x). \tag{6.33}$$

Generally, a coordinate transformation may be described as a rotation in space. Hence, the rotation to the new coordinates (or basis state) defined by

$$\varphi_{2i}(x) = \sum_j S_{ij}\varphi_{1i}(x). \tag{6.34}$$

This expression may be inverted to yield

$$\varphi_{1i}(x) = \sum_j S_{ij}^{-1}\varphi_{2i}(x), \tag{6.35}$$

hence,

$$\psi(x) = \sum_{i,j} a_i S_{ij}^{-1}\varphi_{2i}(x) = \sum_i b_i\varphi_{2i}(x), \tag{6.36}$$

where (6.28) has been added as an expansion in the second basis directly. Thus, one finds that

$$\begin{aligned} [a] &= [S][b], \\ [b] &= [S^{-1}][a]. \end{aligned} \tag{6.37}$$

The operator A of (6.32) may be used to develop the relationship between the different basis sets to show that

$$[A_a]a] = [S][A_b]b]. \tag{6.38}$$

In the last equation, the operator acts upon the individual basis functions and hence will move through the matrix sum. Since the basis functions are changed on the right-hand side, the result is different and leads to there being different matrix elements, which have been denoted by the subscripts corresponding to the old and new sets of basis functions. By using (6.37), we can relate the two matrix representations (in the new and old basis sets) via

$$[A_a] = [S][A_b][S]^{-1} \tag{6.39}$$

where the last term is the inverse of the transformation matrix. Equation (6.39) defines a *similarity* transformation. An important property of such transformations is that they satisfy

$$\sum_j S_{jk}^* S_{ji} = \delta_{ik}. \tag{6.40}$$

This means that the implied matrix product produces a unit matrix, and

$$[S]^{\dagger} = [S]^{-1}. \tag{6.41}$$

The last expression is the relationship defining a *unitary matrix*. The similarity transformations from one coordinate (orthonormal basis) set to another must be unitary transformations and thus are generalized rotations.

6.3.3 The Eigenvalue Problem

Sometimes, a difficulty arises in carrying out the above prescriptions, especially in respect to the eigenvalues. In previous sections, it was assumed that one could find a basis set in which an operator returned the same basis function multiplied by a c-number constant, as in (6.21). In general, one would have to be quite prescient to choose a basis set properly for this to occur for all, or even a significant fraction, of the operators of interest. That is, it is quite unusual to choose a basis set and find that it directly provided eigenvalues of some operators. How then does one find the eigenvalues of operators? We know from the arguments of Section 6.3.1 that, if an operator produces two different values when operating upon a set of functions, these latter functions must be orthogonal to one another. This means, in general, that a selected basis set can be found that will lead to a diagonal matrix for any selected operator, but it is unlikely to be diagonal for other operators unless the latter commute with the selected operator. This is a very special situation, which is not found in general.

On the other hand, it has been asserted above that the choice of the basis set is not particularly important to the physics but is generally made based upon other considerations such as the ease of solving the boundary condition problem. Thus, in general, one should expect that some linear combination of the chosen basis states, much like a Fourier series representation, will describe the basis states for any operator. The connection between the selected basis states and those that yield the diagonal representation for an operator is given by the similarity transformations described in the previous section. In fact, given the solutions in using one basis set, which does not produce an eigenvalue equation, it is possible to diagonalize the system through a coordinate rotation defined by the similarity transformations. It is this ability that will be demonstrated in this section.

The first step for setting up the solution is to choose a basis set of wave functions $\{\varphi_i(x)\}$ that will be applied to the Hamiltonian of the problem. These functions are properly orthogonalized, so the inner product of any distinct pair is zero, and each is normalized. From this selection of a basis set, it is possible to determine the expectation value of a general operator in such a way that the operator lies in a space defined by these operators. It may be asserted that there actually does exist a proper choice of basis functions, in which the operator is directed solely along one of the basis functions, and the task is to find these functions. So, with the selection of the basis set determined by the set mentioned above, the wave function is expanded in the usual way, according to (6.33), except that it is not at all clear that the coefficients a_i are known.

The next step is to operate on the basis set with an operator. One of the more common operators is the Hamiltonian itself as a method to determine the set of energies. However, here the operator A is used, as it is truly arbitrary. This operation is assumed to produce a set of equations

$$\sum_i a_i A\varphi_i(x) = \sum_i \lambda_i a_i \varphi_i(x), \tag{6.42}$$

where the λ_i are unknown; here they remain to be found. If we now multiply by the complex conjugate of any one of the original basis functions and integrate, it gives one row of the matrix equation

$$[A][a] = [\lambda][a], \tag{6.43}$$

where $[\lambda]$ is a diagonal matrix, according to (6.42). This last equation is the *eigenvalue equation.* Since one does not expect the vector $[a]$ to be a null vector, the only solution of (6.43) is the choice

$$det([A]-[\lambda]) = 0. \tag{6.44}$$

This produces an algebraic equation, whose roots are the desired values of λ. If, for example, the dimension of the above matrices were n, this last algebraic equation would be of order n. These roots are the eigenvalues of the operator A. Once these n values are known, then $n-1$ can be substituted, in turn, into (6.43) to find the coefficients that enter the similarity transformation to rotate the original basis set into one for which an eigenvalue equation results from the operation with A.

The above method is the mathematical description of the solution method, but on a computer, one follows a different method, in which (6.44) is rewritten as

$$[A-\lambda][a]=0. \tag{6.45}$$

There exists a large number of routines that diagonalize the first matrix (recall that the $[\lambda]$ is a diagonal matrix to begin with). In the process, the similarity matrix is developed at the same time, as the proper set of coefficients results in the diagonalization method, and this is used to transform the group of a's to the new set in the diagonal basis.

The presence of eigenvalues raises an important issue with regard to the uncertainty discussed above. *An operator that satisfies an eigenvalue equation has zero uncertainty.* This is especially important for the time-independent Schrödinger equation. This equation is an eigenvalue equation for the total energy. On the left-hand side is the Hamiltonian, or energy operator, while the energy value is on the right-hand side. Since this is an eigenvalue equation, there is no uncertainty in the energy that results from its solutions. Each energy value is well described and subject to no variation that can be connected to an uncertainty principle.

6.4 ENTANGLEMENT

In Chapters 3 and 4, the idea of Cooper pairs—two electrons with opposite spin that are bound together to form a composite particle—was discussed. Suppose one of these Cooper pairs is taken and illuminated with a photon of sufficient energy to break up the pair, in which case the two electrons begin to repel and move off in opposite directions. Which spin went in which direction? One really doesn't know the answer to this question. As a result, if a wave function is desired to describe these two electrons and their spins, it will have to be one of unknown validity; that is, it will describe the possibility of any choice.

The obvious is that if the first electron goes to the left, it will have either a spin up or a spin down, where the choice up/down describes the arbitrary direction of the spin vector. The second electron will have the opposite spin as it moves to the right. One possible wave function can be written (properly normalized) as

$$\psi(L,R)=\frac{1}{2}\big(\varphi(L,\uparrow)\varphi(R,\downarrow)\pm\varphi(L,\downarrow)\varphi(R,\uparrow)\big). \tag{6.46}$$

Here, the φ describe the individual electrons and tell us that that particular electron went to the left (or right) and has spin up (or down). So, the left-going electron is either spin up or down, and the right-going electron has the opposite spin.

Now, one could describe a simple basis set for, e.g., the left-going electron as representing either spin up or down, such as

$$\psi(L)=a\varphi(L,\uparrow)+b\varphi(L,\downarrow), \tag{6.47}$$

where a and b can be either 0 or 1, but cannot both be the same. A similar wave function can be defined for the second electron. Now, note that if we combine these two functions in an attempt to describe the pair of electrons that emanated from the Cooper pair, it would look like

$$\begin{aligned}\psi(L)\psi(R)&=a_La_R\varphi(L,\uparrow)\varphi(R,\uparrow)+a_Lb_R\varphi(L,\uparrow)\varphi(R,\downarrow)\\&\quad+b_La_R\varphi(L,\downarrow)\varphi(R,\uparrow)+b_Lb_R\varphi(L,\downarrow)\varphi(R,\downarrow).\end{aligned} \tag{6.48}$$

This result is not at all the result (6.46) and, in fact, contains some terms that are not permitted to appear, even though they are certainly part of the product of the two different basis sets. The point is that (6.46) *cannot be decomposed* into two separate simple basis sets.

The wave function (6.46), which represents the uncertainty as to which electron has spin up, is an *entangled* wave function. The fact that the two electrons must have opposite spins is a result of the constraint that existed in the Cooper pair. It is not a simple boundary condition. Rather, it is a constraint that was placed on the system when the Cooper pair was formed and is properly a result of the interaction between the electrons at this prior time. The memory of this interaction, and its constraint, remains with the two particles after the pair is broken up.

Entanglement is always produced by an interaction that connects two separate states or particles. They become joined and must be described by a new two-body wave function (or larger when many particles are interacting simultaneously). Schrödinger (1935a, 1935b) stated that entanglement was the most important single property of quantum mechanics. It has also been described as the strangest property of quantum mechanics and the least understood property of quantum mechanics. Einstein called it "spooky action at a distance."[2] Nevertheless, entanglement is an important property and has come to be essential in the new fields of quantum information and quantum computing, which will be discussed in Chapter 11.

6.4.1 The Einstein–Podolsky–Rosen (EPR) Problem

It is fair to say that, from the very beginning, there has been controversy in quantum mechanics. This controversy centers upon the interpretation, or philosophical basis, that one wants to assign to quantum mechanics. This controversy has divided the field into two relatively distinct camps. On one hand, there is the far larger group that is based upon Heisenberg, von Neumann, and especially Bohr. On the other hand there is a much smaller group based upon de Broglie, Schrödinger, and especially Einstein.

We know Ernst Mach primarily from his giving us the "Mach number," or velocity with respect to the speed of sound. However, he was also known at the end of the nineteenth century for his philosophical views (and was one of a group of scientists proposing such a view), which were that the reality in science is limited to what one can observe and sense themselves. This very strict concept of reality precludes the presence of atoms and molecules, simply because they are not observable (at least, they were not observable for much of the nineteenth century). This means that the understanding of science is limited to the results of physical measurements. When Bohr developed his atomic theory, and even later when his protege Heisenberg brought forth his new quantum mechanics, this phenomenological view of reality was forced onto the understanding of quantum mechanics. This leads to the view that reality only exists in quantum mechanics following a measurement of the desired variables. That is, measurement establishes the conditions for which reality can exist for the examined system.

The above view on reality through measurements is a condemnation of the idea of predicting the results of measurement from an underlying theoretical basis. Importantly, this also does away with determinism and causality in science, and this bothered many, including the second group. De Broglie, Schrödinger, and Einstein were leading theorists, so it is not unexpected that they would contest such a view. However, this does not condemn their view merely because they were theorists. It is also the case that they did not accept the end of causality and determinism. Indeed, the world of classical physics, and the statistical physics that is part of this world, have a solid theoretical basis that comes from centuries of work and is supported by an enormous number of actual measurements, which infer a cause and effect in physical processes. It is this theory that allows predictions to be made and calculations performed to allow such predictions.

Einstein, Podolsky, and Rosen (EPR, 1935) created a major challenge to the views of the Bohr camp. First, they asserted that, for a *complete* theory, every element of the physical reality must have a counterpart in the physical theory. Then, if, without disturbing the real system, one could predict with certainty the value that a physical quantity would take, there has to be an element of reality corresponding to this quantity. The implication of these assertions was that Bohr's interpretation of quantum mechanics could not constitute a complete theory, a result that Bohr would admit a couple of years later (Bohr, 1937). However, the idea of predicting a result insures determinism and causality remain important.

EPR proposed a situation to illustrate the issues. They considered that two systems existed, termed I and II, which were allowed to interact for a time $0 \le t \le T$, after which they moved apart and no longer interacted (the two electrons above interacted only for a very short T during which the Cooper pair was formed and then when it was broken into two distinct parts). They also considered that the two systems were fully known at the initial time. They asserted that, with the use of the Schrödinger equation, they should be able to calculate the state of the combined system at any future time (after the interaction). However, due to uncertainty, quantum mechanics would only allow them to know with certainty one-half of the variables. The other half would be unknowable due to the uncertainty principle (if one of a conjugate pair is known with certainty, the other is completely uncertain). For example, after the interaction, only the position or the momentum could be accurately determined on system I. Thus, EPR concluded that this asserted that either the quantum mechanical description of reality was not complete or else one of the two quantities (position or momentum) did possess reality, depending upon which measurement was made.

The result of using the measurement to bring reality to the system proposed by EPR created a problem. In the interaction, for example, the momentum of the two particles is completely conserved, and afterward is split between the two outgoing systems. Under Bohr's vision, a measurement of the momentum of one of the systems would immediately determine the momentum of the second. However, this knowledge had to jump from the first system to the second faster than the speed of light, a result that was intolerable to Einstein. It was this that he referred to as "spooky action at a distance." So, the controversy settles around whether the two momenta are determined at the time of the interaction or jump into their values at the time of measurement. The latter is the view of Bohr and his followers and came to be known as part of the "Copenhagen view." It is this view that Feynman probably had in mind when he remarked that "... nobody understands quantum mechanics ..." (Feynman, 1965).

6.4.2 Further Insights

In the EPR example, the two systems become entangled during the time of the interaction. They remain entangled even after separating until something else acts on one or the other in a manner to destroy the entanglement. For example, in solid state, it is generally assumed that a scattering process destroys a significant part of a particle's "memory" of its state prior to the collision. In classical statistical transport, Boltzmann (1868, 1872) assumed complete loss of this memory during a scattering event.

As mentioned above, Schrödinger considered the entanglement property the most important aspect of quantum mechanics. When the EPR systems have completed their interaction, they are no longer two separate systems. Rather, they constitute a single system, but one with entanglement, as discussed through (6.47). As an entangled system, they have correlated properties whose reality exists through the entanglement. However, this assumes that reality exists prior to the measurement, as suggested by Schrödinger, who also pointed out that no classical physicist of the time would ever have dared to believe that all possible states of a classical system could ever be measured. That is, most complex systems, whether classical or not, contain unmeasurable states, and this remains true in modern systems and control theory where there are both observable and non-observable states (Kalman, 1960). So, even in classical systems, one cannot in general observe all of the possible states, and yet there remains reality in these systems without everything being observed or measured. So, it is natural to believe, as Schrödinger did, that to jump to a view that only what was measured constituted the reality of a quantum was quite unsupportable, and he described a great many possible situations where the measurements could not be fully meaningful.

Of course, there is the nonsense of Schrödinger's cat, a paradox he proposed in these 1935 papers. The cat is in a box with a vial of poisonous gas, and the question is whether the cat is dead or alive when the observer opens the box. The standard view is that these are the only two results: (1) the vial is broken and the cat is dead, or (2) the vial is unbroken and the cat is alive. One can find many YouTube videos that express just this pair of results. However, this is nonsense, any good engineer knows that there are many more possible results. The nuclear decay that triggers the hammer to break the vial is not a sudden event. It occurs over time. Suppose, in fact, the observer opens the box, and then the hammer allows the vial to break, allowing the cat to escape but killing the observer. Has a measurement been made when an observer cannot communicate his observations? Other scenarios are possible as well (Ferry, 2019).

Schrödinger also pointed out a paradox in the Copenhagen view, since the latter would say that the measurement collapsed the system into a new state governed by the value of the measurement. This means that you cannot repeat a measurement because the original system has been destroyed (collapsed into the new system). Hence, you need to make copies of the original system in order to repeat the measurement. However, you do not have sufficient information on the system with which to make copies, because the measurement has not yet been done. If one cannot make copies, you cannot make additional measurements. As Schrodinger points out, we are in a contradictory loop, which suggests the Copenhagen view just doesn't fit with the natural world.

It is important to keep in mind, however, that none of the interpretation problems affect the importance and existence of entanglement. This is a viable, and real, consequence of quantum interactions and exists without concern over just where reality arises in the process. For the systems involved reality occurs with the interaction and remains with the entanglement.

6.5 SUPERSYMMETRY

When treating the harmonic oscillator in Chapter 5, a method of splitting the Hamiltonian was introduced. This method divided the Hamiltonian into a pair of operators, which were termed the creation and annihilation operators. In this special case, it was the quadratic potential in x that allowed the use of the operator splitting of the Hamiltonian. The use of these operators greatly simplified the solution of the Schrödinger equation for the harmonic oscillator.

In fact, this quadratic potential is one of a group of potentials that allow such a splitting of the Hamiltonian (Cooper *et al.*, 1995; Sukumar, 1985). The general name given to this approach (splitting the Hamiltonian) is supersymmetric quantum mechanics (SUSYQM), or simply supersymmetry. The concept of supersymmetry has usually been discussed in reference to quantum field theory and describes transformations between fermions and bosons. However, Witten (1981) described a quite simple supersymmetric system consisting of a single spin-1/2 particle moving in one dimension. This latter work had no connection to quantum field theory and provides the basis of what is now known as SUSYQM, and this involves the potentials mentioned above. Just as the harmonic oscillator was simplified by operators, Hamiltonians containing these special potentials also benefit from Hamiltonian splitting via the operators.

In general, we know that a second-order (in space) differential equation, such as the Schrödinger equation, has two types of solutions. In the harmonic oscillator, these are the even and odd solutions. In more complicated situations, they may be, for example, Bessel functions of the first and second kind which occur in spherical coordinates. To address the current SUSYQM, suppose we have a one-dimensional Schrödinger equation of the form

$$H\psi(x) = -\frac{\hbar^2}{2m}\frac{d^2}{dx^2}\psi(x) + V_1(x)\psi(x) = E\psi(x). \tag{6.49}$$

This is, of course, just the normal time-independent Schrödinger equation that comes from (1.18). Here, however, it is assumed that the potential $V_1(x)$ is one that admits to splitting of the Hamiltonian. Then, one can proceed to define two operators that satisfy (see Section 5.1.2)

$$H_1 = A^+ A^- + E, \tag{6.50}$$

where E is a constant (the ground state energy, for example), and H_1 is the Hamiltonian in (6.49). The two operators may be defined arbitrarily as

$$A^{\pm} = W(x) \pm \frac{\hbar}{\sqrt{2m}} \frac{d}{dx}, \tag{6.51}$$

where $W(x)$ will be shown to be connected to the potential energy term of (6.49) and the derivative comes properly from the kinetic energy term.

These two operators of (6.51) have been given the fancy name "super-charge operators," and $W(x)$ is usually referred to as the *super-potential.* The form (6.51) is often encountered in operator treatments, such as the harmonic oscillator of Chapter 5, where the two super-charges are known as creation (A^+) and annihilation (A^-) operators, or raising and lowering operators. The relation between the potential $V_1(x)$ and the super-potential $W(x)$ may be found by direct substitution of (6.51) into (6.49), where

$$V_1(x) = W^2(x) - \frac{\hbar}{\sqrt{2m}} \frac{dW(x)}{dx} + E. \tag{6.52}$$

If the ground state wave function is known, as in the case of the harmonic oscillator, where the product of the annihilation operator and the wave function gave a first-order equation that was easily solved, then the super-potential may be found as

$$W(x) = -\frac{\hbar}{\sqrt{2m}} \frac{1}{\varphi_0(x)} \frac{d\varphi_0}{dx}. \tag{6.53}$$

For the second solution, we can define a partner Hamiltonian with a partner potential via the description (note that the two operators are reversed in their order)

$$H_2 = A^- A^+ + E, \tag{6.54}$$

where

$$V_2(x) = W^2(x) + \frac{\hbar}{\sqrt{2m}} \frac{dW(x)}{dx} + E. \tag{6.55}$$

At this point, it is worthwhile to examine the two operators a little closer. For example, the natural product of them becomes

$$\begin{aligned}
A^+ A^- \psi &= \left[W(x) + \frac{\hbar}{\sqrt{2m}} \frac{d}{dx} \right] \left[W(x) - \frac{\hbar}{\sqrt{2m}} \frac{d}{dx} \right] \psi. \\
&= W^2(x)\psi + \frac{\hbar}{\sqrt{2m}} \frac{d}{dx}(W\psi) - W \frac{\hbar}{\sqrt{2m}} \frac{d\psi}{dx} - \frac{\hbar^2}{2m} \frac{d^2\psi}{dx^2}. \\
&= -\frac{\hbar^2}{2m} \frac{d^2}{dx^2}\psi + \left[W^2(x) - \frac{\hbar}{\sqrt{2m}} \frac{dW}{dx} \right] \psi \\
&= -\frac{\hbar^2}{2m} \frac{d^2}{dx^2}\psi + V_1(x)\psi(x) - E\psi(x)
\end{aligned} \tag{6.56}$$

This, of course, just returns (6.49), as it should. Now, the reverse order of the operators yields

$$\begin{aligned} A^-A^+\psi &= \left[W(x) - \frac{\hbar}{\sqrt{2m}}\frac{d}{dx}\right]\left[W(x) + \frac{\hbar}{\sqrt{2m}}\frac{d}{dx}\right]\psi. \\ &= W^2(x)\psi - \frac{\hbar}{\sqrt{2m}}\frac{d}{dx}(W\psi) + W\frac{\hbar}{\sqrt{2m}}\frac{d\psi}{dx} - \frac{\hbar^2}{2m}\frac{d^2\psi}{dx^2} \\ &= -\frac{\hbar^2}{2m}\frac{d^2}{dx^2}\psi + \left[W^2(x) + \frac{\hbar}{\sqrt{2m}}\frac{dW}{dx}\right]\psi \\ &= -\frac{\hbar^2}{2m}\frac{d^2}{dx^2}\psi + \left[V_2(x) - E\right]\psi \end{aligned} \tag{6.57}$$

This is the partner Hamiltonian for the second potential. Quite often, the energy scale (and the potential) is uniformly shifted so that $E = 0$, and this simplifies the mathematics.

In general, the commutator relationship is complicated. If one subtracts (6.57) from (6.56), the result is

$$\left[A^+, A^-\right] = A^+A^- - A^-A^+ = \frac{2\hbar}{\sqrt{2m}}\frac{dW}{dx}. \tag{6.58}$$

That is, the commutator does not give a simple result. However, let us give an example using the harmonic oscillator. In this case, one obtains

$$V_1 - E = \frac{m\omega^2}{2}x^2, \tag{6.59}$$

where the energy scale has been rescaled. This leads to the result that

$$W(x) = \sqrt{\frac{m}{2}}\omega x, \tag{6.60}$$

and

$$\left[A^+, A^-\right] = \hbar\omega. \tag{6.61}$$

Hence, if the energy is renormalized into units of $\hbar\omega$, the commutator relations would be the simple ones obtained in the harmonic oscillator in Chapter 5.

To proceed to the wave functions, it is usually the case that the ground state lies in the first solution of the Hamiltonian (6.49), and these solutions will be indicated with the subscript "1." That is, it is assumed that there actually is a ground state, as was done in Chapter 5, so that

$$A^-\varphi_{1,0} = 0. \tag{6.62}$$

From (6.51), this leads to

$$\varphi_{1,0} = C_{1,0}exp\left[-\frac{\sqrt{2m}}{\hbar}\int^x W(x')dx'\right]. \tag{6.63}$$

This also leads to the result of the expectation value

$$\left\langle \varphi_{1,0} \middle| A^+A^- \middle| \varphi_{1,0} \right\rangle = 0, \tag{6.64}$$

which implies that the energy scale has been renormalized so that the ground state energy is zero.

As it was assumed that this ground state lies in the chain of the first solution, then it may be assumed that the lowest energy of the second solution will be greater than zero. To investigate this, assume that there is an energy E_2 that satisfies

$$A^-A^+\psi_2 = E_2\psi_2. \tag{6.65}$$

This means that the reverse order of the two operators also produces an eigenvalue of the second solution. Now, let us pre-multiply (6.65) with A^+ to give

$$A^+A^-\left(A^+\psi_2\right) = E_2\left(A^+\psi_2\right). \tag{6.66}$$

If $A^+\psi_2 \neq 0$, as must be the case if the eigenvalues are greater than zero, then this state must be an eigenfunction of H_1 as well and E_2 is also an eigenvalue of H_1. This is an amazing result: all eigenvalues of H_1, other than the ground state, are also eigenvalues of H_2, and all eigenvalues of H_2 are also eigenvalues of H_1. This means, for example, that the lowest energy of solution 2 must correspond to the first excited state of solution 1, and the remaining energy levels of both series align in a corresponding manner. In addition, the wave function for the first excited state of the solution must also be a wave function of solution 2 with the relationship

$$\psi_{1,1} = \frac{1}{\sqrt{E_2}}A^+\psi_{2,0} \tag{6.67}$$

if the energy E_2 is the lowest energy level of the second solution.

Hence, the set of basis functions $\{\varphi_{1,i}\}$ that are used for ψ_1 are precisely the same set of basis functions, except for the normalizations, that are used for ψ_2. These eigenvalues form a pair of ladders of energy that perfectly line up between the two solutions, with the exception of the ground state of the first set. The result is that the simplicity of the harmonic oscillator results can be extended directly to a more extensive set of potentials, which include the Coulomb, Morse, Rosen–Morse, extended Pöschl–Teller, Eckart, and Scarf potentials (Cooper *et al.*, 1995). This approach provides a relatively universal method of solution to a series of potentials that can occur in the Schrödinger equation. More importantly, some of the solutions given above allow us to recognize and prove that the Schrödinger equation is closely affiliated with the Fokker–Planck equation which is a fundamental equation of *classical* statistical mechanics, with the import that quantum mechanics is not that different from classical mechanics, as suggested years ago by Madelung (1926) and Kennard (1928) (see Chapter 1).

6.6 SCHRODINGER AND FOKKER–PLANCK PROBABILITY

Some years ago, Feynman (1982) pointed out that quantum mechanics is actually based upon a probabilistic equation (Born, 1926), and that there may be more complications in this type of computation especially as the system size is reduced to nanometers. Feynman suggested that solving quantum problems may become exceptionally difficult. Quantum evolution is governed by the Schrödinger equation, and one should ask whether the solution of this equation is any more difficult than the solution of a classical equation for a probabilistic system, particularly when the details become complicated by the size. It can be said that the arguments raised in terms of the system size by Feynman are not unique to a quantum system and are just as applicable to a classical system of comparable size. Hence, the question is whether quantum probability is more difficult than classical probability.

As mentioned above, the Fokker–Planck equation appears in classical statistical physics and yields a distribution function as (Risken, 1984)

$$\frac{\partial f(x,t)}{\partial t} = D\frac{\partial^2 f(x,t)}{\partial x^2} + \frac{\partial}{\partial x}\left[\frac{\partial u(x)}{\partial x} f(x,t)\right], \tag{6.68}$$

where D is a diffusion coefficient that is considered constant here and the first term in the square brackets is a mobility-type term (u is often called the Fokker–Planck potential). The right-hand side of (6.68) may be referred to in many cases as a probability current in analogy to the continuity equation in electromagnetics. This equation can be solved exactly in some cases, but in other cases it does not happen. However, it has been suggested that there is a connection to the Schrödinger equation in that solvable problems for this latter equation also represent solvable problems for the Fokker–Planck equation (Risken, 1984). This connection is strong since the potential $u(x)$ will be shown to be related to the ground state of a quantum system. This remarkable connection arises through the use of the supersymmetry of the previous section in formulating the solutions. Thus, for a wide class of potentials, these equations seem to be strongly related and can be solved exactly.

To examine this connection between the two equations, one begins with (6.68) with the general assumption that $D = 1$, which provides a normalization of the remaining terms. Debosscher (1991) has shown that, in this case, the Fokker–Planck potential provides a stationary solution to (6.68) in the form

$$f_0(x) \sim e^{-u(x)}, \tag{6.69}$$

and this distribution may be normalized to provide a stable potential. This gives the steady-state probability distribution function (PDF). For a time-dependent solution of (6.68), one needs the function $p(x,t|x_0)$ of the stochastic process, which has an eigenfunction expansion given by

$$\begin{aligned} p(x,t\,|\,x_0) &= \sum_k f_k(x) g_k(x_0) e^{-\lambda_k t} \\ &= e^{-u(x)} \sum_k g_k(x) g_k(x_0) e^{-\lambda_k t} \\ &= e^{-[u(x)-u(x_0)]/2} \sum_k \psi_k(x) \psi_k(x_0) e^{-\lambda_k t} \end{aligned}, \tag{6.70}$$

where f_k, g_k, and ψ_k are solutions, respectively, to the Fokker–Planck eigenvalue equation

$$-\lambda f(x) = \frac{\partial^2 f(x)}{\partial x^2} + \frac{\partial}{\partial x}\left[\frac{\partial u(x)}{\partial x} f(x)\right], \tag{6.71}$$

the adjoint Fokker–Planck equation

$$-\lambda f(x) = \frac{\partial^2 g(x)}{\partial x^2} - \frac{\partial}{\partial x}\left[\frac{\partial u(x)}{\partial x} g(x)\right], \tag{6.72}$$

and the imaginary-time Schrödinger equation

$$\frac{\partial^2 \psi(x)}{\partial x^2} + [\lambda - V(x)]\psi(x) = 0. \tag{6.73}$$

According to Debosscher, the three problems have the same eigenvalue spectrum and the three solutions are inter-related through the relations

$$f_k(x) = e^{-u(x)} g_k(x) = e^{-u(x)/2} \psi_k(x). \tag{6.74}$$

The above statements are strong, and have to be shown. To begin, we shift the energy levels as in the last section, so that the Schrödinger equation may be written as

$$H_1 \psi(x) = -\frac{\hbar^2}{2m}\frac{d^2}{dx^2}\psi_0(x) + V_1(x)\psi_0(x) = 0. \tag{6.75}$$

The Hamiltonian has been written as (6.50) with $E = 0$. For the lowest energy level, the ground state has the form from (6.63)

$$\psi_0(x) = C_0 exp\left[-\frac{\sqrt{2m}}{\hbar}W(x)\right]. \tag{6.76}$$

From the Fokker–Planck equation, there exists an invertible transformation (Schulze-Halberg and Jimenez, 2009)

$$P\left[f(x)\right] = exp\left(u - \frac{Et}{2}\right)f(x), \tag{6.77}$$

where E is an arbitrary constant. Then, one can equate the wave function $\psi(x) = P[f(x)]$, which leads to the equation

$$\frac{d^2\psi}{dx^2} + \left[E + \frac{d^2u}{dx^2} - \left(\frac{du}{dx}\right)^2\right]\psi = 0. \tag{6.78}$$

While this is a normalized version, one may still recognize the connection that

$$W(x) = \frac{du(x)}{dx} \tag{6.79}$$

so that (6.78) becomes the Schrödinger equation.

While the approach may be somewhat vague in its complications, it remains clear that the Fokker–Planck equation of classical statistical physics and the Schrödinger equation of quantum mechanics are of equal degrees of difficulty, and they are very closely related. Quantum mechanics is no more difficult, even in the probability realm, for which to find solutions than for comparable classical problems and may actually be easier to understand.

6.7 THE DENSITY MATRIX

In many applications, some prefer to have their quantum system defined in a manner that gives more relevance to the probability density (the magnitude squared of the wave function). This can be achieved in many different ways, but one particular method—the density matrix—will be discussed in this section. The density matrix arises directly from the wave function and particularly from the product of the basis functions. As described above, one can solve the Schrödinger equation by assuming an expansion of the wave function in a suitable basis set so that each basis function is an energy eigenfunction according to

$$H\varphi_i = E_i\varphi_i, \tag{6.80}$$

in which E_i is the energy level corresponding to the particular basis function φ_i. Then, the total wave function can be written as

$$\psi(x,t) = \sum_i c_i\varphi_i(x)e^{-iE_it/\hbar}, \tag{6.81}$$

where the specific time variation of each basis function has been retained. Using this wave function (6.81) and its adjoint, the density matrix can be defined as

$$\begin{aligned}\rho(x,x',t) &= \psi^\dagger(x,t)\psi(x',t). \\ &= \sum_{n,m} c_{nm}\varphi_n^\dagger(x)\varphi_m(x')exp\left[i(E_n - E_m)t/\hbar\right].\end{aligned} \tag{6.82}$$

The diagonal terms are obviously the probability of that particular basis state. The off-diagonal terms are going to oscillate rapidly at a frequency given by the difference in the two energies. However, this is the case for which the energies are entirely real. In the non-equilibrium state, in which interactions affect the wave functions, the energies can obtain imaginary components. In this case, the off-diagonal parts may have long-lived components, and these represent the entanglement between the various states.

As mentioned, the diagonal terms are the probability for the individual basis states, and one can say that

$$Tr\{\rho\} = \sum_n c_{nn} |\varphi_n|^2 = 1. \tag{6.83}$$

Obviously, here $c_{nn} = c_n^* c_n$ is a real, positive number, and this last equation follows precisely from the requirement that

$$\int \rho(x,x)\,dx = \int |\psi(x)|^2 dx = 1. \tag{6.84}$$

The above implies that one can use the density matrix quite generally to evaluate the expectation values of various operators. For example, this expectation value can be written as

$$\begin{aligned} \langle A \rangle &= \int \psi^\dagger(x) A \psi(x)\,dx \\ &= \sum_{n,m} c_{nm} \int \varphi_n^\dagger(x) A \varphi_m(x)\,dx \\ &= \sum_{n,m} c_{nm} A_{nm} = \sum_{n,m} A_{nm} \rho_{mn} \\ &= Tr\{A\rho\} = Tr\{\rho A\} \end{aligned} \tag{6.85}$$

The last two terms in (6.85) are important. The properties of matrix multiplication tell us that the order of the two summations is not important, since one is summing over both indices. This means that one can move the functions (or operators) in the trace by any cyclic permutation; in this case moving the ρ from the last position around to the first position. For multiple operators, this operation may be written as

$$Tr\{AB\rho\} = Tr\{\rho AB\} = Tr\{B\rho A\}. \tag{6.86}$$

As the density matrix satisfies other quantum equations, we can write the time dependence as

$$i\hbar \frac{d\rho}{dt} = [H,\rho] = \left[-\frac{\hbar^2}{2m}\left(\frac{\partial^2}{\partial x^2} - \frac{\partial^2}{\partial x'^2} \right) + V(x) - V(x') \right] \rho(x,x',t). \tag{6.87}$$

This latter equation is known as the *Liouville equation*, or sometimes as the Liouville–von Neumann equation. Note that, this equation, like the Schrödinger equation has no dissipative terms included, so is reversible. This reversibility will disappear when interactions appear as in the real world.

6.8 THE WIGNER FUNCTION

One problem that is always raised with respect to the density matrix, and also the wave function itself, is that they are a function only of time and the position variables. In classical physics, one has the Boltzmann equation which is a function of position, momentum, and time, so that one can generate a phase-space formulation of the transport problem. None of the quantum functions that have been discussed so far have this convenient phase-space formulation. One either uses the wave functions in position space or in momentum space, but not both. Nevertheless, it is often very convenient to describe the system with a phase-space formulation. And, in fact, such a formulation exists (Wigner, 1932). To produce Wigner's phase-space function, the center-of-mass and difference coordinates are first produced through the transformations

$$X = \frac{1}{2}(x + x'), \quad x = X + \frac{s}{2},$$
$$s = (x - x'), \quad x' = X - \frac{s}{2}. \tag{6.88}$$

Here, X is obviously the average position of the two wave functions, and s is the difference between the two positions. The Wigner distribution function may now be generated as a Fourier transform in the off-diagonal elements (for which there is a difference in position) of the density matrix as (in one dimension)

$$f_W(X, p, t) = \frac{1}{2\pi\hbar} \int_{-\infty}^{\infty} ds \rho\left(X + \frac{s}{2}, X - \frac{s}{2}, t\right) e^{ips/\hbar}. \tag{6.89}$$

This function is now similar to the Boltzmann distribution function, although this one is not generally positive definite. Normally, the ground state Wigner function (when we use a basis function expansion) is positive definite, because it represents the ground state. The ground state must be positive definite. However, excited states can have negative excursions, and these negative parts, as well as a non-Gaussian shape, are signals of the quantum nature of the system (Ferry and Nedjalkov, 2018; Weinbub and Ferry, 2018) and particularly the entanglement in the system.

The Wigner function has certain normalizations, which arise from the wave functions themselves. If it is integrated over the position, one obtains

$$\begin{aligned} \int f_W(X, p, t) dX &= \frac{1}{2\pi\hbar} \int_{-\infty}^{\infty} ds \int dX \psi^\dagger\left(X + \frac{s}{2}\right) \psi\left(X - \frac{s}{2}\right) e^{ips/\hbar} \\ &= \frac{1}{2\pi\hbar} \int_{-\infty}^{\infty} ds \delta(s) e^{ips/\hbar} = 1. \end{aligned} \tag{6.90}$$

Similarly, if each wave function is Fourier transformed as in (1.30), and then integrated over the momentum, the result will be the same. Hence, the Wigner function maintains the orthonormality of the wave function and its basis functions in the phase space.

Incorporating the coordinate transformations (6.88) into the Liouville Equation (6.87) gives a new equation of motion for the density matrix, which appears as

$$i\hbar \frac{\partial \rho}{\partial t} = \left[-\frac{\hbar^2}{2m} \frac{\partial^2}{\partial X \partial s} + V\left(X + \frac{s}{2}\right) - V\left(X - \frac{s}{2}\right) \right] \rho(X, s, t). \tag{6.91}$$

Fourier transforming (on the variable s as above) this equation leads to a momentum-space version of the equation of motion as

$$\frac{\partial f_W}{\partial t} + \frac{p}{m} \frac{\partial f_W}{\partial X} - \frac{1}{i\hbar} \left[V\left(X + \frac{i\hbar}{2} \frac{\partial}{\partial p}\right) - V\left(X - \frac{i\hbar}{2} \frac{\partial}{\partial p}\right) \right] f_W(X, p, t) = 0. \tag{6.92}$$

In Chapter 4, the concept of a displacement operator was introduced in (4.17) as

$$|j\pm 1\rangle = e^{\pm ika}|j\rangle \rightarrow \psi(x\pm a) = e^{\pm ipa/\hbar}\psi(x). \tag{6.93}$$

Using these operators to shift the position of the Wigner function within the integral allows one to rewrite (6.91) as

$$\frac{\partial f_W}{\partial t} + \frac{p}{m}\frac{\partial f_W}{\partial X} - \frac{1}{2\pi\hbar}\int dp' W(X,p') f_W(X,p+p',t) = 0, \tag{6.94}$$

where the effective Wigner potential has been rewritten using

$$W(X,p') = \int ds sin\left(\frac{p's}{\hbar}\right)\left[V\left(X+\frac{s}{2}\right) - V\left(X-\frac{s}{2}\right)\right]. \tag{6.95}$$

One can see immediately, that if the actual potential is of quadratic order or less, only the first-order derivatives appear, and the result is simply the classical Boltzmann equation. Moreover, only odd orders of X appear in the expansion of the square bracket in (6.92) or (6.95). The minus sign assures that all even orders cancel. Like the Boltzmann equation, one cannot simply solve (6.94), or even the Liouville equation. This is because the time derivative tells us we need an initial condition in order to solve these equations.

As an illustrative example, suppose we have a wave function that is composed of two Gaussians, such as those in (2.70). One Gaussian is taken to be moving in the positive x-direction (and the argument will be written as $x - vt$), while the second Gaussian is moving in the negative x-direction (and the argument will be written as $x + vt$). We are also going to write these wave functions with an exponent propagator so that

$$\psi_{1,2} = \frac{1}{\left(2\pi\sigma^2\right)^{1/4}} exp\left[\frac{\left(x\mp vt\right)^2}{4\sigma^2} + ik_0 x\right], \tag{6.96}$$

where the last term in the exponential represents the velocity and is related to the wave number by $v = \hbar k_0/m$. If these two wave functions are assumed to be a single combined wave function initially as

$$\Psi(x,0) = \frac{1}{\sqrt{2}}\left[\psi_1(x,0) + \psi_2(x,0)\right]. \tag{6.97}$$

When the Wigner function of this composite wave function is created, three terms are produced. Two of these terms are simply of the same form as (6.74), but the third term involves an off-diagonal term in the density matrix that has both X and X' in it. If we plot the Wigner function, it will appear as in Figure 6.1. If we had gone to two beginning independent wave functions, we would have had only the two Gaussians of the form of (6.90). The third, mixed term represents the fact that we are really only using a single total wave function, and it represents the *entanglement* between the two beginning wave functions.

The concept of entanglement surfaced originally in the above sections, particularly in the discussion of the Einstein–Podolsky–Rosen (1935) thought experiment, in which two classical particles interacted with one another and then went on their way. As discussed above, the term entanglement comes from Schrödinger (1935a), who pointed out that, once the two particles had interacted, they were a single entangled quantum entity. Even when they separated and moved away, they remained a single quantum mechanical entity until some other process broke up the entanglement. In essence, this entanglement is a memory of the nature of the interaction that occurred between

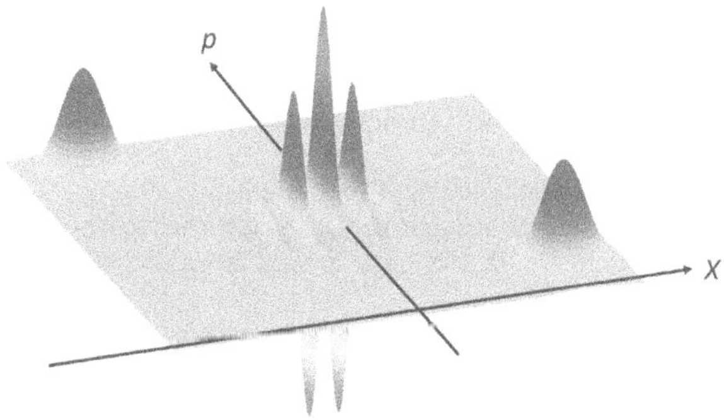

FIGURE 6.1 A composite wave function of two Gaussian packets produces a Wigner function with three components. The central one is the entanglement created between the two original wave packets by the definition (6.97).

the two original particles (this was created here by the definition of the single wave function). The entanglement in Figure 6.1 arises from the off-diagonal terms in the density matrix, so it is not unusual that these terms are referred to as the entanglement, and this entanglement arises from the fact that a single wave function was initially created in (6.97).

NOTES

1 The idea of non-orthogonal, or fuzzy, sets has been extended to quantum mechanics, particularly in the case of uncertain measurements (Prugovecki, 1974).
2 Although the quote is universally attributed to Einstein, it is not at all clear where and when he said it. It certainly is often associated with the EPR paper, discussed in the next section.
3 Page numbers refer to the reprinted papers in Boltzmann, L. 1909. *Wissenschaften Abhandlungen* (Leipzig: Barth).

REFERENCES

Bohr, N. 1937. Causality and complementarity. *Phil. Sci.* **4** 289–298.

Born, M. 1926. Zur Quantenmechanik der Stossvorgänge. *Zeit. Phys.* **37** 863–867.

Boltzmann, L. 1868. Studien über das Gleichgewicht der Lebendigen Kraft zwischen bewegten materiellen Punkten. *Wiener Ber.* **58** 517–560.[3]

Boltzmann, L. 1872. Weitere Studien über das Wärmegleichgewicht unter Gasmolekulen. *Wiener Ber.* **66** 275–370*.

Cooper, F., Khareb, A. and Sukhatme, U. 1995. Supersymmetry and quantum mechanics. *Phys. Repts.* **251** 267–385.

Debosscher, A. 1991. Class of one-dimensional Fokker-Planck and Schrödinger equations with various potentials. *Phys. Rev. A* **44** 908–915.

Einstein, A., Podolsky, B. and Rosen, N. 1935. Can quantum-mechanical description of reality be considered complete? *Phys. Rev.* **47** 777–780.

Ferry, D. K. 2019. *The Copenhagen Conspiracy* (Singapore: Pan Stanford Publishing).

Ferry, D. K. and Nedjalkov, M. 2018. *The Wigner Function in Science and Technology* (Bristol: IOP Publishing).

Feynman, R. 1965. *The Character of Physical Law* (Cambridge: The MIT Press) p. 129.

Feynman, R. P. 1982. Simulating physics with computers. *Intern. J. Theor. Phys.* **21** 467–488.

Heisenberg, W. 1925. Über quantentheoretische Umdeutung kinematischer und mechanischer Beziehunger. *Z. Phys.* **33** 879–893; Tr. in B. L. van der Warden. 1967. *Sources of Quantum Mechanics* (Amsterdam: North-Holland) p. 261–276.

Kalman, R. E. 1960. On the general theory of control systems. *IFAC Proc.* **1** 491–502.

Kaschner, S. and Russell, A. 2023. *Linear Transformations on Vector Spaces; an Introduction to Linear Algebra* (Indianapolis, IN: Private Academic Library Network of Indiana).

Kennard, E. H. 1928. On the quantum mechanics of a system of particles. *Phys. Rev.* **31** 876–890.
Madelung, E. 1926. Quantentheorie in hydrodynamischer Form. *Z. Phys.* **40** 322–326.
Morse, P. M. and Feshbach, H. 1953. *Methods of Theoretical Physics* (New York: McGraw-Hill).
Prugovecki, E. 1974. Fuzzy sets in the theory of measurement of incompatible observables. *Found. Phys.* **4** 9–18.
Risken, H. 1984. *The Fokker-Planck Equation* (New York: Springer).
Schrödinger, E. 1935a. Die gegenwärtige Situation in der Quantenmechanik. *Naturwissenschaften* **23** 807–812, 823–828, 844–849; Tr. Trimmer, J. D. 1980. The Present Situation in Quantum Mechanics. *Proc. Am. Phil. Soc.* **124** 323–338.
Schrödinger, E. 1935b. Discussion of probability relations between separated systems. *Math. Proc. Cambridge Phil. Soc.* **31** 555–563, 32 446–452.
Schulze-Halberg, A. and Jimenez, J. M. C. 2009. Supersymmetry of generalized Schrödinger equations in (1+1) dimensions. *Symmetry* **1** 115–144.
Sukumar, C. V. 1985. Supersymmetric quantum mechanics of one-dimensional systems. *J. Phys. A: Math. Gen.* **18** 2917–2936.
von Neumann, J. 1932. *Mathematische Grundlagen der Quantummechanik* (Berlin: Springer).
Weinbub, J. and Ferry, D. K. 2018. Recent advances in Wigner function approaches. *Appl. Phys. Rev.* **5** 041104.
Wigner, E. 1932 On the quantum correction for thermodynamic equilibrium. *Phys. Rev.* **40** 749–759.
Witten, E. 1981. Dynamical breaking of supersymmetry. *Nucl. Phys. B* **188** 513–555.
Zadeh, L. 1965. Fuzzy sets. *Informat. Cont.* **8** 338–353.

PROBLEMS

1. A dynamic system has $H = p^2 + Ax + Bpx$. Calculate the equations of motion for the operators x,p; that is, compute the time derivatives of the expectation values for these operators.
2. A particular operator A has been used to evaluate a dynamical property in an orthonormal set containing only three basis functions. When its expectation value is found, the resulting matrix for the operator is given by

$$\left[A\right] = \begin{bmatrix} \sqrt{2} & \sqrt{2} & 0 \\ -\sqrt{2} & \sqrt{2} & 0 \\ 0 & 0 & 2 \end{bmatrix}$$

   Determine the eigenvalues of this operator.
3. Compare the different dynamics that results from Hamiltonians that are written as $2x^2p^2$, $x^2p^2 + p^2x^2$ and $2(xp)^2$. What can you conclude about the proper symmetrization of operators?
4. Two operators, A and B, satisfy [A,B] = C. Compute the difference $A^3B^3 - (AB)^3$. Assume that C is a c-number and not an operator.
5. A particular projection operator $P_j = |j\rangle\langle j|$ is defined on the basis set

$$\psi_k\left(x\right) = \sqrt{\frac{2}{a}}\sin\left(\frac{k\pi x}{a}\right)$$

   for $0 \le x \le a$ (k,i,j are integers). Expand the function $V(x)$ in the basis set $|k\rangle$ if $V(x) = 1$ for $0 \le x \le a$, and $V(x) = 0$ elsewhere; that is, determine the coefficients c_j in the expansion

$$V\left(x\right) = \sum_k c_k \left|k\right\rangle.$$

6. Consider a computational scheme in which the real axis (the position coordinate) is discretized to points $x = na$, where a is a "lattice" constant. Instead of the continuous eigenvalue characteristics, each lattice site must be characterized by a function that allows the orthogonality to be retained. Determine a function localized at each lattice site that satisfies

normalization and is orthogonal with all of its neighbors. (The functions will yield an integrated product that is one of the generalized distribution forms of the delta function.)

7. Consider a potential well which is described by the potential $V(x) = Fx$ for $0 < x < a$, and $V(x) \rightarrow \infty$ for $x < 0$ and $x > a$. Using the wave functions of the infinite potential well

$$\psi_n(x) = \sqrt{\frac{2}{a}}\sin\left(\frac{n\pi x}{a}\right),$$

compute the Hamiltonian matrix for the lowest five energy levels. Take $F = 0.02$ eV nm^{-1} and $a = 5$ nm. Then, diagonalize the matrix to determine the energy eigenvalues and their wave functions. (While a proper approach would use an infinite set, approximate results can be found from these lowest five wave functions.) Compare the results with those of Chapter 2.

8. Consider now a representation in which the wave function is represented as a series of harmonic oscillator wave functions from Chapter 4. Calculate the elements of the matrix A for the following operators:

$$(a)\,A = x + i\frac{p}{m\omega_0},$$

$$(b)\,A = x - i\frac{p}{m\omega_0},$$

$$(c)\,x,$$

$$(d)\,p.$$

9. Find the eigenfunctions $\psi(x)$ and the nature of the eigenvalues of the operator:

$$O = \frac{d^2}{dx^2} + \frac{2}{x}\frac{d}{dx},$$

such that $O\psi = \lambda\psi$. Hint: consider the wave function $u = x\psi$, where ψ is, e.g., an arbitrary wave function with representation over the entire x-axis.

10. A system is initially in the mixed state

$$|\psi\rangle = \left(\sqrt{2}|\psi_1\rangle + |\psi_2\rangle + \sqrt{2}|\psi_3\rangle\right)/\sqrt{5},$$

where $|\psi_n\rangle$ are eigenstates of the system's Hamiltonian that satisfies the eigenvalue equation

$$H|\psi_n\rangle = n^2 E_0 |\psi_n\rangle.$$

Using the initial state of the system, consider the following questions:

a. If the energy is measured, what values can be obtained?

b. Consider an operator defined by

$$A|\psi_n\rangle = (n+1)a_0|\psi_n\rangle;$$

if we measure A, what values can be obtained?

c. Suppose that a measurement of the energy yields the value $4E_0$. If we measure A immediately afterward, what value is obtained?

d. Calculate the probabilities with which the individual energies in (a) can be obtained. Use the relation $P(E_n) = |\psi_n|\psi|^2$ and normalize the initial wave function prior to using this approach.

7 Stationary Perturbation Theory

It is generally preferable to solve a problem exactly. However, this is often not possible, and some approximation scheme must be applied. Examples of this were the finite potential well of Section 2.4.2 and the triangular potential well of Section 2.5. In the former example, the solution could not be obtained in closed form, and graphical solutions, or available software packages for simulation on a computer, were used to find the eigenvalues of the energy. In the latter example, complicated special functions were required to solve Schrödinger's equation. These are examples of more common problems. In Chapter 6, however, it was pointed out that one could choose an arbitrary basis set of functions that formed a complete orthonormal set as a defining linear vector space. The only requirement on these functions (other than their properties as a set) was that they made the problem easy to solve. In the two previous examples mentioned, what choice would we have made for the set? One choice would have been to use the basis functions that arise from an infinitely deep potential well, but these are not really defined outside the range of the well itself. Consequently, it is often found that it is difficult to find a proper basis set of functions for which the problem can be easily solved.

When the problem is not calculated easily with a direct method, one must then resort to these approximate methods. One such method is *perturbation theory.* This approach is useful when the Hamiltonian can be split into two parts

$$H = H_0 + H_1, \tag{7.1}$$

in which the first part H_0 can be solved directly, and the second part H_1 is small with respect to the first part. When this can be done, then the second part can be initially ignored, and the problem solved exactly in terms of a natural basis function set. After this is done, the second term in (7.1) is treated as a perturbation, and the eigenvalues and eigenfunctions are then developed in a perturbation series in terms of the natural basis function set.

In this chapter, the general methodology of the perturbation technique will be developed. Then several examples of the technique will be considered to illustrate the applications. Finally, a totally different approach, the variational method, will be introduced. The latter is a useful approximation when the split of the Hamiltonian into two parts, as in (7.1), does not lead to a small perturbing term.

7.1 THE PERTURBATION SERIES

As was discussed above, the basic approach revolves around being able to solve the quantum mechanical problem with only a portion of the total Hamiltonian, with the remainder being a small term. For this, (7.1) is rewritten as

$$H = H_0 + \lambda V, \tag{7.2}$$

where λ is a small parameter (if V is small compared with H_0, one can eventually let λ go to unity without introducing any inconsistency). The introduction of this parameter is solely to help develop the proper terms in the perturbation series. The term $V (= H_1)$ is the perturbing part of the Hamiltonian and contains all of the extra terms that have been removed from H to produce H_0. Now, the assumption is that the solutions to H_0 are easily obtained in terms of a set of basis functions for which

$$H_0 \,|\, k\rangle_0 = E_k^{(0)} \,|\, k\rangle_0, \tag{7.3}$$

DOI: 10.1201/9781003717195-7

where the subscripts and superscripts indicate the zero-order solution to the reduced Hamiltonian H_0. Here, the reduced Hamiltonian has provided a set of zero-order basis functions, that in this case are energy eigenfunctions where k is an integer denoting the term in the set. If the deviation of the actual basis functions and eigenvalues from these zero-order ones is small, then we can write the entire value of the basis functions as

$$|k\rangle = |k\rangle_0 + \lambda |k\rangle_1 + \lambda^2 |k\rangle_2 + \cdots. \tag{7.4}$$

Similarly, the energies can be written as

$$E_k = E_k^{(0)} + \lambda E_k^{(1)} + \lambda^2 E_k^{(2)} + \cdots. \tag{7.5}$$

These last two equations provide the starting point for the perturbation series. To develop this series, these two equations are inserted into the time-independent Schrödinger equation using (7.2) as the Hamiltonian. That is, the three expansions (7.2), (7.4), and (7.5) are inserted into

$$H|k\rangle = E_k|k\rangle. \tag{7.6}$$

One now has to recognize that all the terms with the same coefficient λ^n may be combined into a single equation together (these terms have to satisfy the equation on their own, and this is true for each power of λ). The idea is that all terms with the same power of λ must satisfy the equation independently of the other terms. The three lowest equalities give

$$\begin{aligned}
\lambda^0 &\Rightarrow H_0|k\rangle_0 = E_k^{(0)}|k\rangle_0, \\
\lambda^1 &\Rightarrow H_0|k\rangle_1 + V|k\rangle_0 = E_k^{(1)}|k\rangle_0 + E_k^{(0)}|k\rangle_1, \\
\lambda^2 &\Rightarrow H_0|k\rangle_2 + V|k\rangle_1 = E_k^{(2)}|k\rangle_0 + E_k^{(1)}|k\rangle_1 + E_k^{(0)}|k\rangle_2.
\end{aligned} \tag{7.7}$$

The first equation of (7.7) is obviously just the appropriate (7.3) for the unperturbed solutions. The second equation of (7.7) gives the basis for *first-order* perturbation theory. Let us examine this equation in more detail.

The initial corrections to the basis functions and to the eigenvalues are found by working with the second equation of (7.7). This equation may be rearranged, to give a more direct approach to the solution, as

$$\left(H_0 - E_k^{(0)}\right)|k\rangle_1 = \left(E_k^{(1)} - V\right)|k\rangle_0. \tag{7.8}$$

The basic assumption is that the perturbed basis functions are only small deviations from the unperturbed ones, so that it is fruitful to expand the deviations in terms of the unperturbed functions as

$$|k\rangle_1 = \sum_j c_j |j\rangle_0. \tag{7.9}$$

Introducing this assumption of an expansion into (7.8) leads to

$$\sum_j c_j\left(H_0 - E_k^{(0)}\right)|j\rangle_0 = \sum_j c_j\left(E_j^{(0)} - E_k^{(0)}\right)|j\rangle_0 = \left(E_k^{(1)} - V\right)|k\rangle_0. \tag{7.10}$$

Due to (7.3), the left-hand side vanishes for the term $j = k$, which will leave c_j undetermined. The right-hand side then gives us

$$E_k^{(1)} = V_{kk} = {}_0\langle k \,|\, V \,|\, k_0. \tag{7.11}$$

The first-order deviation of the wave function $|k\rangle_1$ is made up of small admixtures of the other members of the unperturbed basis set. Yet, we want the new wave function to remain normalized. To achieve this, we will require that $c_k = 0$ by definition. Then, using (7.3), Equation (7.10) can be rewritten as

$$\sum_{j \neq k} c_j \left(E_j^{(0)} - E_k^{(0)}\right) |\, j\rangle_0 = \left(E_k^{(1)} - V\right) |\, k\rangle_0. \tag{7.12}$$

In order to evaluate the change in the wave function, multiply both sides of (7.12) by the adjoint function ${}_0\langle i|$, and perform the implied integral. The resulting Kronecker delta functions allow us to perform the summation and rearrange (7.12) to give

$$\sum_{j \neq k} c_j \left(E_j^{(0)} - E_k^{(0)}\right) \delta_{ij} = -V_{ik}. \tag{7.13}$$

Doing the summation using the δ-function produces a single term

$$c_i = -\frac{V_{ik}}{E_i^{(0)} - E_k^{(0)}}. \tag{7.14}$$

The choice $c_k = 0$ is now seen to also avoid an inconvenient zero in the denominator of this equation. Thus, to first order, the eigenvalues and eigenfunctions are ($\lambda = 1$)

$$\begin{aligned} E_k &= E_k^{(0)} + V_{kk}, \\ |\, k\rangle &= |\, k\rangle_0 + \sum_{j \neq k} \frac{V_{jk}}{E_k^{(0)} - E_j^{(0)}} |\, j\rangle_0. \end{aligned} \tag{7.15}$$

Now consider the second-order corrections to the eigenvalues and the basis functions. The third equation of (7.7) can be rearranged as

$$\left(H_0 - E_k^{(0)}\right) |\, k\rangle_2 = \left(V_{kk} - V\right) |\, k\rangle_1 + E_k^{(2)} \,|\, k\rangle_0. \tag{7.16}$$

Here, (7.12) already has been introduced for the first-order perturbation of the energy levels. So, the first-order correction on the wave function can also be inserted to give

$$\left(H_0 - E_k^{(0)}\right) |\, k\rangle_2 = \left(V_{kk} - V\right) \left[\, k\rangle_0 + \sum_{j \neq k} \frac{V_{jk}}{E_k^{(0)} - E_j^{(0)}} \,|\, j\rangle_0 \right] + E_k^{(2)} \,|\, k\rangle_0. \tag{7.17}$$

Again, the second-order perturbation of the wave function is expanded in terms of the unperturbed basis set as

$$|\, k\rangle_2 = \sum_{s \neq k} b_s \,|\, i\rangle_0. \tag{7.18}$$

The term b_k (= 0) is omitted since the perturbation is small. Now, (7.18) in (7.17) gives

$$\sum_{s\neq k} b_s\left(H_0 - E_k^{(0)}\right)|s\rangle_0 = \left(V_{kk} - V + E_k^{(2)}\right)|k\rangle_0 + \left(V_{kk} - V\right)\sum_{j\neq k}\frac{V_{jk}}{E_k^{(0)} - E_j^{(0)}}|j\rangle_0 . \tag{7.19}$$

In order to evaluate the change in the eigenvalue, let us multiply both sides of (7.19) by the adjoint function ${}_0\langle k|$, and perform the implied integral. The resulting Kronecker delta functions allow us to perform the summation, which gives the result

$$E_k^{(2)} = \sum_{j\neq k}\frac{V_{kj}V_{jk}}{E_k^{(0)} - E_j^{(0)}}. \tag{7.20}$$

To recover the coefficients for the wave function expansion, (7.19) is pre-multiplied by an arbitrary adjoint function ${}_0\langle i|$ and the implied integration performed. This leads to

$$b_i\left(E_i^{(0)} - E_k^{(0)}\right) = \sum_{i,j\neq k}\frac{V_{jk}\left(V_{kk}\delta_{ji} - V_{ij}\right)}{E_k^{(0)} - E_j^{(0)}}, \tag{7.21}$$

or

$$b_i = \sum_{j\neq k}\left\{\frac{V_{jk}V_{ij}}{\left(E_k^{(0)} - E_j^{(0)}\right)\left(E_k^{(0)} - E_i^{(0)}\right)} - \frac{V_{jk}V_{kk}}{\left(E_k^{(0)} - E_j^{(0)}\right)^2}\delta_{ij}\right\},\quad i\neq k. \tag{7.22}$$

Thus, to second order in the perturbation, the energy and wave function may be written using (7.20) and

$$\begin{aligned} E_k &= E_k^{(0)} + V_{kk} + \sum_{j\neq k}\frac{V_{kj}V_{jk}}{E_k^{(0)} - E_j^{(0)}}, \\ |k\rangle &= |k\rangle_0 + \sum_{j\neq k}\frac{V_{jk}}{E_k^{(0)} - E_j^{(0)}}|j\rangle_0 \\ &\quad + \sum_{\substack{j\neq k\\ i\neq k}}\frac{V_{jk}\left(V_{ij} - V_{kk}\delta_{ij}\right)}{\left(E_k^{(0)} - E_j^{(0)}\right)\left(E_k^{(0)} - E_i^{(0)}\right)}|i\rangle_0. \end{aligned} \tag{7.23}$$

This procedure can obviously be extended to many higher orders of perturbation. In general, each higher order adds another summation over an intermediate state. For example, in (7.15), the first-order correction just couples two states, with a summation over all other states that couple to the kth one. At second order, however, this coupling generally goes through an intermediate state, with an additional ratio of a matrix element to an energy difference (in the denominator). Thus, at third order, it would be expected that an additional summation would appear as the connection moves through two intermediate states. In the world of perturbation theory, it is of course quite important to be sure that the correction terms are indeed quite small. If they are not, it is possible that the series does not converge, which means the entire approach violates the assumptions and is invalid. It is usual to check the results at an order above the desired one to make sure that the solution is stable. Otherwise other approaches must be pursued.

In the above discussion, it also has been assumed that the eigenvalues $E_k^{(0)}$ are all discrete, or at least all different if the spectrum is continuous. This may not always be the case, and there may be several members of the set of wave functions that have the same energy. Then, the approaches described above must be modified to "lift" this degeneracy prior to proceeding with the perturbation series in order to avoid unwanted zeros in the denominators of the summations. Let us consider the case where there are several different wave functions that all have the same energy value $E_k^{(0)}$. The correct wave functions in the zero-order approximation are then some linear combinations of these wave functions

$$c_k \mid k\rangle_0 + c_{k'} \mid k'\rangle_0 + c_{k''} \mid k''\rangle_0 \cdots \tag{7.24}$$

For the constants c_k, it suffices to take their zero-order values as the solutions of the process. We now will write out a set of equations with $i = k, k', k''$, etc. and substitute in them, as the first approximation, $\mathcal{E}_k = \mathcal{E}_k^{(0)} + \mathcal{E}_k^{(1)}$ obtained from the $H \mid k\rangle_0$ term in (7.8), which then gives

$$E_k^{(1)} c_k = \sum_{k'} V_{kk'} c_{k'}. \tag{7.25}$$

Here, k and k' take on all the possible values that span the set of degenerate wave functions. This leads to a system of homogeneous equations

$$\sum_{k'} \left(V_{kk'} - E_k^{(1)} \delta_{kk'} \right) c_{k'} = 0. \tag{7.26}$$

This system of homogeneous equations has solutions for the various values of the coefficients c_k if the determinant vanishes:

$$det \left| V_{kk'} - E_k^{(1)} \delta_{kk'} \right| = 0. \tag{7.27}$$

If there are n degenerate eigenvalues, then Equation (7.27) is of order n. The n roots of the equation then give the values of $E_k^{(1)}$. Equation (7.27) is called the *secular equation*. The sum of the various new eigenvalue corrections is given by the sum of the diagonal matrix elements, a general result in degenerate perturbation theory. This process can be continued to second order if the degeneracy is not fully lifted by the first-order treatment.

7.2 SOME EXAMPLES OF PERTURBATION THEORY

In this section, a few examples of perturbation theory will be discussed in order to illustrate the stationary perturbation technique. The first of these will consider a trapezoidal potential well, which is a well with a linear potential term. Then a shifted harmonic oscillator, again obtained as a result of adding a linear potential to the harmonic oscillator, will be considered. We then turn to a pair of interacting potential wells, and finally a form of electron scattering by impurities is considered.

7.2.1 The Stark Effect in a Potential Well

The system to be considered is shown in Figure 7.1, where we have an infinitely deep potential well (one with infinitely high barriers, corresponding to Section 2.4.1), but with a linear potential existing within the well. Here, we take the unperturbed Hamiltonian as the quantity

$$H_0 = -\frac{\hbar^2}{2m}\frac{d^2}{dx^2} + V_0, \tag{7.28}$$

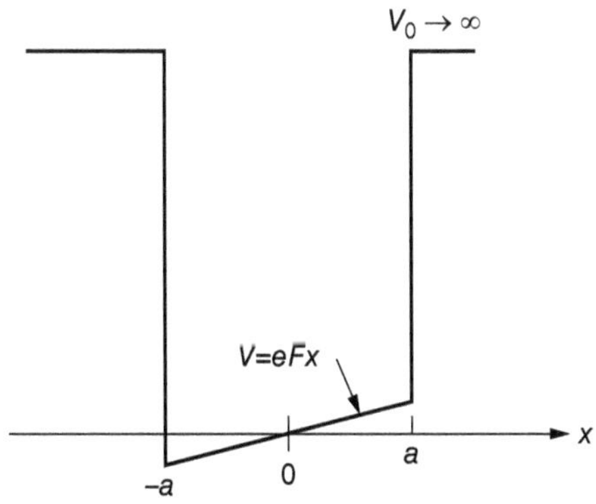

FIGURE 7.1 When an electric field is added to the infinite potential well, a trapezoidal well arises. The slope at the bottom of the well is determined by the value of the electric field that is added.

where

$$V_0 = \begin{cases} 0, & |x| < a, \\ \infty, & |x| > a. \end{cases} \tag{7.29}$$

From Section 2.4.1, the eigenvalues and eigenfunctions are

$$\begin{aligned} \psi_n(x) &= \frac{1}{\sqrt{a}} sin\left[\frac{n\pi}{2a}(x+a)\right], \\ E_n &= \frac{n^2\pi^2\hbar^2}{8ma^2}. \end{aligned} \tag{7.30}$$

Of course these exist only for $|x| < a$. For the perturbing potential, a linear potential corresponding to adding an electric field F as (to avoid the double use of E) is used as

$$V = eEx. \tag{7.31}$$

The entire problem is reduced to determining the matrix elements of this potential between the various basis states. There are two sets to be determined, the diagonal ones and the off-diagonal ones. First, the diagonal matrix elements are

$$\begin{aligned} V_{nn} &= \langle ne \,|\, Fx \,|\, n \rangle \\ &= \frac{eF}{a} \int_{-a}^{a} x sin^2\left[\frac{n\pi}{2a}(x+a)\right] dx \\ &= \frac{eF}{2a} \int_{-a}^{a} x\left\{1 - cos\left[\frac{n\pi}{2a}(x+a)\right]\right\} dx = 0. \end{aligned} \tag{7.32}$$

This result is a particular result of choosing the potential to be symmetrical about the center of the well, so that the average under the perturbing potential is zero. Thus, there is no first-order shift in the energy levels themselves. In other words, the particular choice of the potential results in the center of the well not moving, which means that, on average, the energy levels are not shifted to first order.

The off-diagonal terms, however, are nonzero, and these are evaluated in the same manner. The result is

$$
\begin{aligned}
V_{nm} &= \langle n \mid eFx \mid m \rangle \\
&= \frac{eF}{a}\int_{-a}^{a} x\sin\left[\frac{n\pi}{2a}(x+a)\right]\sin\left[\frac{m\pi}{2a}(x+a)\right]dx. \\
&= \frac{eF}{2a}\int_{-a}^{a} x\left\{\cos\left[\frac{(n-m)\pi}{2a}(x+a)\right]-\cos\left[\frac{(n+m)\pi}{2a}(x+a)\right]\right\}dx \\
&= \begin{cases} -\dfrac{16eFa}{\pi^2}\dfrac{nm}{\left(n^2-m^2\right)^2}, & (i\pm j)odd, \\ 0, & (i\pm j)even. \end{cases}
\end{aligned}
\tag{7.33}
$$

The last result tells us that the odd potential mixes only those states that have different parity. Thus, the odd potential produces matrix elements only between an even-parity state and an odd-parity state. (The plus/minus sign just reminds us that if the difference between two integers is odd, then the sum of these same two integers is also odd, and thus the two terms that result from the two integrals in (7.33) have been combined in the final result.)

Normally, the addition of an electric field produces a shift of the energy levels, an effect called the Stark effect. Here, because the original potential has been taken to be a symmetric one, the linear shift in the energy levels vanishes, and there is no first-order Stark effect. However, this is merely because of the choice of the reference level for the potential. If the zero of energy had been taken to lie at the point $x = -a$, then all the energy levels would have been pushed upward by the value of the potential at the center of the well. To see this, let us shift the origin to $-a$, so that the well goes from 0 to $2a$. Now, the basis functions are given by

$$
\psi_n(x) = \frac{1}{\sqrt{a}}\sin\left(\frac{n\pi x}{2a}\right), \quad 0 \le x \le a. \tag{7.34}
$$

Here, the symmetry of the perturbing potential has been changed. The diagonal elements now become

$$
\begin{aligned}
V_{nn} &= \langle n \mid eFx \mid n \rangle \\
&= \frac{eF}{a}\int_{0}^{2a} x\sin^2\left[\frac{n\pi x}{2a}\right]dx \\
&= \frac{eF}{2a}\int_{0}^{2a} x\left\{1-\cos\left[\frac{n\pi x}{2a}\right]\right\}dx = eFa
\end{aligned}
\tag{7.35}
$$

The energy levels are all shifted to first order by the same amount, *eFa*, which is the potential at the center of the well, as suggested above. This is the linear Stark shift of the energy levels. However, the off-diagonal elements do not change from (7.33), which is a reflection of the fact that both energy levels shift, but their difference does not (the linear Stark shift is the same for all energy levels). It may be noted that the denominator of (7.33), other than for associated constants, is the difference (squared) of two energy levels.

Thus, whether or not we see a uniform shift of the energy levels in a perturbation approach often depends upon the reference level chosen for the overall potential. However, this does not appear in the off-diagonal matrix elements. This latter is significant, as it is only the off-diagonal matrix elements that appear in the summations for the various orders of perturbation theory (the term in (7.20) involving the difference between two diagonal matrix elements is identically zero due to the fact that all of these diagonal elements are equal).

7.2.2 The Shifted Harmonic Oscillator

As a second example of the application of perturbation theory, consider a harmonic oscillator that is subjected to a linear potential that arises from an applied electric field. Then, the Hamiltonian can be written as

$$\begin{aligned} H &= \frac{p^2}{2m} + \frac{1}{2} m\omega^2 x^2 + eFx \\ &= \hbar\omega\left(a^\dagger a + \frac{1}{2}\right) + eFx. \end{aligned} \tag{7.36}$$

In the last line, the operator results of Chapter 5 have been used. The application of the linear potential still produces a harmonic oscillator, but one that is shifted in both position and energy. This can be seen, as the first line of (7.36) can be rewritten as

$$H = \frac{p^2}{2m} + \frac{1}{2} m\omega^2 (x + x_0)^2 - \frac{e^2F^2}{2m\omega^2}, \tag{7.37}$$

where

$$x_0 = \frac{eF}{m\omega^2}. \tag{7.38}$$

The last term in (7.37) is the downward shift of the quadratic potential in energy, while (7.38) is the shift of the center of the potential well in position. This is shown in Figure 7.2. There are now two ways to approach the solution of the new Hamiltonian in (7.36). In the first approach, the linear potential is taken to be the perturbing potential, and the wave functions of the harmonic oscillator centered at $x = 0$ are used. In the second approach, the perturbing potential is the uniform shift of the harmonic oscillator downward by the constant energy term, which is the last term on the right-hand side of (7.37). The wave functions are those corresponding to the harmonic oscillator centered at $x = -x_0$. We take the last approach first.

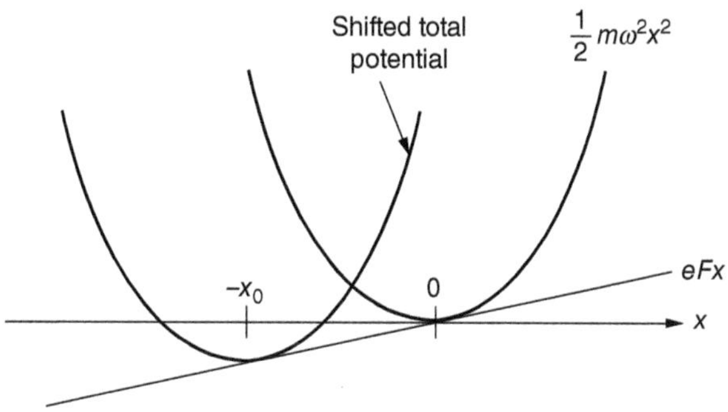

FIGURE 7.2 The combination of a quadratic potential and a linear potential produces a shifted quadratic potential, that is, a shifted harmonic oscillator.

If at the harmonic oscillator is treated directly as a shifted harmonic oscillator with the minimum centered at $x = -x_0$, then the wave functions are merely shifted Hermite polynomials of (5.28)

$$\psi_n \sim \left(a^\dagger\right)^n \psi_0 \sim \left(\frac{m\omega}{2\hbar}\right)^{n/2} \left(x + x_0 - i\frac{p}{m\omega}\right)^n \psi_0. \tag{7.39}$$

The matrix elements are

$$V_{ij} = \langle i | \frac{e^2F^2}{2m\omega^2} | j \rangle = \frac{e^2F^2}{2m\omega^2}\delta_{ij}, \tag{7.40}$$

and *only the diagonal elements are nonzero.* A constant shift of the Hamiltonian provides only a uniform and constant shift of all energy levels. It provides no mixing of the states. Thus, when the linear potential is applied to the harmonic oscillator, the only result is a shift downward of all energy levels by the shift in the minimum, and a shift of the centroid of the wave functions to the new center position of the harmonic oscillator.

Using perturbation theory should yield a similar answer. For this, the matrix elements can be computed using the linear potential as the perturbation. Then, the matrix elements can be computed using the position x given by the operators as

$$\begin{aligned} V_{ij} &= \langle i | eFx | j \rangle = eF\langle i | \sqrt{\frac{\hbar}{2m\omega}}\left(a^\dagger + a\right) | j \rangle \\ &= eF\sqrt{\frac{\hbar}{2m\omega}}\left[\sqrt{i}\delta_{j+1,i} + \sqrt{i+1}\delta_{j-1,i}\right] \end{aligned} \tag{7.41}$$

Surprisingly, there is no first-order shift of the energy, but the second-order shift involves only two terms in the summation. In each term, the product of matrix elements connects a given level only with its neighbors just above and just below, so only one of the two delta functions in (7.41) appears in each matrix element. We compute the second-order shift in the diagonal energy to be

$$E_n = E_n^{(0)} - \frac{e^2F^2}{m\omega^2}. \tag{7.42}$$

This produces the full shift of the energy levels given by (7.37), and it also is independent of the value of the energy level under consideration. Thus, higher-order corrections to the energy levels should vanish, and this can be checked by examining the fourth-order term, which is the next one expected. However, the off-diagonal corrections are not given in the exact theory above. These should not exist and are an artifact of using perturbation theory.

It cannot always be ensured that higher-order terms vanish, but rational thinking for this problem suggests that they should sum to zero, when all orders are included. The corrections to the wave functions are just those that try to move the centroid to the new center of the harmonic potential. This is a perfect example of a case where low-order perturbation theory does not produce a result that is close to the exact answer (at least for the wave functions), which we know from being able to solve the problem exactly. The first-order shift in the wave functions can be obtained by using the matrix elements above. For the lowest energy level, this correction is given as

$$|0\rangle^{(1)} = -\frac{eF}{\hbar}\sqrt{\frac{\hbar}{2m\omega}}\,|1\rangle^{(0)}, \tag{7.43}$$

and for the higher levels

$$|n\rangle^{(1)} = -\frac{eF}{\hbar}\sqrt{\frac{\hbar}{2m\omega}}\left(\sqrt{i}\,|\,i-1\rangle^{(0)} + \sqrt{i+1}\,|\,i+1\rangle^{(0)}\right). \tag{7.44}$$

Equations (7.43) and (7.44) give different values for the shifts of the wave functions, which is different than the exact approach above (which gives exactly the same shift for each wave function). Perhaps if the perturbation theory were evaluated to much higher orders, the proper solutions might appear.

7.2.3 Multiple Quantum Wells

The next example is the consideration of coupled quantum wells, which will each be taken to be quadratic in nature. This treatment provides a discussion of *degenerate perturbation theory*, a process in which we decide how two levels that otherwise would have the same energy interact to cause a small splitting. This was discussed at the end of Section 7.1. Consider, for example, two harmonic wells that are displaced from one another by a distance d, as shown in Figure 7.3. Here, the unperturbed wave functions in each well are at the same energy, so the degeneracy is between wave functions centered in each of the two wells. The interaction between the two potentials causes only a small barrier between the two wells, so the wave function of an electron localized in one of the wells actually extends somewhat into the other well. Thus, the two wave functions overlap with each other and there is an interaction between them. It is this interaction that causes a modification of the two wave functions, lifts the degeneracy, and leads to two new levels, which are then separated slightly in energy. One of the levels has a wave function composed of the constructive interference of the two individual wave functions, while the second (the higher energy state) has a wave function that is anti-symmetric in the combined wells and thus arises from the destructive interference of the two individual wave functions.

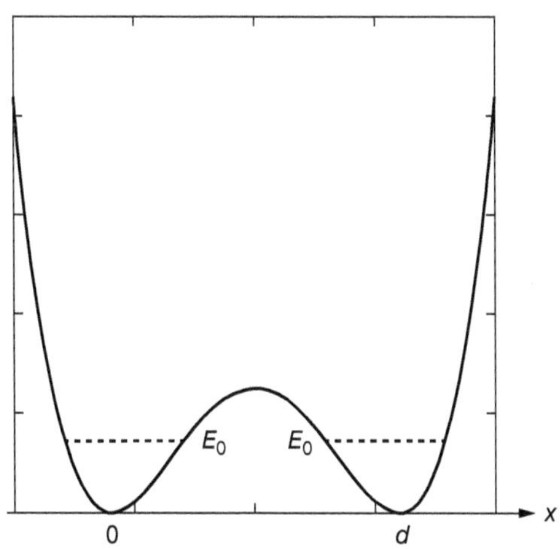

FIGURE 7.3 When two quadratic potentials are close, their overlap produces only a small potential in the region where they overlap. This is much like the modification of the atomic potential in Chapter 4. Now, the wave functions in each potential can overlap slightly and produce an interaction which modifies the energy levels as well as the wave functions.

In each of the two wells, the unperturbed problem is just that of a harmonic oscillator, centered at the particular well, so the Hamiltonian and lowest wave function of the well at $x = 0$ and $x = d$ are given by (and denoted by a different subscript)

$$\begin{aligned} H_{0,1} &= \frac{p^2}{2m} + \frac{1}{2}m\omega^2 x^2, \\ H_{0,2} &= \frac{p^2}{2m} + \frac{1}{2}m\omega^2 (x-d)^2. \end{aligned} \tag{7.45}$$

The two wave functions are taken to be

$$|0\rangle_0^{(0)}, |0\rangle_d^{(0)}, \tag{7.46}$$

where the subscript indicates the center position of the wave function. These two also have the same ground state energy $E_0^{(0)}$. Both wave functions, one centered in each well, have the same eigenvalues for the unperturbed energy. Thus, these two eigenvalues are said to be *degenerate*. The task here is to use perturbation theory to lift this degeneracy.

The interest is not so much in how the interaction with higher energy levels changes the lowest levels as with how these two wave functions interact with each other to change the lowest energy levels. Although the treatment focuses on the lowest energy level, this is not a limitation, and one could work with any of the levels of the isolated well that lie below the central peak between the two wells (for those levels above this peak, perturbation theory is not appropriate). To proceed, the interest is in the matrix element

$$V_{00} = {}_0^{(0)}\langle 0 | 0\rangle_d^{(0)} = \int_{-\infty}^{\infty} \psi_0^{\dagger(0)}(x) V(x) \psi_0^{(0)}(x-d)\,dx. \tag{7.47}$$

Here, the potential is the actual total potential in the problem, but the main part is that part of the potential that lies between the two wells. Equation (7.47) is the *overlap* integral defining the *mixing* of the two wave functions.

The composite wave functions should be able to be written as a sum of the two individual wave functions, as

$$|0\rangle = A|0\rangle_0^{(0)} + B|0\rangle_d^{(0)}. \tag{7.48}$$

If this last equation is pre-multiplied with the conjugate of the wave function of the left-hand well, and then again with the conjugate of the wave function of the right-hand well, two equations are produced. This is done by treating the variation of the potential difference from the single well for the diagonal terms only with the interaction potential of (7.47). The process produces the resulting matrix equation

$$\begin{bmatrix} E_0 & V_{00} \\ V_{00} & E_0 \end{bmatrix} \begin{bmatrix} A \\ B \end{bmatrix} = \begin{bmatrix} E \\ E \end{bmatrix} \begin{bmatrix} A \\ B \end{bmatrix}. \tag{7.49}$$

If this is to have a solution, then the determinant must vanish, or

$$det \begin{vmatrix} (E_0 - E) & V_{00} \\ V_{00} & (E_0 - E) \end{vmatrix} = 0. \tag{7.50}$$

This leads to the new energy levels

$$E = E_0 \pm V_{00}. \tag{7.51}$$

Thus, one eigenstate moves up slightly in energy while the other moves down in energy. The lowest of the two states is the bonding state with $A = B$, while the upper state is the antibonding state with $B = -A$. The total energy splitting between these two states is $2V_{00}$.

7.3 AN ALTERNATIVE APPROACH—THE VARIATIONAL METHOD

There are often times when the perturbation approach just does not work, because the basis set cannot be determined consistently with (7.3). An example is the triangular potential well of Section 2.4. The problem with this particular barrier is that the solution of the Schrödinger equation is complicated, and the wave functions are special functions that, in general, are evaluated numerically. Often, however, the exact eigenvalues and eigenfunctions are not required; rather, only good approximations to them are required. To accomplish this, another approximation technique is generally used—the Rayleigh–Ritz method, or *variational* method. Here, it may be assumed that we cannot begin with (7.3) or any equivalent form. Nevertheless, an approximate wave function may be generated as an expansion in a convenient basis set as

$$\psi(x) = \sum_i a_i \varphi_i(x). \tag{7.52}$$

If this is an orthonormal basis set, it may be assumed that the expectation value of the Hamiltonian is given by

$$\langle H \rangle = \sum_i E_i |a_i|^2. \tag{7.53}$$

Here, the expectation value has been constructed by pre-multiplying by the conjugate of (7.52), post-multiplying by (7.52), integrating, and using the orthogonality properties of the basis set. An inequality can be created by assuming that *only the lowest energy level* is used in the expansion (instead of the actual energy level for each of the eigenstates), so that the real expectation value of the Hamiltonian is

$$\langle H \rangle \geq E_0 \sum_i |a_i|^2 = E_0. \tag{7.54}$$

The sum over all the basis states is unity to ensure the normalization of the wave function in (7.52).

The principle of the Rayleigh–Ritz method is that a trial wave function can be adopted, which has some parameters in it, and these parameters are adjusted to minimize the energy calculated with this wave function. This energy will still be above the actual lowest energy level E_0. This (hopefully) gives a good approximation to the actual energy level. The next higher state can be developed by using a second parametrized basis function, which is made orthogonal to the first. The energy level of this wave function is then found by the same adjustment technique. To illustrate this, the actual triangular-potential-well problem of Section 2.4 will be used.

To begin, it may be noted that the wave function vanishes at the origin (where the potential is infinite). The potential is typically defined by

$$V(x) = \begin{cases} \infty, & x \leq 0, \\ eFx, & x > 0, \end{cases} \tag{7.55}$$

where F has been used for the field to avoid conflicting with the energy. Now, a trial function that vanishes at $x = 0$, and also vanishes for large x, where the energy lies at a lower level than the potential, is necessary. It should not be a difficult function, because it is going to be integrated. For this, a useful form that fits these requirements is

$$|0\rangle = Axe^{-bx}. \tag{7.56}$$

The prefactor coefficient A is determined by normalization to be

$$\langle 0|0\rangle = 1 = A^2\int_0^\infty x^2 e^{-2bx}\,dx = \frac{A^2}{4b^3}, \tag{7.57}$$

or

$$A = 2b^{3/2}. \tag{7.58}$$

Now, this wave function can be used with the Hamiltonian to compute the expectation value as

$$\begin{aligned}\langle H\rangle &= 4b^3\int_0^\infty xe^{-bx}\left(-\frac{\hbar^2}{2m}\frac{d^2}{dx^2}+eFx\right)xe^{-bx}dx\\ &= \frac{\hbar^2b^2}{2m}+\frac{3eF}{2b}.\end{aligned} \tag{7.59}$$

This value is now above the actual lowest energy level according to (7.54), and one can minimize this value by setting the derivative with respect to b to zero. This then leads to (we take the first level as the lowest level in keeping with the general results of Section 2.4)

$$b = \left(\frac{3eFm}{2\hbar^2}\right)^{1/3}. \tag{7.60}$$

The actual lowest variational energy level is found by inserting (7.60) into the last line of (7.59) to give

$$E_0 \approx \frac{3\hbar^2}{2m}\left(\frac{3eEm}{2\hbar^2}\right)^{2/3}. \tag{7.61}$$

This result should be compared with (2.65). The present result differs from (2.66) only by a factor of $3(2/3\pi)^2/3 \simeq 1.07$, or about a 7% error, which is quite small considering the approximations. As expected, the right-hand side of (7.61) lies above the actual value, given by (2.66), but one should remember that this last value was itself an approximation using the values of the zeroes of the Airy function.

To compute the next level, a wave function that is orthogonal to (7.56) must be created. The wave function of (7.56) has a single maximum, decaying away to zero at the origin and for large values of x. The next level should have a wave function with two maxima but also decay away for large values of x and at the origin. For this purpose, assume the first excited–state wave function may be approximated as

$$|1\rangle = C\left(x - Dx^2\right)e^{-fx}. \tag{7.62}$$

This function must now be made orthogonal to the zero state. It must also be normalized, and, following this, it may be adjusted in minimizing the energy level that corresponds to this state. The first of these tasks gives

$$\begin{aligned}\langle 0 \mid 1\rangle = 0 &= 2b^{3/2}C\int_0^\infty xe^{-bx}\left(x - Dx^2\right)e^{-fx}dx \\ &= 2b^{3/2}C\left[\frac{2}{(b+f)^3} - \frac{6D}{(b+f)^4}\right]\end{aligned} \quad (7.63)$$

This may be satisfied by choosing $D = (b + f)/3$. The second task is to normalize the wave function as

$$\langle 1 \mid 1\rangle = 1 = \frac{C^2}{4f^5}\left(f^2 - 3Df + 3D^2\right). \quad (7.64)$$

Now, inserting the value for D, normalization gives

$$C^2 = \frac{12f^2}{f^2 - bf + b^2}. \quad (7.65)$$

Now, the last factor is to minimize the expectation value of the energy for this level. The expectation value of the energy is

$$\begin{aligned}\langle H\rangle = &\frac{\hbar^2 f^2}{6m}\left(1 + \frac{15f^2}{f^2 - bf + b^2}\right) \\ &+ \frac{eF}{2f}\left(1 + \frac{3b^2}{f^2 - bf + b^2}\right)\end{aligned}. \quad (7.66)$$

Since B is known from (7.60), this last expression can be minimized to yield the value of f. While the tendency might be to ignore the second terms in each of the large parentheses, this would be a mistake, as it would lead to an energy level below the lowest level. These terms are actually the major terms. Unfortunately, this leads to a complicated, high-order algebraic equation for f in terms of b. This can be solved most easily by numerical calculations, which lead to the result $f \approx 0.73b$, which in turn leads to $E_1 \approx 2.4E_0$.

A similar procedure can now be used to generate a variational wave function for the next higher eigenstate. A wave function, probably with a cubic term in the prefactor, could be assumed and the coefficients adjusted first to make it orthogonal to each of the two lower eigenfunctions. Then, the energy is minimized to provide an estimate of the energy level. Of course, each successive higher level becomes more complicated to determine, and most efforts using this method are addressed only to the lowest energy level.

PROBLEMS

1. In an infinite square well, in which $V(x) \to \infty$ for $x < 0$ and $x > a$, $V(x) = 0$ for $0 \le x \le a$, the basis states are given by $|n\rangle = \sqrt{2/a}\sin(n\pi x/a)$, and $E_n = \hbar^2\pi^2 n^2/(2ma^2)$. For the introduction of a perturbing potential of $V_1 = (x - a)^2/2$, calculate the changes in the wave functions up to first order and in the energy levels up to second order.
2. Determine the third-order correction to the energy in perturbation theory.

3. Except for certain accidents, degeneracy usually does not arise in a one-dimensional problem. Degeneracy is usually found when the dimension is raised. Consider the two-dimensional infinite square well

$$V(x,y)=\begin{cases}0, & |x|\le a/2, |y|\le a/2,\\ \infty, & |x|\ge a/2, |y|\ge a/2.\end{cases}$$

Using this potential, separate the Schrödinger equation into two one-dimensional equations and solve for the allowed energy values using the fact that the potential can be written as the sum of two one-dimensional potentials. The solutions can then be written down using the results of Chapter 2. Consider the application of the simple perturbation $V_1(x,y) = V_0$ for $|x| \le a/4$ and $|y| \le a/4$ and is zero elsewhere (the perturbing potential exists only in the center of the well). Compute the splitting of the lowest two unperturbed energy levels.
4. Using the harmonic oscillator raising and lowering operators, compute directly the matrix element for $\langle n|x^4|n\rangle$. Then, introducing a complete set of states (as a resolution of the delta function) break this into two products x^2x^2, and compute the matrix element from

$$\langle n|x^4|n\rangle=\sum_j\langle n|x^2|j\rangle\langle j|x^2|n\rangle.$$

5. A one-dimensional harmonic oscillator is perturbed by a term αx^3. Calculate the change in each energy level to second order.
6. Consider the simple harmonic oscillator Hamiltonian

$$H=-\frac{\hbar^2}{2m}\frac{d^2}{dx^2}+\frac{C_0}{2}x^2.$$

Let a perturbation $W = C_1x^2/2$ be applied to this system and determine the energy eigenvalues and compare them with the exact solution.
7. Consider an infinite potential well that exists in the region $0 \le x \le b$. Now, add a perturbing potential of the form

$$V_1(x)=\lambda sin\left(\frac{\pi x}{b}\right).$$

Determine the energy shift for the various energy levels to first order.
Calculate the lowest energy level of a harmonic oscillator using the Gaussian wave function as a variational wave function. If the Gaussian is defined from e^{-bx^2}, use the factor b as the variational parameter.

8 Time-Dependent Perturbation Theory

In many physical systems, the interest is in small time-dependent changes in the state. The approach used in the last chapter cannot deal with this situation, and perturbation theory must be developed further, particularly with the inclusion of a time variation. This will be done in the present chapter. In fact, the approach will be developed twice, first for the Schrödinger representation and then for the Heisenberg representation in a linear vector space. This double approach is followed for two reasons. First, it is important to grasp the concept of the time-dependent perturbation approach in a direct method, without the encumbrance of the mathematics that accompanies the latter approach—the interaction representation. Second, it is important to learn the mathematics of the interaction representation, but in doing so it is quite helpful to understand just where one is heading. The double approach achieves both objectives.

In time-dependent perturbation theory, we continue to assume that the Hamiltonian can be split into two parts

$$H = H_0 + H_1(t), \tag{8.1}$$

in which the first part H_0 can be solved directly and the second part H_1 is small. It is this second part that is assumed to have a time variation. When this can be done, the second part can be initially ignored, and the problem is solved directly to give the wave function, or its expansion in a natural basis function set, typically the energy eigenstates. After this is done, the second term in (8.1) is treated as a perturbation, and the eigenvalues and eigenfunctions are then developed in a time-dependent perturbation approach in terms of the natural basis function set. But the formulation (8.1) remains a closed system, even though it may be considered that the perturbation arises from an interaction from outside the system (commonly referred to as the *bath*), such as from phonons when we are considering the properties of the electrons.

In this chapter, the approach in the Schrödinger representation will be developed first and an example due to a harmonic potential will be considered. Then the interaction representation will be developed, and the perturbation theory redeveloped in the Heisenberg representation. Our attention will then be turned to an extended approach in which the initial state of the system has an exponential decay. It is this latter approach that allows us to return once again to the idea of an energy–time relationship as a result of the lifetime of the initial state. Finally, the scattering matrix approach (the T-matrix) is developed. One note of importance is that it will be assumed that the term H_0 is a time-independent quantity, while the perturbation term H_1 will be allowed to have an oscillatory nature. In the final sections, an important application, that of optical absorption in a semiconductor, is taken as an example of the time-dependent perturbation theory.

8.1 THE PERTURBATION SERIES

If the Schrödinger equation is used with a time-independent and unperturbed Hamiltonian H_0, for which it may readily be solved, the solution may be written from (6.5) as (the position coordinate is suppressed)

$$\psi^{(0)}(t) = \sum_n c_n e^{-i\omega_n t} \varphi_n, \tag{8.2}$$

DOI: 10.1201/9781003717195-8

where

$$H_0\varphi_n = E_n\varphi_n, \quad \omega_n = E_n/\hbar, \tag{8.3}$$

are the usual assumptions, and

$$c_n = \int_{-\infty}^{\infty} \varphi_n^\dagger \psi^{(0)}(0)dx \tag{8.4}$$

is the projection of the total wave function onto the nth basis state.

In the presence of the perturbation H_1, a solution is sought that has the same general form as (8.2)–(8.4), and this can be achieved by making the assumption that the perturbation $H_1 = V$ is small and produces a slow variation in the coefficients c_n. Then, we may write the new solution as

$$\psi(t) = \sum\nolimits_n c_n(t)e^{-i\omega_n t}\varphi_n, \tag{8.5}$$

in which the time variation of the coefficient is very slow compared to any of the frequencies arising from the various energy levels. This leads to the only problem, that is, determining the time variation of these coefficients. To begin, the complete Schrödinger equation becomes

$$i\hbar\frac{\partial\psi}{\partial t} = (H_0 + V)\psi. \tag{8.6}$$

To proceed, the assumed solution (8.5) is introduced into the Schrödinger equation which leads to

$$\sum_n \left(c_n E_n + \frac{\partial c_n}{\partial t}\right)e^{-i\omega_n t}\varphi_n = \sum_n (E_n + V)c_n e^{-i\omega_n t}\varphi_n. \tag{8.7}$$

In this expansion, (8.3) has been used to insert the energies of the eigenfunctions. The first term on the left and the first term on the right thus cancel. This last equation is now pre-multiplied by an arbitrary basis function $\varphi_k^\dagger$, and this result is integrated over all space, using the orthonormality of these functions. It will be further assumed that $V = V_0 e^{-i\omega_0 t}$. This leads to

$$i\hbar\frac{\partial c_k}{\partial t}e^{-i\omega_k t} = \sum_n V_{kn}c_n e^{-i(\omega_n - \omega_0)t}, \tag{8.8}$$

where

$$V_{kn} = \langle k|V_0|n\rangle \tag{8.9}$$

is the matrix element connecting the two states with the *time-independent* part of the perturbing potential. To make life easy, we introduce the reduced frequency

$$\omega_{kn} = \frac{E_k - E_n}{\hbar} = \omega_k - \omega_n, \tag{8.10}$$

so that (8.8) can be rewritten as

$$i\hbar\frac{\partial c_k}{\partial t} = \sum_n V_{kn}c_n e^{-i(\omega_{nk} - \omega_0)t}. \tag{8.11}$$

What has appeared is a complicated set of equations for the coefficients and their time dependence. In general, this can be solved by matrix techniques when the matrix elements are known.

The approach to be followed first, however, assumes that the perturbation is very small. Moreover, it will also be assumed that the system is in a single eigenstate at the initial time, which will be taken as $t = 0$, which gives the initial conditions

$$\begin{aligned} c_s(0) &= 1, \\ c_k(0) &= 0, k \neq s. \end{aligned} \tag{8.12}$$

To be consistent with this assumption, it also will be assumed that the perturbation V is zero for $t < 0$. With the above approximations, (8.11) now becomes a simple equation

$$i\hbar \frac{\partial c_k}{\partial t} = V_{ks} e^{i(\omega_{ks} - \omega_0)t}, \tag{8.13}$$

and the solution becomes

$$c_k(t) = -\frac{V_{ks}}{\hbar(\omega_{ks} - \omega_0)} \left[e^{i(\omega_{ks} - \omega_0)t} - 1 \right]. \tag{8.14}$$

The change in the occupancy of the state is mainly determined by the change in the magnitude squared of the wave function. Thus, the quantity of interest is the magnitude squared of (8.14), which becomes

$$|c_k(t)|^2 = \frac{4|V_{ks}|^2}{\hbar^2(\omega_{ks} - \omega_0)^2} sin^2\left(\frac{(\omega_{ks} - \omega_0)t}{2}\right). \tag{8.15}$$

At this point, it should be noted that any decay of the initial state has been ignored, under the assumption that the perturbation is very small. This initial decay will be treated in Section 8.3. The result (8.15) is now the probability that density moves from the initial state s to the state k. Our interest is in the rate at which this probability transfers, or the transition rate, which is the probability per unit time, or

$$\Gamma_{ks} = \frac{|c_k(t)|^2}{t} = \frac{4|V_{ks}|^2}{\hbar^2(\omega_{ks} - \omega_0)^2 t} sin^2\left(\frac{(\omega_{ks} - \omega_0)t}{2}\right). \tag{8.16}$$

In the long-time limit this becomes (Landau and Lifshitz, 1958)

$$\lim_{t \to \infty} \frac{sin^2(at)}{a^2 t} \to 2\pi\delta(a). \tag{8.17}$$

When $a \neq 0$, the limit is zero and the transition rate vanishes (a is associated with $(\omega_{ks} - \omega_0)/2$). The delta function serves to ensure that the transition rate is nonvanishing only for $a = 0$. Thus, the final transition rate can be written as

$$\Gamma_{ks} = \frac{2\pi}{\hbar} |V_{ks}|^2 \delta(\hbar\omega_{ks} - \hbar\omega_0). \tag{8.18}$$

The last form in (8.18) is often referred to as *the Fermi golden rule*. The transition rate is given by a numerical factor times the squared magnitude of the matrix element and a delta function that conserves the energy in the transition. The latter is important, as it ensures that time-dependent variations arise only from those states in which the transition can conserve energy, a result that is required for the system to remain in equilibrium with no gain or loss of energy. This does not preclude the *transfer* of energy to or from another part of the system. In fact, here the energy is transferred to or from that part of the closed system that provided the perturbing potential V. This means, for example, that the wave function (8.5) does not represent the entire system, as there is another part that provided V, which is often referred to as the *bath*, as mentioned above. While not represented in (8.5), it certainly has been assumed to exist in (8.1).

8.2 THE INTERACTION REPRESENTATION

As mentioned in the introduction, there is another method of getting at the perturbation expression through a more formal approach, which will give rise to all orders of the perturbation (although no one really goes past first order without moving to more extensive formulations). This approach is based upon the Heisenberg representation introduced in Chapter 6. To review this approach, the wave function is expanded in a linear vector space of basis functions, just as in the approach that leads to (8.3), but it is assumed that the basis set is not time-varying, at least in the unperturbed state. Rather, the time variation is attached to the operators themselves, which rotate in the fixed linear vector space (which forms our coordinate system) as a function of time according to (6.7) as

$$\frac{\partial V(t)}{\partial t} = \frac{i}{\hbar}\left[H_0, V\right], \qquad V_H(t) = e^{iH_0 t/\hbar} V e^{-iH_0 t/\hbar}. \tag{8.19}$$

This relationship is the basis of the Heisenberg representation. It is quite possible that the operator A actually varies with time, and, if this is the case, the final form of (8.19) leaves that time variation intact (as one can show by adding the partial derivative of A with respect to time to the right-hand side of the first equation, and then integrating the system).

In the presence of a time-varying perturbation, however, it will now be assumed that the perturbation introduces a slow variation in the basis set itself. Just as we earlier assumed that the coefficients of the expansion were now time-varying, it will be assumed that this slow variation is entirely due to the perturbing potential $V(t)$ defined according to (8.19). It will also be assumed that there is an explicit time variation to this perturbation (such as might arise if the perturbation were due to an interaction with a harmonic oscillator). Hence, there is a slow variation of the wave function in the linear vector space that is introduced by V, while all other operators still respond to the unperturbed Hamiltonian according to (8.19). This mixed representation is termed the interaction representation.

Before proceeding, it is useful to consider further the time variation in general. Earlier, the time variation of a wave function was written as (6.2) the following:

$$\psi(x,t) = e^{-iHt/\hbar}\psi(x,0). \tag{8.20}$$

This assumed that the time of interest began at $t = 0$ when (6.1) was integrated. But suppose it began at t_0. Then, (8.20) would be rewritten as

$$\psi(x,t) = e^{-iH(t-t_0)/\hbar}\psi(x,t_0). \tag{8.21}$$

This suggests that a time-translation operator can be defined as

$$T(t,t_0) = exp\left[-\frac{i}{\hbar}H_0(t-t_0)\right]. \tag{8.22}$$

Now, (8.21) and (8.22) can be used to rewrite (8.20) as

$$\psi(x,t) = T(t,t_0)\psi(x,t_0). \tag{8.23}$$

With (8.19), a new operator can be defined as (Merzbacher, 1970)

$$U(t,t_0) = e^{iH_0t/\hbar}T(t,t_0)e^{-iH_0t/\hbar}. \tag{8.24}$$

When the Hamiltonian is Hermitian, both $T(t,t_0)$ and $U(t,t_0)$ will be unitary operators. Using the Schrödinger equation on this time-translation operator gives the latter the variation as

$$i\hbar\frac{\partial U}{\partial t} = V_H(t)U(t,t_0), \tag{8.25}$$

where $V_H(t)$ has the Heisenberg form of (8.19). The operator U has some interesting properties. First, it must be unity when the two times are the same

$$U(t,t) = 1, \tag{8.26}$$

because no change is occurring in the wave function (or any other entity such as an operator). It must also have the sequential property

$$U(t,t_0) = U(t,t_2)U(t_2,t_0). \tag{8.27}$$

Now, let us return to the differential equation for the time-translation operator (8.24) and integrate it over time as

$$U(t,t_0) = 1 - \frac{i}{\hbar}\int_{t_0}^{t} V_H(t')U(t',t_0)dt'. \tag{8.28}$$

This can be solved by iteration in the following manner (the goal is to find a more useful formulation for the solution, and this arises from the iteration process)

$$\begin{aligned} &U_0(t,t_0) = 1, \\ &U_1(t,t_0) = 1 - \frac{i}{\hbar}\int_{t_0}^{t} V_H(t')U_0(t',t_0)dt' = 1 - \frac{i}{\hbar}\int_{t_0}^{t} V_H(t')dt', \\ &U_2(t,t_0) = 1 - \frac{i}{\hbar}\int_{t_0}^{t} V_H(t')dt' + \left(\frac{i}{\hbar}\right)^2 \int_{t_0}^{t} V_H(t')\int_{t_0}^{t'} V_H(t'')dt''dt', \end{aligned} \tag{8.29}$$

and so on. This iteration can be continued and suggests that the summation can be resummed as a known integral in the form

$$U(t,t_0) = exp\left[-\frac{i}{\hbar}\int_{t_0}^{t} V(t')dt'\right]. \tag{8.30}$$

This form is obtained by noting that the multiple integrations can all be changed to integrate over the full temporal range and this adds the 1/numerical factor that gives the exponential series.

Using the above result, the matrix element and the higher-order terms in the perturbation series between two states can be written as

$$\begin{aligned} \langle k | U(t,t_0) | s \rangle = \langle k | s \rangle &- \frac{i}{\hbar}\int_{t_0}^{t} \langle k | V_H(t') | s \rangle\, dt' \\ &+ \left(\frac{i}{\hbar}\right)^2 \int_{t_0}^{t}\int_{t_0}^{t'} \langle k | V_H(t')V_H(t'') | s \rangle\, dt''dt' + \cdots \end{aligned} \tag{8.31}$$

Obviously, the first term on the right-hand side vanishes by the principle of orthogonality, while the second term is essentially the integral of (8.13) if one recognizes that the exponential is buried in the time dependence of the perturbing potential. We note that the left-hand side is the coefficient c_k. The result here tells us that the result obtained earlier is just the first term in a perturbation series expansion, which may be terminated only so long as the terms decrease sufficiently rapidly in order. The rest of the development follows Section 8.1 through to the Fermi golden rule.

8.3 EXPONENTIAL DECAY

In the preceding sections, it has been assumed that the initial state remains at unit amplitude. In fact, this is really unlikely, because it does not maintain a constant probability for the wave functions—the initial state cannot remain constant at unity if probability is being transferred to the other states via scattering by the Fermi golden rule. It is now desirable to see how this transfer of probability leads to the decay of the initial state, which in turn affects the transition probability. Each approach yields the same form for the transition rate, so we will use the direct form of (8.13)–(8.14). These results can then be used to determine the decay of the initial state by summing over all the other states as (recall that s was the original state, with $s = 1$)

$$i\hbar\frac{\partial c_s}{\partial t} = \sum_{k\neq s} V_{ks} c_k(t) e^{i\omega_{ks}t}. \tag{8.32}$$

The value for $c_k(t)$ is now taken from (8.13) to give (here, $c_s(t)$ has been reinserted, since it is no longer expected to remain at unity)

$$\frac{\partial c_s}{\partial t} = -\frac{1}{\hbar^2}\sum_{k\neq s}\int_0^t |V_{ks}|^2 c_s(t') e^{i(\omega_{ks}-\omega_0)t'} dt'. \tag{8.33}$$

This last term is a convolution integral of this initial state amplitude and the exponential function. While this complicates the behavior, it is easy to handle with a Laplace transformation whose variable we take as z. Transforming (8.33) now leads us to

$$zC_s(z) - 1 = -\frac{1}{\hbar^2}\sum_{k\neq s}|V_{ks}|^2 C_s(z)\frac{1}{z - i(\omega_{ks}-\omega_0)}, \tag{8.34}$$

which leads to

$$C_s(z) = \left\{z + \frac{1}{\hbar^2}\sum_{k\neq s}|V_{ks}|^2\frac{1}{z - i(\omega_{ks}-\omega_0)}\right\}^{-1}. \tag{8.35}$$

Rationalizing the last term in the curly brackets and separating the real and imaginary parts gives

$$C_s(z) = \left\{z + \frac{1}{\hbar^2}\sum_{k\neq s}|V_{ks}|^2\frac{z}{z^2 + (\omega_{ks}-\omega_0)^2} + \frac{i}{\hbar^2}\sum_{k\neq s}|V_{ks}|^2\frac{(\omega_{ks}-\omega_0)}{z^2 + (\omega_{ks}-\omega_0)^2}\right\}^{-1}. \tag{8.36}$$

As in Section 8.1, the long-time limit is going to be imposed, which may be found by taking $z \to 0$ limit of the expression. The same result will be found if the normal expansion

$$\int \frac{1}{z - i\omega} \to P + i\pi\delta(\omega) \tag{8.37}$$

is used to evaluate (8.35). The last term in the curly brackets of (8.35) becomes one-half of (8.18) for the normal scattering rate, although this time it is the total scattering rate out of the state s. It would be easy to set the second term in (8.36) to zero, as this is the principal part of the integral (or summation in this case) (8.37), but this leads to an energy shift of the initial state. In general the real and imaginary terms together furnish the self-energy of the state, so that (8.2) becomes

$$\psi(x,t) = \sum_n c_n e^{i(E_n - \Sigma_n)t/\hbar} \varphi_n(x), \tag{8.38}$$

where the self-energy is

$$\sum\nolimits_n = \Delta E_n + i\frac{\Gamma_n}{2}. \tag{8.39}$$

The first term on the right-hand side of this last equation is the shift in the energy of the state and the last term is the scattering out of the state which leads to a decay of the amplitude of that state. In general, (8.13) can now be rewritten as

$$c_k = -\frac{i}{\hbar}\int_0^t V_{ks} e^{-\frac{\Gamma_s t'}{2} + i(\Omega_{ks} - \omega_0)t'} dt', \tag{8.40}$$

where

$$\Omega_{ks} = \omega_{ks} - i\Delta E_s/\hbar. \tag{8.41}$$

If the long-time limit is now taken for the actual magnitude squared of the coefficient, the result is

$$\begin{aligned} |c_k|^2 &= \lim_{t\to\infty} \frac{|V_{ks}|^2}{\hbar^2} \frac{1 - 2e^{-\frac{\Gamma_s t}{2}} cos((\Omega_{ks} - \omega_0)t) + e^{-\Gamma_s t}}{(\Omega_{ks} - \omega_0)^2 + (\Gamma_s/2)^2} \\ &\to \frac{|V_{ks}|^2}{\hbar^2\left[(\Omega_{ks} - \omega_0)^2 + (\Gamma_s/2)^2\right]}. \end{aligned} \tag{8.42}$$

Rather than dividing by the time as done in Section 8.1, this resulting Lorentzian is used to replace the delta function, and the scattering rate is added to the numerator to assure integration to unity. This gives the new scattering rate as

$$\Gamma_{ks} = \frac{|V_{ks}|^2}{\hbar^2} \frac{\Gamma_s}{(\Omega_{ks} - \omega_0)^2 + (\Gamma_s/2)^2}. \tag{8.43}$$

The result is both a shift of the resonance and a broadening of this resonance arising from the perturbing process. Here, the delta function has been broadened into a Lorentzian line shape. One often assumed that the half-width of the Lorentzian, given by Γ_s, to be something like a time uncertainty. However, this width is not fundamental, as it depends upon the scattering process and the interaction strength. In this sense, it is more like a classical limit on the ability to measure the energy of the resonance term.

8.4 THE LIPPMANN–SCHWINGER EQUATION

In treating the Fermi golden rule transition rate of the last few sections, it was assumed that the perturbing potential was turned on at $t = 0$. This was done simply to avoid facing the evaluation of some of the integrals at $t \to -\infty$. The latter limit is problematic when the excitation is taken to be a simple sinusoid or a constant function. This has allowed us to use the Laplace transforms where necessary. A somewhat different approach is often used, in which Fourier transforms are utilized as necessary. In order to avoid the problem of the limit at large negative times, a different strategy is adopted. In this approach, we generate what is called the T-matrix, which in turn is used in the Lippmann–Schwinger equation to describe the scattered states.

8.4.1 The Scattering State T-Matrix

In order to describe negative time, it will be assumed that the perturbation is slowly turned on over a very-long time period, but in such a manner that the matrix element can be written as

$$V_{ks} = T_{ks} e^{\alpha t} \tag{8.44}$$

with

$$\begin{aligned} \alpha &> 0, \quad when\ t < 0, \\ \alpha &= 0, \quad when\ t > 0. \end{aligned} \tag{8.45}$$

In general, this factor can be set to zero at that point in the calculation in which all integrals have been evaluated at the lower limit of large, negative time. With this approach, (8.13) and (8.14) become

$$\begin{aligned} c_k &= -\frac{iT_{ks}}{\hbar} \int_{-\infty}^{t} e^{i(\omega_{ks}-\omega_0)t' + \alpha t'} dt' \\ &= \frac{T_{ks}}{\hbar} \frac{e^{i(\omega_{ks}-\omega_0)t' + \alpha t'}}{i\alpha - (\omega_{ks} - \omega_0)} + \delta_{ks} \end{aligned} \tag{8.46}$$

The caveat here is that the last form is only for times $t \ll 1/\alpha$, so the growing exponential is not to be considered as a major factor, but we don't want to set α to zero just yet. Now, (8.15) becomes

$$|c_k|^2 = \frac{|T_{ks}|^2}{\hbar^2} \frac{e^{2\alpha t}}{\alpha^2 + (\omega_{ks} - \omega_0)^2}, \tag{8.47}$$

and

$$\begin{aligned} \Gamma_{ks} &= \lim_{\alpha \to 0} \frac{\partial |c_k|^2}{\partial t} = \lim_{\alpha \to 0} \frac{|T_{ks}|^2}{\hbar^2} \frac{2\alpha e^{2\alpha t}}{\alpha^2 + (\omega_{ks} - \omega_0)^2} \\ &\to \frac{2\pi}{\hbar^2} |T_{ks}|^2 \delta(\omega_{ks} - \omega_0). \end{aligned} \tag{8.48}$$

Here, only the derivation of the exponential term has been treated, and the result is just (8.18), the Fermi golden rule (the extra factor of $\hbar$ goes with the delta function to produce a delta function in energy rather than radian frequency).

The matrix of quantities T_{ks} is usually called the T-matrix, and it was introduced in the above discussion in a rather *ad hoc* manner. Let us pursue this a little more deeply and find out how it is really related to the matrix elements V_{ks}. To proceed, (8.44) is inserted into the earlier (8.11) for the coefficients of the expansion series. This leads to

$$i\hbar\frac{\partial c_k}{\partial t} = \sum_n V_{kn}T_{ns}e^{i(\omega_{kn}+\omega_{ns}-\omega_0)t+\alpha t}, \tag{8.49}$$

which then leads to

$$c_k = T_{ks}e^{i(\omega_{ks}-\omega_0)t} = \frac{1}{\hbar}\sum_n V_{kn}T_{ns}\frac{e^{i(\omega_{kn}+\omega_{ns}-\omega_0)t+\alpha t}}{i\alpha-(\omega_{ns}-\omega_0)}. \tag{8.50}$$

The two frequencies in the exponential reduce to ω_{ks} in keeping with the left-hand side of the equation where it relates the initial and final states of the T-matrix. For all practical purposes, we can now set α to 0 and the exponential terms drop out of the equation. This result now can be used in (8.48) to once again regain the Fermi golden rule.

8.4.2 Gaining the Lippmann-Schwinger Equation

The T-matrix can now be used to define scattering states that arise from the perturbation of an incoming wave by any scattering center. This incoming wave gives rise to the scattered outgoing wave $\varphi_s^{(+)}$ as

$$T_{ks} = \int \varphi_k^\dagger V(x,t)\varphi_s^{(+)}dx = \sum_n \int \varphi_k^\dagger V(x,t)\varphi_n dx \cdot \langle \varphi_n^\dagger \mid \varphi_s^{(+)}\rangle, \tag{8.51}$$

where the resolution of the delta function has been inserted via the relationship

$$\sum_n \varphi_n^\dagger(x')\varphi_n(x) = \delta(x-x'). \tag{8.52}$$

Using this last equality led to the summation in (8.51). The plus sign in that equation refers to an incremental positive time to assure causality. With (8.51), the form of (8.50) can be expressed as

$$\langle \varphi_k|V|\varphi_s^{(+)}\rangle = V_{ks} + \sum_n \frac{\langle \varphi_k|V|\varphi_n\rangle\langle \varphi_n|V|\varphi_s^{(+)}\rangle}{\hbar\left(i\alpha-(\omega_{kn}+\omega_{ns}-\omega_0)\right)}. \tag{8.53}$$

The diagonal term follows this same form as

$$\langle \varphi_s|V|\varphi_s^{(+)}\rangle = V_{ss} + \sum_n \frac{\langle \varphi_s|V|\varphi_n\rangle\langle \varphi_n|V|\varphi_s^{(+)}\rangle}{\hbar\left(i\alpha-(\omega_{nk}-\omega_0)\right)}, \tag{8.54}$$

which will be recognized as being just the correction to the energy at a second-order time-independent perturbation, especially when $\alpha \to 0$. The correction to the wave function follows in a similar manner as

$$\varphi_s^{(+)} = \varphi_s + \sum_n \varphi_n \frac{\langle \varphi_n|V|\varphi_s^{(+)}\rangle}{\hbar\left(i\alpha-(\omega_{nk}-\omega_0)\right)}, \tag{8.55}$$

which is recognized as the correction arising at first order in the perturbation. One can now back out the resolution of the delta function in (8.52) to give

$$\varphi_s^{(+)} = \varphi_s + \frac{1}{\left(i\hbar\alpha - \left(H_0 - E_s - \hbar\omega_0\right)\right)} V\varphi_s^{(+)}, \tag{8.56}$$

where the frequency has been replaced with the appropriate energies (prior to the last operation). This is the Lippmann–Schwinger equation, and it provides a self-consistent evaluation of the outgoing scattered wave function at the same energy level (Merzbacher, 1970). As in Section 8.3, the inverse energy operator can be expanded in a series that will give rise to the perturbation series, and this will give the exact solution for the scattered wave in many cases. Equation (8.56) can be rewritten as (with $\alpha = 0$), with the approximation that the $\hbar\omega_0$ term can be canceled on the two sides,

$$\left(H_0 - E_s\right)\varphi_s^{(+)} = \left(H_0 - E_s\right)\varphi_s - V\varphi_s^{(+)}. \tag{8.57}$$

Since the first term on the right-hand side vanishes by (8.3), this can be rewritten as

$$\left(H_0 + \mathrm{V}\right)\varphi_s^{(+)} = E_s\varphi_s^{(+)}, \tag{8.58}$$

which tells us that the outgoing waves are solutions to the total Hamiltonian, and because of this $\varphi_s^{(+)}$ will differ from its value at the initial time (it evolves in time as it propagates).

8.4.3 Orthogonality of the Scattering States

The Lippmann–Schwinger equation can now be used to show that the scattering states remain normalized and orthogonal. Before doing this, though, we diverge to illustrate another representation of the equation. If the resolvent operator, which is the fraction in the center of (8.56), is defined by this fraction as

$$G\left(E_s\right) = \frac{1}{E_s - H_0 + i\hbar\alpha}, \tag{8.59}$$

then (8.56) can be rewritten as

$$\varphi_s^{(+)} = \frac{1}{1 - G\left(E_s\right)V\varphi_s}. \tag{8.60}$$

This latter form is termed Dyson's equation (Fetter and Walecka, 1971) and clearly illustrates how the perturbation series is obtained by expanding the denominator in a power series. In fact, the form (8.60) is often termed the resummed perturbation series.

Another useful result is obtained by a rearrangement (8.56), in which all the terms are pre-multiplied by the denominator of the last term to give

$$\left(E_s - H_0 + i\hbar\alpha\right)\varphi_s^{(+)} = \left(E_s - H_0 + i\hbar\alpha\right)\varphi_s + V\varphi_s^{(+)}, \tag{8.61}$$

which, of course, is just (8.57) with the temporal decay term retained. This can be rearranged to give

$$\left(E_s - H_0 - V + i\hbar\alpha\right)\varphi_s^{(+)} = \left(E_s - H + V + i\hbar\alpha\right)\varphi_s, \tag{8.62}$$

where the term H_0 on the right-hand side has been replaced using (8.1). Once again, this leads to

$$\varphi_s^{(+)} = \varphi_s + \frac{V}{E_s - H + V + i\hbar\alpha}\varphi_s^{(+)}. \tag{8.63}$$

Finally, we want to show that the scattering states retain their orthonormality. To begin, we use (8.63) to describe just one of the scattering states, to the lowest order in the scattering, according to

$$\begin{aligned} \langle \varphi_r^{(+)} | \varphi_s^{(+)} \rangle &= \langle \varphi_r + \frac{V}{E_r - H + i\hbar\alpha}\varphi_r^{(+)} | \varphi_s^{(+)} \rangle. \\ &\sim \langle \varphi_r | \varphi_s^{(+)} \rangle + \frac{V}{E_r - H - i\hbar\alpha}\varphi_s^{(+)} \rangle \\ &= \langle \varphi_r | \varphi_s^{(+)} \rangle + \frac{V}{E_r - E_s - i\hbar\alpha}\varphi_s^{(+)} \rangle. \end{aligned} \tag{8.64}$$

Here, we have used (8.58) to replace the total Hamiltonian with the energy of the scattered state. We can now factor out the minus sign of the denominator, and since the fraction is now a *c*-number, we can reverse the order of the terms multiplying the last scattering state. We then take the total operator *back* to the adjoint term, where the E_r-term becomes the operator H_0. Then, returning this operator to the position shown in (8.64), this latter equation can be written as

$$\langle \varphi_r^{(+)} | \varphi_s^{(+)} \rangle = \langle \varphi_r | \varphi_s^{(+)} \rangle + \langle \varphi_r | \frac{V}{E_r - H_0 - i\hbar\alpha}\varphi_s^{(+)} \rangle. \tag{8.65}$$

Finally, replacing $\varphi_s^{(+)}$ in the first term on the right by (8.63) and rationalizing the result, one obtains

$$\langle \varphi_r^{(+)} | \varphi_s^{(+)} \rangle = \langle \varphi_r | \varphi_s \rangle = \delta_{rs}. \tag{8.66}$$

This establishes the orthonormality of the scattering states.

8.5 AN EXAMPLE—OPTICAL ABSORPTION IN SEMICONDUCTORS

Excess carriers in semiconductors that are generated by optical absorption provide some of the more interesting applications in devices today. These range from the growing field of solar cells to advanced lasers found in most scanning devices used by commercial vendors. In this section, the study of optically created, non-equilibrium carriers is discussed. In particular, the direct absorption transition of electrons from the valence band to the conduction band, which creates an electron-hole pair, is treated.

At frequencies near to, or in the optical region, where $\hbar\omega > E_G$, the electromagnetic field can induce electrons from the valence band into the conduction band. Here, $\hbar\omega$ is the energy of the photon and E_G is the energy between the top of the valence band and the bottom of the conduction band, as shown already in Figure 4.5, where the gap is at $k = 0$. This photon absorption creates a new electron-hole pair that contributes to the conduction process. For this to occur, the lower energy state (top of the valence band) must be occupied and the upper energy state (bottom of the conduction band) empty. Of interest in this section are the direct transitions, which occur, e.g., in GaAs at Γ, and similarly in most direct-gap III–V semiconductors.

8.5.1 The Light-Matter Interaction

The first step of the process is to create an electron-hole pair by the absorption and subsequent annihilation of a photon. This step is no different in a semiconductor from the same process to excite an electron between energy shells in an atom. Some believe that the electron just "jumps" into the conduction band via a quantum leap. This is nonsense. The transition proceeds through a well-organized process described by standard probability functions that take a few femtoseconds to occur (Geltman, 1977; Haan and Geltman, 1982). This involves entanglement, as the electromagnetic field creates a correlation between the electron wave function in the valence band and an empty state wave function in the conduction band. This is the first step. Subsequently, the field then breaks this correlation, either by breaking it up, or by finishing the process with the excitation of the electron to the conduction band (Haug, 1988). The ratio of the electron-hole pair production to the photon number is the quantum efficiency. However, this initial process is beyond the level of detail that is considered here, where the interest is in the dependence upon the number of electron-hole pairs produced rather than the first few femtoseconds.

It is the absorption coefficient, describing how the optical electromagnetic wave creates electron-hole pairs, that is of interest. The electromagnetic wave produces a perturbation, which leads to the transition probability for the inter-band transition directly according to Section 8.1. The matrix elements, needed for the transition probability, arise in a natural manner as will be shown. Once the transition probability is determined, the number of electrons that make the transition for a given electromagnetic field power can be determined, and this will lead us to the absorption coefficient. In Figure 8.1, the basic process for direct absorption is shown. This transition is vertical, because the photon momentum is orders of magnitude smaller than the electron or hole momentum.

The transition of electrons from the essentially full valence band to the essentially empty conduction band may be described with the result (8.16) as

$$\Gamma_{12} = \frac{|c_{12}|^2}{t} = \frac{4|H_{12}|^2}{\hbar^2}\frac{\sin^2\left[(\omega_{21}-\omega)t/2\right]}{(\omega_{21}-\omega)^2 t^2}, \tag{8.67}$$

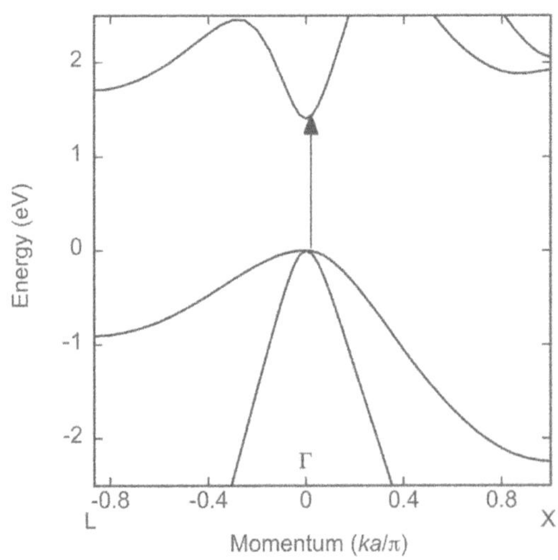

FIGURE 8.1 The band structure of GaAs around the center of the Brillouin Zone. The labels Γ, L, and X indicate the zone center and two zone edges, respectively. The vertical arrow indicates the direct optical transition induced by a photon.

where state 1 is the valence band and state 2 is the conduction band, so that $\omega_{21} = [E_c - E_v]/\hbar$. The matrix element is defined to be

$$H_{12} = 2\langle|H_F|1\rangle = \int \psi_v^\dagger(k_1) H_F \psi_c(k_2) d^3\mathbf{r}, \tag{8.68}$$

and the perturbing element H_F is still to be determined.

The properties of the electromagnetic wave were described in Section 5.3.1 where the fields were quantized. It was the vector potential that was important there, and it is also the important parameter here. In the presence of an electromagnetic wave, the Hamiltonian for the electrons is given by

$$H = \frac{1}{2m_0}\left(-i\hbar\nabla - e\mathbf{A}\right)^2 + V(\mathbf{r}) + \phi(\mathbf{r}). \tag{8.69}$$

Here, the leading terms of importance are

$$H = -\frac{\hbar^2}{2m_0}\nabla^2 + V(\mathbf{r}) + \frac{i\hbar e}{m_c}\mathbf{A}\cdot\nabla. \tag{8.70}$$

The last term in (8.70) is the perturbing potential H_F. The wave functions are normalized Bloch functions for each of the two bands, which may be written as

$$\psi_{c,v}(\mathbf{r}) = \frac{1}{\sqrt{N}} u_{c,v}(\mathbf{r}) e^{i\mathbf{k}\cdot\mathbf{r}}, \tag{8.71}$$

where the $u_{c,v}(\mathbf{r})$ are periodic in the unit cell as described in Chapter 4, and the exponential terms give the wave propagation. We take the vector potential to have the form

$$\mathbf{A} = \frac{1}{2}\mathbf{A}_0 e^{i(\omega t - \mathbf{k}_0\cdot\mathbf{r})} + c.c.. \tag{8.72}$$

With these expressions, the matrix element may now be written as

$$\begin{aligned} H_{12} &= \frac{i\hbar e A_0}{2m_0 N}\int d^3 r u_c^\dagger(\mathbf{r}) e^{-i\mathbf{k}\cdot\mathbf{r}} e^{-i\mathbf{k}_0\cdot\mathbf{r}} \mathbf{a}\cdot\nabla u_v(\mathbf{r}) e^{i\mathbf{k}'\cdot\mathbf{r}} \\ &= \frac{i\hbar e A_0}{2m_0 N}\int d^3 r u_c^\dagger(\mathbf{r}) e^{-i(\mathbf{k}+\mathbf{k}_0-\mathbf{k}')\cdot\mathbf{r}}\left[\mathbf{a}\cdot\nabla u_v(\mathbf{r}) + i\mathbf{a}\cdot\mathbf{k}' u_v(\mathbf{r})\right] \end{aligned}, \tag{8.73}$$

where $\mathbf{a}$ is a unit polarization vector in the direction of the vector potential. Since the Bloch functions are periodic over a unit cell, the integration can be separated into an integral over a unit cell and a summation over the unit cells. The sum over the unit cells provides the closure condition for the Fourier transforms, so that, within a reciprocal lattice vector **G**,

$$\mathbf{k} + \mathbf{k}_0 - \mathbf{k}' = 0. \tag{8.74}$$

The optical wave vector is orders of magnitude smaller than either the valence band or conduction band wave vectors, as was mentioned above, so that it may be ignored, which leads to the vertical transition in Figure 8.1. Using (8.74) and summing over the unit cells in the lattice, the final perturbation term can be found to be

$$H_{12} = \frac{i\hbar e A_0}{2m_0}\int d^3 r u_c^\dagger(\mathbf{r})\left[\mathbf{a}\cdot\nabla u_v(\mathbf{r}) + i\mathbf{a}\cdot\mathbf{k} u_v(\mathbf{r})\right]. \tag{8.75}$$

8.5.2 Allowed Transitions

The first term in the square brackets of (8.75) dominates the matrix element. This term requires that the angular momentum of the electron states differ due to the gradient term (which is the operator for momentum). Since the top of the valence band is usually composed of atomic p-states, and the bottom of the conduction is usually composed of atomic s-states, in direct-gap materials like GaAs, it is this momentum term that is most important and is called the "allowed" transition (Ferry, 2020). The second term in the square brackets is much smaller and the factor $\boldsymbol{k}$ breaks this momentum selection rule. It is termed the "forbidden" transition as it can occur whatever the relationship between the wave functions. These two terms will be treated separately in the following.

As mentioned in the previous paragraph, allowed transitions arise from the first term in the bracket in (8.75). The gradient operator produces a dipole interaction, and this requires a change of angular momentum between the valence wave function and the conduction wave function. The matrix element may be written as

$$H_{12} = \frac{i\hbar e A_0}{2m_0} \int d^3\boldsymbol{r} u_c^\dagger(\boldsymbol{r}) \boldsymbol{a} \cdot \nabla u_v(\boldsymbol{r}) = -\frac{eA_0}{2m_c} \boldsymbol{a} \cdot \boldsymbol{P}_{12}, \tag{8.76}$$

in which a *dipole* contribution $\boldsymbol{P}_{12}$ has been introduced to represent the unknown wave functions and the resulting integral through (recall that $i\hbar\nabla$ is the momentum operator)

$$\boldsymbol{P}_{12} = i\hbar \int d^3\boldsymbol{r} u_c^\dagger(\boldsymbol{r}) \nabla u_v(\boldsymbol{r}). \tag{8.77}$$

This term is basically the expectation value of the momentum operator when evaluated with the cell-periodic part of the conduction and valence band wave functions. In addition, the explicit evaluation of the photon energy needs to account for the momentum of the electron states, as

$$\hbar\omega_{21} = E_c(k) - E_v(k) = E_G + \frac{\hbar^2 k^2}{2}\left(\frac{1}{m_{c,e}} - \frac{1}{m_{c,h}}\right) = E_G + \frac{\hbar^2 k^2}{2m_R}, \tag{8.78}$$

and a *relative* mass m_R has been introduced in the final term.

A sum now must be performed over the available states k that are allowed in the process. This is done in spherical coordinates and involves an angle averaging as well. The first step is to get the resulting angle-averaged dipole moment given by

$$\int d\Omega \left|\boldsymbol{a} \cdot \boldsymbol{P}_{12}\right|^2 \to 4\pi \overline{P_{12}^2}. \tag{8.79}$$

Using this angle-averaged parameter, the integration over momentum may be performed to arrive at

$$\begin{aligned} P(\omega) &= \frac{e^2 A_0^2 t}{m_0^2 \hbar^2} \int d^3\boldsymbol{k} \frac{\left|\boldsymbol{a} \cdot \boldsymbol{P}_{12}\right|^2}{4\pi^3} \frac{sin^2\left[(\omega_{21} - \omega)t/2\right]}{(\omega_{21} - \omega)^2 t^2} \\ &= \frac{e^2 A_0^2}{m_0^2 \hbar^3} \frac{2k\overline{P_{12}^2} m_R}{\pi^2} \int \frac{sin^2 x}{x^2} dx \\ &= \frac{2e^2 A_0^2 k \overline{P_{12}^2} m_R}{\pi m_0^2 \hbar^3}. \end{aligned} \tag{8.80}$$

As there remains a significant number of unknown parameters, many of these are gathered into a term known as the "oscillator strength" that is defined to be the ratio of the kinetic energy in the transition to the photon energy, as

$$f_{osc} = \frac{2\overline{P_{12}^2}}{m_0 \hbar\omega}. \tag{8.81}$$

The factor of k may be replaced from (8.78), and the oscillator strength introduced, both in (8.80), to give

$$P(\omega) = \frac{e^2 A_0^2 (2m_R)^{3/2} \omega f_{osc}}{2\pi m_0 \hbar^3} (\hbar\omega - E_G)^{1/2}. \tag{8.82}$$

The present result must still be normalized with the number of photons arriving, or by the power contained in the electromagnetic wave. The number of photons arriving is given by the incident power (measured by the Poynting vector S of the plane wave) and the photon energy, or

$$F_0 = \frac{\mathbf{S}\cdot\mathbf{a}_k}{\hbar\omega} = \frac{\omega A_0^2}{2\hbar}\sqrt{\frac{\varepsilon_\infty}{\mu_0}} = \frac{\omega A_0^2}{2\hbar} G_\omega, \tag{8.83}$$

where $G_\omega = \sqrt{\varepsilon_\infty / \mu_0}$ is the wave admittance. The absorption coefficient now becomes

$$\alpha = \frac{P(\omega)}{F_0} = \frac{e^2 (2m_R)^{3/2} f_{osc}}{\pi m_0 \hbar^2 G_\omega} (\hbar\omega - E_G)^{1/2}. \tag{8.84}$$

The mass in the numerator is the reduced mass accounting for both the valence and conduction states. The mass in the denominator, however, arose from the Hamiltonian and should be the free electron mass. Tabulations of the properties of materials, including the optical absorption coefficient (as a function of energy), can be found at (Ioffe Physico-Technical Institute, 2001).

8.5.3 Forbidden Transitions

When the dipole transition above is forbidden by symmetry of the wave functions, the absorption can still occur via the last term in the bracket of (8.75). This is true for silicon where the conduction band extrema, located near the X points in the Brillouin zone, and the valence band maximum, located at the Γ point, are both formed primarily from atomic p-states. Hence, the dipole term is expected to vanish. In this case, the forbidden absorption term will be the significant driver for optical absorption. Here, the matrix element is written from (8.75) as

$$H_{12} = -\frac{e\hbar A_0}{2m_0} \mathbf{a}\cdot\mathbf{k} \int u_c^\dagger u_v d^3\mathbf{r}. \tag{8.85}$$

The process basically now follows that of the previous section. An effective "oscillator strength" is defined from the integral in (8.85) as

$$f' = \left| \int u_c^\dagger u_v d^3\mathbf{r} \right|^2. \tag{8.86}$$

Then, doing the angle averaging, the power absorbed is given as

$$P(\omega)=\frac{4\pi e^2 A_0^2 f'}{3m_0^2}\int\frac{k^4 t}{4\pi^3}\frac{\sin^2\left[(\omega_{21}-\omega)t/2\right]}{(\omega_{21}-\omega)^2 t^2}dk=\frac{e^2 A_0^2 k^3 m_R f'}{6\pi^2 m_0^2\hbar}. \tag{8.87}$$

An important difference is that this form depends upon the cube of k, rather than the linear dependence of the last section. When this is replaced, through (8.78), the result is

$$P(\omega)=\frac{e^2 A_0^2 (2m_R)^{5/2} f'}{12\pi^2 m_0^2\hbar^4}(\hbar\omega-E_G)^{3/2}. \tag{8.88}$$

Now, using (8.83), the absorption coefficient is found to be

$$\alpha=\frac{e^2 (2m_R)^{5/2} f'}{6\pi^2 m_0^2\hbar^2 G_\omega}\frac{(\hbar\omega-E_G)^{3/2}}{\hbar\omega}. \tag{8.89}$$

This absorption is considerably weaker than the allowed version, but it rises much faster with photon energy due to the higher power in the last term (while this term is raised to the same power, the difference in the numerator term $(\hbar\omega-E_G)^{3/2}$ leads to a faster rise).

8.6 AN EXAMPLE—PHONON SCATTERING IN SEMICONDUCTORS

Our final example is the scattering of an electron in a semiconductor by the acoustic modes of the lattice vibrations, the latter of which were discussed in Section 5.4. The acoustic modes are the low-energy set of phonons discussed there, and scattering is normally from those phonons in very low $\boldsymbol{q}$ modes, and primarily from the longitudinal mode. The assumption is that these vibrations cause small shifts in the energy bands, and this shift is the scattering potential. This scattering potential will be used in a straight-forward application of the Fermi golden rule (8.18).

In this process, electrons are scattered out of one state $\boldsymbol{k}$ and into another state $\boldsymbol{k}'$, while either absorbing or emitting a phonon of wave vector $\boldsymbol{q}$. Each of the different processes, or interactions, leads to a different "matrix element," where the matrix element is the key ingredient in the perturbation theory. The scattering potential δE leads to a matrix element

$$M(\boldsymbol{k},\boldsymbol{k}')=\langle\psi_{\boldsymbol{k},\boldsymbol{q}}|\delta E|\psi_{\boldsymbol{k}',\boldsymbol{q}}\rangle, \tag{8.90}$$

where each wave function is a composite of the electron wave function (with momentum $\boldsymbol{k}$ or $\boldsymbol{k}'$) and the phonon wave function (with momentum $\boldsymbol{q}$).

This perturbing potential is different for each different type of phonon. As mentioned, the longitudinal acoustic phonon modes will be treated here. Normally, the electronic wave functions are Bloch functions (already used in the previous section) that exhibit the periodicity of the lattice. In addition, the matrix element usually contains the momentum conservation condition. Here this conservation condition leads to

$$\boldsymbol{k}-\boldsymbol{k}'\pm\boldsymbol{q}=\boldsymbol{G}, \tag{8.91}$$

which tells us that the momentum is conserved up to including a reciprocal lattice vector $\boldsymbol{G}$, which is a result of the periodic boundary conditions introduced for the lattice discussed in Chapter 4. The upper sign in (8.91) corresponds to the absorption of a phonon by the electron. The lower sign leads to the emission of a phonon by the electrons. The straight-forward application of the Fermi golden rule (8.18) produces

$$P(\boldsymbol{k},\boldsymbol{k}') = \frac{2\pi}{\hbar}\left|M(\boldsymbol{k},\boldsymbol{k}')\right|^2 \delta\left(E_k - E_{k'} \pm \hbar\omega_q\right), \tag{8.92}$$

where the two signs correspond to those of (8.89).

The scattering rate out of the state with momentum $\boldsymbol{k}$ and energy E_k is obtained by integrating (8.92) over all final states. Because of the momentum conservation condition (8.91), the integration can be over either $\boldsymbol{k}'$ or $\boldsymbol{q}$ with the same result. In this example, the integration will be over the final state wave vector $\mathbf{k}'$, and

$$\Gamma(\boldsymbol{k}) = \frac{2\pi}{\hbar}\sum_{\boldsymbol{k}'}\left|M(\boldsymbol{k},\boldsymbol{k}')\right|^2 \delta\left(E_k - E_{k'} \pm \hbar\omega_q\right). \tag{8.93}$$

Two major assumptions are traditionally made for the acoustic modes. The first is that $E_{k'}E_{k'} \gg \hbar\omega_q$, so that the phonon energy is neglected in (8.93). The second is that M is independent of the phonon wave vector, so that the matrix element can be removed from the summation. With these two approximations, the summation in (8.93) just leads to the density of final states $n_{3D}(E)$, which is given by (Ferry, 2020)

$$n_{3d}(E) = \frac{V}{2\pi^2}\left(\frac{2m_c}{\hbar^2}\right)^{3/2} E^{1/2}. \tag{8.94}$$

To proceed, the matrix element must now be evaluated. A long-wavelength longitudinal acoustic wave moving through the lattice causes a local strain in the crystal that perturbs the energy bands, which leads to a small change in energy defined as (Bardeen and Shockley, 1950)

$$\delta E = \Xi_1 \Delta = \Xi_1 \nabla \cdot \boldsymbol{u}_q, \tag{8.95}$$

where Ξ_1 is the *deformation potential* and Δ is the *dilation* of the lattice caused by the phonon. The phonon itself is defined by $\boldsymbol{u}_q$ which is a displacement of the harmonic oscillator for that mode. This displacement is given by the position through the operators of Chapter 5 as

$$\boldsymbol{u}_q = \left(\frac{\hbar}{2\rho_m V\omega_q}\right)^{1/2}\left[a_q e^{i\boldsymbol{q}\cdot\boldsymbol{r}} + a_q^\dagger e^{-i\boldsymbol{q}\cdot\boldsymbol{r}}\right]\boldsymbol{e}_q e^{-i\omega_q t}, \tag{8.96}$$

where ρ_m is the mass density, V is the volume, and $\mathbf{e_q}$ is the polarization vector for the phonon plane wave. The matrix element is calculated by considering the sum over both the lattice and the electronic wave functions. Consider the second term in the square brackets of (8.96), which accounts for emission of a phonon by the carrier. This term leads to the matrix element squared, as

$$\left|M(k,q)\right|^2 = \frac{\hbar\Xi_1^2 q^2}{2\rho_m V\omega_q}\left(N_q + 1\right)I_{k,q}^2, \tag{8.97}$$

where N_q is the Bose–Einstein distribution (5.58), and

$$I_{k,q} = \int_\Omega u_{k-q}^\dagger u_k d^3\boldsymbol{r} \tag{8.98}$$

is the *overlap* integral between the cell portions of the Bloch functions. For the elastic process considered here, this integral is unity. If the first term in the square brackets of (8.96) had been used, the result (8.97) would be the same except for the omission of the factor of 1 in the parentheses. Since the phonons are quite dense in the acoustic modes, the differences in these two processes are small, and the two can be added together by replacing the term in parentheses by a factor of $2N_q \sim 2k_BT/\hbar\omega_q \sim 2k_BT/\hbar qv_s$, where v_s is the velocity of the phonons, which is the velocity of sound in the crystal. Putting all of this together, the matrix element becomes

$$\left|M(k)\right|^2 \sim \frac{\Xi_1^2 k_B T}{\rho_m v_s^2 V}. \tag{8.99}$$

This final form of the matrix element in (8.99) is independent of the wave vector q of the phonons, so that the assumptions above may be used. For electrons in a simple, spherical energy surface and parabolic bands, this leads to

$$\begin{aligned}\Gamma(k) &= \frac{2\pi}{\hbar}\frac{\Xi_1^2 k_B T}{\rho_m v_s^2 V}\frac{V}{4\pi^2}\left(\frac{2m_c}{\hbar^2}\right)^{3/2} E^{1/2} \\ &= \frac{\Xi_1^2 k_B T\left(2m_c\right)^{3/2}}{2\pi\rho_m v_s^2\hbar^4} E^{1/2}\end{aligned}. \tag{8.100}$$

Although most of the parameters easily may be obtained for a particular semiconductor (Ioffe Physico-Technical Institute, 2001), it is found that the conduction band deformation potential itself is almost universally of the order of 7–10 eV for nearly all group IV and III–V semiconductors. The result obtained is merely the evaluation of the Fermi golden rule for a particular matrix element and material (through the deformation potential and density of states). It is perhaps noteworthy that this approach is almost universally used for all scattering processes in semiconductors, with only changes in these parameters due to different materials and perturbing potentials.

REFERENCES

Bardeen, J. and Shockley, W. 1950. Deformation Potentials and Mobilities in Non-Polar Crystals. *Phys. Rev.* **80** 72–80.
Ferry, D. K. 2020. *Semiconductors: Bonds and Bands, 2nd Ed.* (Bristol: IOP Publishing).
Fetter, A. L. and Walecka, J. D. 1971. *Quantum Theory of Many-Particle Systems.* (New York: McGraw-Hill).
Geltman, S. 1977. Ionization of a Model Atom by a Pulse of Coherent Radiation. *J. Phys. B.* **10** 831–840.
Haan, S. L. and Geltman, S. J. 1982. Time Development of Resonant Multiphoton Ionization of Sodium. *J. Phys. B.* **15** 1229–1242.
Haug, H. 1988. *Optical Nonlinearities and Instabilities in Semiconductors.* (San Diego: Academic).
Ioffe Physico-Technical Institute. 2001. *New Semiconductor Materials, Characteristics and Properties.* http://www.ioffe.ru/SVA/
Landau, L. D. and Lifshitz, E. M. 1958. *Quantum Mechanics.* (London: Pergamon).
Merzbacher, E. 1970. *Quantum Mechanics, 2nd Ed.* (New York: John Wiley).

PROBLEMS

1. The propagator $U(t, t_0)$ satisfies the integral equation (8.25). If $V(t) = V_0\delta(t - t_1)$, show that

$$U(t,t_0) = \frac{1}{1+(iV_0/\hbar)\vartheta(t-t_1)},$$

where $\vartheta(x)$ is the Heaviside step function, $\vartheta(x) = 1$ for $x > 0$, and $\vartheta(x) = 0$ for $x < 0$. (Hint: break up the integral into segments $0 < t < t_1 - \delta$, $t_1 - \delta < t < t_1 + \delta$, $t_1 + \delta < t < \infty$, where δ is a very small parameter that is assumed to vanish, and use the property (8.23).)

2. Verify (8.30). (Hint: the multiple time integrals in each term must be transformed all to have the same limits of integration. This introduces a factorial in the prefactor and also changes the multiple integrals from a nested set to a product of individual integrals. The series is then summed.)
3. At $t < 0$, an electron is assumed to be in the $n = 3$ eigenstate of an infinite square potential well, which extends from $-a/2 < x < a/2$. At $t = 0$, an electric field is applied, with the potential $V = Ex$. The electric field is then removed at time τ. Determine the probability that the electron is in any other state at $t > \tau$. Do not make any assumptions about the relative size of $\omega_{k_3}\tau$. What are the differences for the cases in which this latter quantity is small or is large?
4. Using Fermi's Golden rule described in class, calculate the total scattering rate out of the state k for the case of a delta-function perturbing potential, i.e., $V(\boldsymbol{r}) = V_0\delta(\boldsymbol{r})$. In your calculations consider a three-dimensional crystal and plane waves as basis functions.

9 Motion in a Centrally Symmetric Potential

The problem of interacting particles can usually be reduced in quantum mechanics to that of one particle, as can be done in classical mechanics. Normally, this problem is one of an electron orbiting around, or being affected by, a positive atomic core or a scattering center with the interaction governed by the Coulomb potential. In general, this is a multi-dimensional problem, and not one of the simpler one-dimensional problems with which we have been concerned in the previous chapters. Once we begin to treat multiple dimensions, then state degeneracy begins to arise more frequently, and the most common problem treated is that of the hydrogen atom vs. the helium atom. These extra dimensions provide more degrees of freedom and more complexity. In the hydrogen atom, there is only a single electron, and the problem is solvable. On the other hand, adding a second electron for the helium atom makes the problem extremely difficult to solve, as it has become a three-body problem in which the interactions are difficult to handle.

In this chapter, the motion of a charged particle in a centrally symmetric potential will be treated. First, the degeneracies that may arise, and what they mean, will be discussed in terms of a two-dimensional harmonic oscillator. The manner in which the degeneracies are split by a magnetic field will then be addressed (this has already been introduced in a previous chapter in terms of the electron spin). Following this, we will discuss the hydrogen atom with its three-dimensional potential. Finally, the energy levels that arise in atoms more complex than the hydrogen atom will be discussed for real, but non-Coulombic potentials. Finally, the discussion will turn to hydrogenic impurities in semiconductors.

9.1 THE TWO-DIMENSIONAL HARMONIC OSCILLATOR

The harmonic oscillator in one dimension was discussed in Chapter 5. This simple problem of a particle in a quadratic potential energy is one of the typical problems of quantization. In two (or more) dimensions, the problem is more interesting, but not particularly more complicated. Only two dimensions will be discussed in this section in order first to understand better just how the degeneracies arise. However, as will be seen, the two-dimensional problem is of particular interest for electrons at semiconductor surfaces or at interfaces between two dissimilar semiconductors. Here, this typical two-dimensional problem, similar to the one that will be considered, can arise when electrons are confined in an inversion layer at the interface between, for example, silicon and silicon dioxide or at the interface of a heterostructure or in a potential well in the direction normal to that considered here (see Section 2.4 for the first example). In any case, it is assumed that there is no z-motion, due to the confinement of the carriers in this dimension. Furthermore, it will be assumed for simplicity that centrally symmetric properties of the potential require that the "spring" constants of the harmonic potential are the same in the two free coordinates.

9.1.1 Rectangular Coordinates

To begin, consider that the two dimensions are aligned along two perpendicular axes in a normal rectangular coordinate system. The Hamiltonian for the two-dimensional harmonic oscillator may be obtained by expanding (5.9) to two dimensions, which leads to

$$-\frac{\hbar^2}{2m}\left(\frac{\partial^2}{\partial x^2}+\frac{\partial^2}{\partial y^2}\right)\psi+\frac{m\omega^2}{2}\left(x^2+y^2\right)\psi=E\psi. \tag{9.1}$$

DOI: 10.1201/9781003717195-9

In order to simplify the results, the creation and annihilation operators from (5.12) will be used, but with one set for the *x*-coordinates (denoted with a subscript *x*) and one set for the *y*-coordinates (with a subscript *y*). With the introduction of these operators, the Hamiltonian corresponding to (9.1) may be expressed simply as

$$H = \hbar\omega\left(a_x^\dagger a_x + \frac{1}{2}\right) + \hbar\omega\left(a_y^\dagger a_y + \frac{1}{2}\right) = \hbar\omega\left(a_x^\dagger a_x + a_y^\dagger a_y + 1\right). \tag{9.2}$$

Introducing the number operators from below (5.14), this last expression may be written as

$$H = \hbar\omega\left(n_x + n_y + 1\right). \tag{9.3}$$

We note from (9.2), or (9.3), that the lowest energy level E_0 is just $\hbar\omega$, which arises for $n_x = n_y = 0$. This level arises from just the zero-point motion of the two harmonic oscillators, and there is just one possible state that contributes to this level (the state where both number operators are identically zero). Thus, there is no degeneracy in this lowest energy level. On the other hand, the next highest energy level E_1 ($= 2\hbar\omega$) is doubly degenerate, since it can arise from either $n_x = 1$ and $n_y = 0$, as well as $n_x = 0$ and $n_y = 1$. In a similar manner, the third energy level is triply degenerate, since it arises from the three ways in which we can have $n_x + n_y = 2$. This discussion can be continued to higher energy levels, from which it is found that the energy value of the *n*th level is

$$E_n = \left(n + 1\right)\hbar\omega. \tag{9.4}$$

This level has an *n*-fold degeneracy in which the two number operators add up to $n_x + n_y = n$.

How does one determine the *x*- and *y*-axes? The problem has no central property that allows us to determine uniquely the orientation of either of these axes. Thus, the selection of some arbitrary direction for the *x*-axis is one of convenience, but not one of basic physics. This means that the properties of the harmonic oscillator are not those associated with these axes. Rather, the various degeneracies arise from the radial and angular motion of the particle, with the energy level set by the radial motion (in cylindrical coordinates). To illustrate this better, a change of variables to cylindrical coordinates will be made, and this is done in the next section.

9.1.2 Polar Coordinates and Angular Momentum

If one specifies the oscillator merely by its energy Hamiltonian (9.3), a complete set of commuting observables is difficult to define (other than those of the individual coordinates), but these miss important points of the two-dimensional problem. There are a number of combinations of *x*- and *y*-oscillator states that can combine into any single value of *n*. This leaves things dangerously undefined. However, if the total Hamiltonian is transformed into polar coordinates (cylindrical coordinates with no *z*-variation), the resulting Schrödinger equation can be separated into the radial and the angular parts. It will be seen that the radial part generally determines the energy level, and the degeneracies that existed above go into the various angular variations that will result. For example, for the $n = 2$ level of (9.4), one of the solutions arises for $n_x = n_y = 1$. Thus, each of the two one-dimensional harmonic oscillators is in the first excited state. How do we adjust the phase difference between the two oscillations? This phase difference can make the total oscillation actually rotate in the (x, y) plane as a phasor. The rotation of the oscillation amplitude corresponds to angular momentum, and a momentum operator arises from the angular parts of the Hamiltonian in polar coordinates. By studying this angular momentum, the overall motion of the two-dimensional harmonic oscillator in polar coordinates may be better understood.

Angular momentum in classical mechanics arises from the motion of a body around some center of rotation. This is defined by the vector:

$$\boldsymbol{L} = \boldsymbol{r} \times \boldsymbol{p}. \tag{9.5}$$

In quantum mechanics, the variables $\boldsymbol{r}$ and $\boldsymbol{p}$ are operators, but for the two-dimensional harmonic oscillator, the only component of $\boldsymbol{L}$ is aligned along the z-axis as there is no z-component of either position or momentum in the problem. This is given by

$$L_z = xp_y - yp_x. \tag{9.6}$$

To introduce this into the operator notation used above, the operators can be defined by inverting (5.12) for each coordinate as

$$\begin{aligned} x &= \sqrt{\frac{\hbar}{2m\omega}}\left(a_x + a_x^\dagger\right), \qquad y = \sqrt{\frac{\hbar}{2m\omega}}\left(a_y + a_y^\dagger\right), \\ p_x &= -i\sqrt{\frac{\hbar m\omega}{2}}\left(a_x - a_x^\dagger\right), \quad p_y = -i\sqrt{\frac{\hbar m\omega}{2}}\left(a_y - a_y^\dagger\right). \end{aligned} \tag{9.7}$$

These can be used to identify the angular momentum representation of (9.6) to be

$$\begin{aligned} L_z &= -i\hbar\left[\left(a_x + a_x^\dagger\right)\left(a_y - a_y^\dagger\right) - \left(a_y + a_y^\dagger\right)\left(a_x - a_x^\dagger\right)\right] \\ &= -i\hbar\left(a_x^\dagger a_y - a_y^\dagger a_x\right). \end{aligned} \tag{9.8}$$

In arriving at (9.8), the fact that operators with different subscripts always commute, while operators with the same subscript satisfy (5.16), has been used. Moreover, another commutation result is

$$\begin{aligned} \left[a_x^\dagger a_y, a_x^\dagger a_x + a_y^\dagger a_y\right] &= \left[a_x^\dagger, a_x^\dagger a_x\right]a_y + a_x^\dagger\left[a_y, a_y^\dagger a_y\right] \\ &= \left(a_x^\dagger a_x^\dagger a_x - a_x^\dagger a_x a_x^\dagger\right)a_y + a_x^\dagger\left(a_y a_y^\dagger a_y - a_y^\dagger a_y a_y\right) \\ &= -a_x^\dagger a_y + a_x^\dagger a_y = 0, \end{aligned} \tag{9.9}$$

where we have used (5.16) in evaluating the various commutators. A similar result holds for the second term in the second line of (9.8) when it is put into a commutator with the energy terms as above. This establishes that the z-component of the angular momentum in this system commutes with the energy and therefore is a constant of the motion. Since the energy (9.4) has a set of degeneracies, these must correspond to different values of the angular momentum. In the lowest level ($n = 0$), there is only one state which must correspond to a single angular momentum value, which will be seen to be zero. Because there is automatically a degeneracy between positive and negative angles of rotation (e.g., for every state with positive angular momentum, there must be a state with the opposite negative value of angular momentum, as there is nothing in the problem to give a preferred rotation direction), this single level must have $L_z = 0$. In the next energy level ($n = 1$), there are two degenerate states, one for each of the x- and y-axes. These must also correspond, in polar coordinates, to a nonzero value of angular momentum, with one state for each direction of rotation. This can be continued to the higher energy levels. We shall therefore seek to find the basis of the eigenvalues for the total Hamiltonian H and the angular momentum L_z. The approach will basically follow the treatment of Cohen-Tannoudji *et al.* (1977).

In electromagnetic fields, it is useful to decompose linear-polarized waves into left- and right circularly polarized waves. These circularly polarized waves have angular momentum just like the current situation. This suggests that rotating creation and annihilation operators be introduced to correspond to the rotating coordinates and to remove the meaningless Cartesian coordinates. These are done by the new definitions:

$$\begin{aligned} a &= \frac{1}{\sqrt{2}}\left(a_x - ia_y\right), \\ a^\dagger &= \frac{1}{\sqrt{2}}\left(a_x^\dagger + ia_y^\dagger\right), \end{aligned} \tag{9.10}$$

and

$$\begin{aligned} b &= \frac{1}{\sqrt{2}}\left(a_x + ia_y\right), \\ b^\dagger &= \frac{1}{\sqrt{2}}\left(a_x^\dagger - ia_y^\dagger\right). \end{aligned} \tag{9.11}$$

From these definitions, both a and b will produce the result that both of the number operators in (9.3) are reduced as

$$\begin{aligned} a\,|\,n_x,n_y\rangle &= c_1\,|\,n_x-1,n_y\rangle - ic_2\,|\,n_x,n_y-1\rangle, \\ b\,|\,n_x,n_y\rangle &= c_1\,|\,n_x-1,n_y\rangle + ic_2\,|\,n_x,n_y-1\rangle, \end{aligned} \tag{9.12}$$

where the c_i are constants. This means that the operation of a or b produces a linear combination of states that are at the energy level one quantum ($\hbar\omega$) down and so removes one quantum of energy from the system just as the isolated component of an annihilation operator in a single dimension does from that coordinate's harmonic oscillator. Similarly, the adjoint operators (creation operators) increase the energy of the state by one unit of energy $\hbar\omega$. We will see below that this energy reduction, or increase, is accompanied by a corresponding change in the angular momentum of the state.

The rotational operators satisfy a commutation relation that can be obtained from those of the independent ones for each axis, as

$$\left[a,a^\dagger\right] = \left[b,b^\dagger\right] = 1, \tag{9.13}$$

and all other commutators vanish as expected. The definitions (9.10) and (9.11) can be used to find the connections:

$$\begin{aligned} a^\dagger a &= \frac{1}{2}\left(a_x^\dagger a_x + a_y^\dagger a_y - ia_y^\dagger a_x + ia_x^\dagger a_y\right) \\ b^\dagger b &= \frac{1}{2}\left(a_x^\dagger a_x + a_y^\dagger a_y + ia_y^\dagger a_x - ia_x^\dagger a_y\right) \end{aligned}. \tag{9.14}$$

This means that (9.2) and (9.3) can be replaced with

$$H = \hbar\omega\left(a^\dagger a + b^\dagger b + 1\right) = \hbar\omega\left(n_a + n_b + 1\right). \tag{9.15}$$

Similarly, the angular momentum (9.8) can be replaced with

$$L_z = \left(n_a - n_b\right)\hbar. \tag{9.16}$$

The Hamiltonian remains in a form as simple as that for rectangular coordinates, and the angular momentum has been considerably simplified over its representation above.

Using the operators a and b, one can go through the complete arguments of Section 5.1.3 to determine the wave functions. This will lead to the orthonormal states:

$$|n_a, n_b\rangle = \frac{1}{\sqrt{n_a!n_b!}}\left(a^\dagger\right)^{n_a}\left(b^\dagger\right)^{n_b}|0,0\rangle, \tag{9.17}$$

and these states are eigenstates of *both* the Hamiltonian and the angular momentum. The normal parameters, such as the energy index in (9.4), are given by the integers:

$$\begin{aligned} n &= n_a + n_b, \\ m &= n_a - n_b. \end{aligned} \tag{9.18}$$

In the last line of this last equation, the traditional index m has been introduced for the angular momentum. If one uses the operator $a^\dagger$ on the state in (9.17), the energy is raised by one unit (we increase n_a by one unit), and the value of m is also raised by one unit, thus increasing the angular momentum by one unit of Planck's (reduced) constant. Operating with $b^\dagger$ raises the energy but reduces the angular momentum by decreasing m through increasing the index n_b. This tells us that this operator corresponds to the positive (anti-clockwise) direction of rotation, while $a^\dagger$ refers to the negative (clockwise) direction of rotation. The eigenvalues of the angular momentum are thus $m\hbar$. Eigenvalues with a given value of n are $(n + 1)$-fold degenerate, since the various integers satisfy the constraints of

$$\begin{aligned} n_a &= n, & n_b &= 0, \\ n_a &= n-1, & n_b &= 1, \\ n_a &= n-2, & n_b &= 2, \\ &\cdots & &\cdots. \end{aligned} \tag{9.19}$$

This continues until the limit of $n_a = 0$, $n_b = n$ is reached. Hence, m can be a positive or negative integer, since it can range from $n_{a,max}$ to $-n_{b,\,max}$, or from n to $-n$. For $n = 0$, there is only a single level that requires $m = 0$. For the next energy level, $n = 1$, and there are two levels, for which $m = \pm 1$. For the third level, $n = 2$, and there are three levels, where m takes on the values 0 and ± 2. Hence, for a given energy level, the allowed values of m are

$$-n, -n+2, -n+4, \ldots, n-4, n-2, n. \tag{9.20}$$

This points out that any particular state is specified by the integers n and m, and the wave function is defined through

$$\left|n_a = \frac{n+m}{2}, n_b = \frac{n-m}{2}\right\rangle. \tag{9.21}$$

9.1.3 Finding the Wave Functions

In seeking the wave function for the ground state, the polar coordinates need to be introduced. Of course, this is done through the transformation relations:

$$\begin{aligned} x &= r\cos\varphi, & r &\geq 0, \\ y &= r\sin\varphi, & 0 &\leq \varphi \leq 2\pi. \end{aligned} \tag{9.22}$$

With a bit of algebra, (9.11) and (9.12) will lead to the evaluation of the operators in cylindrical coordinates as

$$a = \frac{e^{-i\varphi}}{2}\left[Br + \frac{1}{B}\frac{\partial}{\partial r} - \frac{i}{Br}\frac{\partial}{\partial \varphi}\right],$$
$$b = \frac{e^{+i\varphi}}{2}\left[Br + \frac{1}{B}\frac{\partial}{\partial r} + \frac{i}{Br}\frac{\partial}{\partial \varphi}\right]. \tag{9.23}$$

Similarly, the adjoint operators are given by

$$a^{\dagger} = \frac{e^{+i\varphi}}{2}\left[Br - \frac{1}{B}\frac{\partial}{\partial r} - \frac{i}{Br}\frac{\partial}{\partial \varphi}\right],$$
$$b^{\dagger} = \frac{e^{-i\varphi}}{2}\left[Br - \frac{1}{B}\frac{\partial}{\partial r} + \frac{i}{Br}\frac{\partial}{\partial \varphi}\right]. \tag{9.24}$$

In the above equations, the reduced quantity

$$B = \sqrt{m\omega / \hbar} \tag{9.25}$$

has been introduced to lessen the clutter. Because of the rotating coordinates, *the adjoint operators are not simply complex conjugates of the annihilation operators* but must be carefully calculated from the definitions themselves.

Both a and b will attempt to lower the energy and change the angular momentum. The lowest eigenstate should satisfy both of these operators, as

$$a|0,0\rangle = b|0,0\rangle = 0. \tag{9.26}$$

The two operators a and b differ only by the sign of the imaginary part. Since (9.26) must satisfy both the real and imaginary parts simultaneously, it leads to

$$\frac{1}{Br}\frac{\partial}{\partial \varphi}|0,0\rangle = 0,$$
$$\frac{1}{B}\frac{\partial}{\partial r}|0,0\rangle = -Br|0,0\rangle. \tag{9.27}$$

The first line just leads to a function that is independent of the angle φ. The second of these equations leads to $\exp(-B^2r^2/2)$ behavior, just as for the linear harmonic oscillator. After normalization, the ground state wave function is

$$|0,0\rangle = \frac{B}{\sqrt{\pi}} e^{-B^2r^2/2}. \tag{9.28}$$

The higher-lying levels are now found by repeatedly using (9.17). The angular variation is introduced into the wave function by the operators $a^{\dagger}$ and $b^{\dagger}$ themselves through the exponential prefactors. For example, the wave function for the $n = 1$, $m = 1$ ($n_a = 1, n_b = 0$) state is given by

$$\begin{aligned}|1,0\rangle &= a^{\dagger}|0,0\rangle \\ &= \frac{e^{+i\varphi}}{2}\left[Br - \frac{1}{B}\frac{\partial}{\partial r} - \frac{i}{Br}\frac{\partial}{\partial \varphi}\right]\frac{B}{\sqrt{\pi}}e^{-B^2r^2/2} \\ &= \frac{B^2 r}{\sqrt{\pi}} e^{+i\varphi} e^{-B^2r^2/2}\end{aligned} \tag{9.29}$$

The state $|0,1\rangle$ ($n_a = 0, n_b = 1, n = 1, m = -1$, n = 1) is given by the complex conjugate of this expression. Thus, the two angular momentum states have counter-rotating properties, but both are made up of linear combinations of the linear harmonic oscillators along the two axes. In the next section, an additional potential due to a magnetic field will be used to raise the degeneracy of the two angular momentum states.

9.1.4 Splitting the Angular Momentum States with a Magnetic Field

In Chapter 4, it was pointed out that at the end of the nineteenth century, Pieter Zeeman was studying the optical absorption spectrum of various atomic gases and found that a magnetic field split the absorption line into a pair of lines, whose energy difference was linearly dependent upon the magnetic field (Zeeman, 1897a, 1897b, 1897c). In that discussion, this splitting was due to the electron spin. However, this spin is an angular momentum, so that the magnetic field will also split other degenerate angular momentum states. These latter are the subject here. The magnetic field introduces a particular direction of rotation of the electrons (or other charge carriers), and this will help one direction of angular momentum while hindering the other direction of angular momentum. This is because the magnetic field also produces a harmonic oscillator behavior, as discussed in Chapter 5, where it was shown that the magnetic harmonic oscillator potential couples with the two-dimensional harmonic oscillator potential of (9.1). The first people to study the two-dimensional harmonic oscillator in a magnetic field were apparently Fock (1928) and Darwin (1931), but the operator approach used here follows the treatment of Rössler (1991). The total Hamiltonian can be found by coupling (9.1) with (5.65), and

$$H = \frac{1}{2m}(\boldsymbol{p} + e\boldsymbol{A})^2 + \frac{m\omega^2}{2}(x^2 + y^2). \tag{9.30}$$

Because of the two-dimensional nature of the electrostatic harmonic oscillator, it is more convenient to take the vector potential in the symmetric gauge, rather than the Landau gauge used in Section 5.4. With the symmetric potential, the vector potential is written as

$$\mathbf{A} = \frac{1}{2}\left(-By\boldsymbol{u}_x + Bx\boldsymbol{u}_y\right). \tag{9.31}$$

(Here, $\boldsymbol{u}_x$ and $\boldsymbol{u}_y$ are used as the unit vectors along the Cartesian axes to avoid confusion with the operator a used above.) As previously, the magnetic field remains oriented in the z-direction, normal to the two-dimensional plane of the motion ($\mathbf{B} = \nabla \times \mathbf{A} = Ba_z$). The momentum term can now be expanded as

$$\begin{aligned}(\boldsymbol{p} + e\boldsymbol{A})^2 &= -\hbar^2\left(\frac{\partial^2}{\partial x^2} + \frac{\partial^2}{\partial y^2}\right). \\ &-i\hbar eB\left(x\frac{\partial}{\partial y} - y\frac{\partial}{\partial x}\right) + \frac{e^2B^2}{4}(x^2 + y^2)\cdot\end{aligned} \tag{9.32}$$

The last term can be combined with the electrostatic harmonic oscillator by defining the new oscillator "spring" frequency to be

$$\begin{aligned}\Omega^2 &= \omega^2 + \left(\frac{\omega_c}{2}\right)^2, \\ \omega_c &= \frac{eB}{m_0}.\end{aligned} \tag{9.33}$$

(In this equation, m_0 is the free electron mass and not the index m, which appears in the next equation.) This last frequency is the cyclotron frequency introduced in (5.68). The second term in (9.32), which is linear in the magnetic field, can be rewritten using (9.6) as simply BL_z. With these definitions, the Hamiltonian, with the change in frequency in the definitions of the operators as $\omega \to \Omega$, can then be rewritten as

$$H = \left(n_a + n_b + 1\right)\hbar\Omega + \frac{1}{2} m\hbar\omega_c, \tag{9.34}$$

where the index m is given in (9.18) and relates to the angular momentum. This new Hamiltonian still commutes with the angular momentum, and hence both are simultaneously measurable. However, the energy now has a contribution from the angular momentum directly. This term raises the degeneracy of the previous eigenvalues (energies), by splitting the eigenvalues for each angular momentum state. This last term in (9.34) is known as the *Zeeman energy*.

For small values of the magnetic field, $\omega_c \ll \Omega$, each normal energy level is split into $(n + 1)$ levels as the degeneracy is raised by the magnetic field. Each pair of these levels is separated by the amount $\hbar\omega_c$ since δm was determined in the previous section to be 2. As the magnetic field value is increased, this spread in energies changes significantly. For slightly larger values of the magnetic field, but still with $\omega_c < \Omega$, the uppermost energy level is described by the frequency:

$$\omega_+ = \left(n+1\right)\left(\omega + \frac{\omega_c^2}{4\omega}\right) + \frac{m\omega_c}{2}. \tag{9.35}$$

This is only slightly above the degenerate energy level that exists in the absence of the magnetic field. The lowest energy level is now described by the frequency:

$$\omega_- = \left(n+1\right)\left(\omega + \frac{\omega_c^2}{4\omega}\right) - \frac{m\omega_c}{2}, \tag{9.36}$$

so all the levels begin to show a weak quadratic upward motion away from the linear spreading of the levels for very small magnetic fields.

In the case of very large magnetic fields, $\omega_c \gg \Omega$, the energy levels are quite different. The energy levels are now given by the frequencies:

$$\omega_{n,m} = \left(n+m+1\right)\frac{\omega_c}{2} + \left(n+1\right)\frac{\omega^2}{\omega_c}. \tag{9.37}$$

If the last term is neglected, the Landau levels are recovered, since $n + m = 2n_a$ is an even integer. Moreover, it may be noted that the lowest Landau level, where $n + m = 0$, is formed from one level from each of the electrostatic harmonic oscillator levels (when $B = 0$), as shown in Figure 9.1. Similarly, for the next higher Landau level, one level from each electrostatic harmonic oscillator level for which $n \geq 1$ converges into the Landau level. This continues upward throughout the spectrum, with the ith Landau level being formed from states arising from the levels for which $n \geq i$. Thus, as the magnetic field is increased in size, the energy levels move smoothly from those values associated with the electrostatic harmonic oscillator to those values associated with the Landau levels.

While a given Landau level has contributions from all equal-index and higher-index harmonic oscillator levels, a given harmonic oscillator level contributes to only a fixed number of Landau levels. The lowest harmonic oscillator level, for example, contributes only to the lowest Landau level,

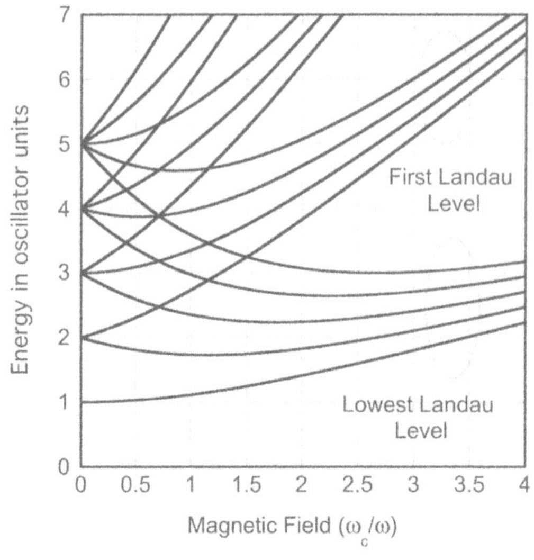

FIGURE 9.1 The lowest five harmonic oscillator levels in the presence of a magnetic field, illustrating the splitting of the angular momentum states and the formation of the Landau levels.

since it has only one non-degenerate state. This is the case for which $n + m = 0$. Similarly, each harmonic oscillator level contributes its lowest level to the lowest Landau level. However, its highest level ($m = n$) goes into the nth Landau level. So the first harmonic oscillator level contributes to the lowest two Landau levels, the second to the lowest three Landau levels, and so on. The spacing of the levels that merge into a particular Landau level is given by the frequency so the electrostatic harmonic oscillator potential splits the degeneracy of the Landau levels. Again, this may be seen in Figure 9.1.

The spectral positions of these "quantum box" levels, shown in Figure 9.1, have been measured by far-infrared absorption measurements for InSb (Sikorski and Merkt, 1989) and via the conductance arising from single-electron charging in GaAs (McEuen *et al.*, 1991) and in GaAs tunneling structures (Tarucha *et al.*, 1996).

9.2 THE HYDROGEN ATOM

The above treatment could easily be extended to a three-dimensional harmonic oscillator, but this would bring little new. A more important problem that has spherical symmetry in the potential is that of the hydrogen atom. This is one of the few problems that can be solved exactly (the motion of two electrons about a nucleus includes the interaction between the two electrons and produces a three-body problem, which is quite complicated and well beyond the scope of this text). For simplicity, only the *relative* motion of the electron about the nucleus is treated, taking the latter as fixed in position. The atom is centrally symmetric in three dimensions, and the interaction between the electron and the nucleus is the Coulomb potential, a $1/r$ potential. Thus, the natural coordinate system is spherical coordinates.

The electron is attracted to the nucleus (which will be assumed to be immensely massive in comparison with the electron) through the Coulomb potential, so the Schrödinger equation is simply

$$H\psi = \left[-\frac{\hbar^2}{2m_0}\nabla^2 - \frac{e^2}{4\pi\epsilon r}\right]\psi = E\psi, \tag{9.38}$$

where m_0 is the free electron mass. To avoid confusion with one of the integers describing the angular momentum in the subsequent material, we will take the mass as the reduced mass $\mu = m_0M/(m_0 + M)$, where M is the nuclear mass. In spherical coordinates, (9.38) can be rewritten as

$$\begin{aligned} E\psi &= -\frac{\hbar^2}{2\mu r^2}\frac{\partial}{\partial r}\left(r^2\frac{\partial\psi}{\partial r}\right) - \frac{e^2}{4\pi\epsilon r}\psi. \\ &-\frac{\hbar^2}{2\mu r^2}\left[\frac{2}{\sin\vartheta}\frac{\partial}{\partial\vartheta}\left(\sin\vartheta\frac{\partial\psi}{\partial\vartheta}\right) + \frac{1}{\sin^2\vartheta}\frac{\partial^2\psi}{\partial\phi^2}\right]. \end{aligned} \tag{9.39}$$

The traditional method of solving this is to assume that $\psi(r,\vartheta,\varphi)$ can be split into two product terms as $R(r)$ and $\psi(\vartheta,\varphi)$, where the latter is also separable into a product $\Theta(\vartheta)\Phi(\varphi)$. By inserting this wave function, and then dividing by the wave function itself, (9.39) can be split into three terms: one a function of r alone, one a function of ϑ alone, and the other a function of φ alone. If this is to be true for all values of the position and angles, each term must be equal to a constant. Thus, the term in the square brackets is a constant, say λ, times the wave function. While this constant is arbitrary, it is known from treatments of spherical harmonics that it is more appropriate to set this separation constant equal to $\ell(\ell + 1)$ (the separation constant for Legendre polynomials which are the known solutions in this type of problem). It is easy to show that there is a one-to-one correspondence to any value of λ and ℓ, so long as $\lambda > 0$, so no generality is lost by making this latter substitution. The solutions will be dealt with separately, and then the combined results will be treated.

9.2.1 The Radial Equation

By eliminating the angular variables with the above constant interpretation, which still must be shown to be valid (and this will be done in subsequent sections below), the radial equation now becomes

$$\frac{1}{r^2}\frac{\partial}{\partial r}\left(r^2\frac{\partial R}{\partial r}\right) - \frac{\ell(\ell+1)}{r^2}R + \frac{2\mu}{\hbar^2}\left[E + \frac{e^2}{4\pi\epsilon r}\right]R = 0. \tag{9.40}$$

To simplify the equation, reduced units are introduced for energy and length through

$$\begin{aligned} \alpha^2 &= \frac{8\mu E}{\hbar^2}, \\ \rho &= \alpha r. \end{aligned} \tag{9.41}$$

In addition, the constant λ is introduced for the set of variables as

$$\lambda = \frac{\mu e^2}{2\pi\epsilon\alpha\hbar^2}. \tag{9.42}$$

With these substitutions, (9.42) may be rewritten as

$$\frac{1}{\rho^2}\frac{\partial}{\partial\rho}\left(\rho^2\frac{\partial R}{\partial\rho}\right) + \left[\frac{\lambda}{\rho} - \frac{1}{4} - \frac{\ell(\ell+1)}{\rho^2}\right]R = 0. \tag{9.43}$$

The choice of ¼ in the square bracket results from the above substitutions but can be considered to be somewhat arbitrary. Nevertheless, one goal was to get this value as it puts this equation into a standard form that is well studied in the field of boundary-value problems.

As has been common for a long time with well-recognized equations, the subsequent procedure is a well-traveled path, and by noting that for sufficiently large values of the normalized radius, the behavior of R is as $\rho^n e^{-\rho/2}$. This follows by retaining just the second-order derivative and the term with ¼ as the prefactor. The factor n will have to be determined, though, but this suggests that solutions of the form $R(\rho) = F(\rho)e^{-\rho/2}$, where $F(\rho)$ will be a polynomial of finite order in ρ, should be pursued. Hence, this substitution leads to

$$\frac{\partial^2 F}{\partial \rho^2} + \left(\frac{2}{\rho} - 1\right)\frac{\partial F}{\partial \rho} + \left[\frac{\lambda - 1}{\rho} - \frac{\ell(\ell+1)}{\rho^2}\right]F = 0, \tag{9.44}$$

where the exponential term has been factored out of this result. The next step, also well known in the field, is to replace $F(\rho)$ with $\rho^s L(\rho)$, so that (9.43) is transformed into

$$\begin{aligned} &\rho^2 \frac{\partial^2 L}{\partial \rho^2} + \rho\left[(2s+1) - \rho\right]\frac{\partial L}{\partial \rho} \\ &+ \left[\rho(\lambda - s - 1) + s(s+1) - \ell(\ell+1)\right]L = 0 \end{aligned} . \tag{9.45}$$

While it may seem that one is wandering around in an equation forest, (9.45) actually leads to some actual results. If $\rho = 0$ is considered and the derivatives are well behaved and are finite at this point, the result is that $s = \ell$ or $-(\ell + 1)$. Since the functions must be finite at $\rho = 0$, only the former value is allowed. With this result, (9.45) becomes

$$\rho \frac{\partial^2 L}{\partial \rho^2} + \left[(2\ell+1) - \rho\right]\frac{\partial L}{\partial \rho} + (\lambda - \ell - 1)L = 0. \tag{9.46}$$

This result is not a well-recognized equation, except in this particular problem. The approach now is to assume that $L(\rho)$ is a power series in ρ itself as

$$L(\rho) = \sum_{v} a_v \rho^v, \tag{9.47}$$

but this power series must be of finite order so that it terminates at some order. If this assumption is used in (9.46), one finds that the coefficients satisfy

$$a_{v+1} = a_v \frac{v + l + 1 - \lambda}{(v+1)(v+2l+2)}. \tag{9.48}$$

For this to terminate, there must be a maximum value of ν, and this maximum value is denoted as n', such that

$$\lambda = n' + \ell + 1. \tag{9.49}$$

Now, λ is set to n, and this becomes the total quantum number (this, and thereby λ, determines the total energy of the level), n' is the radial quantum number, and l is the angular momentum quantum number; all are integers. With this total quantum number, the energy levels are given by (9.42) and (9.41) as

$$E_n = \frac{\mu e^4}{32\pi^2 \epsilon^2 \hbar^2 n^2}. \tag{9.50}$$

Just as in the two-dimensional harmonic oscillator, the energy levels are specified by the total quantum number, except that there is no $n = 0$ level here. This may be used in the first line of (9.41) to show that the effective radius of the ground state is given by $2/\alpha = a_0 = 4\pi\varepsilon\hbar^2/\mu e^2 = 5.3 \times 10^{-9}$ cm. This quantity is called the *Bohr radius*. Note, however, that the actual normalization radius for the wave functions, and the factor α, depends upon the index n.

The coupling between the radial quantum number and the angular momentum quantum number means that, as the angular momentum increases, the radial variations decrease (the order of the polynomial decreases). Furthermore, for the lowest energy level ($n = 1$), both the radial and angular momentum quantum numbers are zero, so the only variation is in the exponential term. In this lowest level, the wave function is spherically symmetric, decaying exponentially away from the center of the atom. For the second level ($n = 2$), ℓ can be either 0 ($n' = 1$) or 1 ($n' = 0$), so there is a spherically symmetric state and one that has preferential directions due to the angular momentum. This continues, with ℓ taking on values $(0, 1, \ldots, n - 1)$, so there are n values for this variable. A few of the lower-level wave functions in terms of the components may be written as $R_{nl}(\rho)$, where n and l are the two eigenvalue integers that we have discussed above. The lowest three energy levels have the unnormalized radial wave functions ($\rho = \alpha r$):

$$\begin{aligned}
&R_{10}(\rho) \sim e^{-\rho/2}, \\
&R_{20}(\rho) \sim (\rho^2 - 2)e^{-\rho/2}, \\
&R_{21}(\rho) \sim \rho e^{-\rho/2}, \\
&R_{30}(\rho) \sim (\rho^2 - 6)(\rho - 1)e^{-\rho/2}, \\
&R_{31}(\rho) \sim \rho(\rho - 4)e^{-\rho/2}, \\
&R_{32}(\rho) \sim \rho^2 e^{-\rho/2}.
\end{aligned} \tag{9.51}$$

We see that the highest-order term is determined by the radial quantum number n', but that the number of lower-order terms is related to the angular momentum quantum number l. Note, again, that the normalization of each wave function will depend upon the level numbers n and l.

9.2.2 Angular Solutions

The angular equation that arises from (9.39) represents the behavior of the wave function with the two spherical angles: ϑ, the polar angle measured from the z-axis (the polar axis), and ϕ, the azimuthal angle measured in the (x,y)-plane from the x-axis. This equation may be written as

$$\frac{1}{\sin\vartheta}\frac{\partial}{\partial\vartheta}\left(\sin\vartheta\frac{\partial\psi}{\partial\vartheta}\right) + \frac{1}{\sin^2\vartheta}\frac{\partial^2\psi}{\partial\phi^2} + \ell(\ell+1)\psi = 0, \tag{9.52}$$

where ψ is the angular version introduced just after (9.39). This equation can be separated further by introducing $\psi = \Theta(\vartheta)\Phi(\varphi)$ and follows the same separation procedure as was done above. This allows us to separate (9.52) into two equations, one for each of the two variables. These are given as

$$\begin{aligned}
&\frac{\partial^2\Phi}{\partial\varphi^2} + m^2\Phi = 0, \\
&\frac{1}{\sin\vartheta}\frac{\partial}{\partial\vartheta}\left(\sin\vartheta\frac{\partial\Theta}{\partial\vartheta}\right) + \left[\ell(\ell+1) - \frac{m^2}{\sin^2\vartheta}\right]\Theta = 0.
\end{aligned} \tag{9.53}$$

The separation constant has been taken to be the square of an integer, m^2, just as was done for the two-dimensional case. In particular, this value of m is expected to be related to the angular motion of the azimuthal angle around the polar axis, and it appears that way in the first of Equation (9.53).

The fact that it is an integer is in recognition that Φ has to be periodic in 2π. Thus, the solution to this equation can easily be written as

$$\Phi(\varphi) = \frac{1}{\sqrt{2\pi}} e^{im\varphi}, \quad |m| \geq 0. \tag{9.54}$$

The last part of this equation denotes the fact that m can take on both positive and negative values, but that Φ must be continuous and have a continuous derivative for any value of (ϕ), and, in addition, and Φ is a constant for $m = 0$.

In addressing the last equation of (9.53), it will be convenient to let $w = \cos\theta$. With this substitution, this equation becomes

$$\frac{d}{dw}\left[\left(1-w^2\right)\frac{d\Theta}{dw}\right] + \left[\ell(\ell+1) - \frac{m^2}{1-w^2}\right]\Theta = 0. \tag{9.55}$$

Since the angle varies from 0 to π, the domain of w is 1 to -1. To begin, a substitution is made for the wave function $\Theta \sim (1-w^2)^2 G(w)$. With this substitution, (9.55) becomes

$$\left(1-w^2\right)\frac{d^2G}{dw^2} - 2w(2s+1)\frac{dG}{dw} + \left[\ell(\ell+1) - 2s(2s+1)\right]G = 0. \tag{9.56}$$

At this point, it is convenient to connect the exponent s to the separation constant through $4s^2 = m^2$, which also says that $2s = |m|$. As a result, it can be asserted that G is a polynomial of finite order (it terminates as a given order), and that this order should be $l - 2s = l - |m|$, just as found in the two-dimensional case. The parity of the polynomial (even or odd) is that of this terminating order. The polynomials G are thus recognized (at least to those cognizant of boundary-value problems in spherical coordinates) as the associated Legendre polynomials $P_l^m(w)$, and this association carries over to $\Theta(w)$.

9.2.3 Angular Momentum Again

The angular momentum, and particularly the z-component of this quantity, has already been discussed in Section 9.1.2. At this point, the total angular momentum of the particle orbiting the central potential in the hydrogen atom needs to be discussed further. By continuation of (9.5), the three components of angular momentum can be written as

$$\begin{aligned} L_x &= -i\hbar\left[y\frac{\partial}{\partial z} - z\frac{\partial}{\partial y}\right], \\ L_y &= -i\hbar\left[z\frac{\partial}{\partial x} - x\frac{\partial}{\partial z}\right], \\ L_z &= -i\hbar\left[x\frac{\partial}{\partial y} - y\frac{\partial}{\partial x}\right]. \end{aligned} \tag{9.57}$$

It is easy to demonstrate that these components obey a set of commutation relationships that may be expressed as

$$\begin{aligned} \left[L_x, L_y\right] &= i\hbar L_z, \\ \left[L_y, L_z\right] &= i\hbar L_x, \\ \left[L_z, L_x\right] &= i\hbar L_y. \end{aligned} \tag{9.58}$$

Since the angular momentum is a well-defined vector quantity, the square of this vector is easily taken to give

$$L^2 = L_x^2 + L_y^2 + L_z^2. \tag{9.59}$$

Moreover, this total angular momentum commutes with each of the components individually. Thus, there are four relatively independent quantities—the three components of the angular momentum and the total angular momentum—that can all be measured simultaneously. Normally, only two of these can be specified independently (there are only two parameters, l and m, that arise in the treatment of the orbital motion). So, some manipulation must be carried out to determine how these various values may be combined.

Consider first L_z, since the second of the parameters, m, is related to the motion about the polar axis, the z-axis, as it were. If this term from (9.57) is transformed to spherical coordinates, with $x = \rho sin\vartheta cos\phi, y = \rho sin\vartheta sin\phi, z = cos\vartheta$, then

$$L_z = -i\hbar\frac{\partial}{\partial\varphi}, \tag{9.60}$$

and this obviously involves only the quantum number m, so that L_z is a good component of the angular momentum to take as one of the independent quantities in the problem. The eigenvalues of the z-component of the angular momentum are then $\pm im\hbar$, with m an integer. From the discussion of (9.56), it is clear that $|m| \leq l$, so the values of m are the sequence:

$$-\ell, -\ell+1, \ldots, -1, 0, 1, \ell-1, \ell. \tag{9.61}$$

Thus, there are $2\ell + 1$ values for the z-component of angular momentum that arise for a given value of ℓ.

One expects that the second independent quantity will be the total angular momentum, and this will be the case. To see this, combinations of the x- and y-components are converted into circular parameters (just as in the operators for the harmonic oscillator, or in circularly polarized photon waves):

$$L_\pm = L_x \pm iL_y = \hbar\left[\pm\frac{\partial}{\partial\vartheta} + icot\vartheta\frac{\partial}{\partial\varphi}\right]. \tag{9.62}$$

The two possible products of these circular operators may be expressed as

$$\begin{aligned} L_+L_- &= \left(L_x + iL_y\right)\left(L_x - iL_y\right) = L_x^2 + L_y^2 + \hbar L_z, \\ L_-L_+ &= \left(L_x - iL_y\right)\left(L_x + iL_y\right) = L_x^2 + L_y^2 - \hbar L_z. \end{aligned} \tag{9.63}$$

Using the results of these products, the total angular momentum may be written as

$$\begin{aligned} L^2 &= L_+L_- - \hbar L_z + L_z^2 \\ &= \left[\frac{2}{sin\vartheta}\frac{\partial}{\partial\vartheta}\left(sin\vartheta\frac{\partial}{\partial\vartheta}\right) + \frac{1}{sin^2\vartheta}\frac{\partial^2}{\partial\phi\varphi^2}\right]. \end{aligned} \tag{9.64}$$

It may be seen that the term in the square brackets is the angular operator from (9.39) for the total wave function. Now, it must be shown that this leads to the factor $l(l+1)$.

The first line of (9.63) gives an important starting point, since this can be rewritten as

$$L^2 - L_z^2 = L_x^2 + L_y^2 \geq 0. \tag{9.65}$$

This guarantees that $L_z^2 \leq L^2$, or that

$$-\sqrt{L} \leq L_z \leq \sqrt{L}. \tag{9.66}$$

This might lead one to jump immediately to a wrong conclusion that $L \sim \ell$, but the angular momentum is more subtle than this. To proceed, consider the commutator relationships (9.58), which lead to

$$L_z\left(L_x \pm iL_y\right) = \left(L_x \pm iL_y\right)\left(L_z \pm 1\right), \tag{9.67}$$

much as in the two-dimensional case. If this result is used to operate on a product of the angular solutions, one obtains

$$L_z\left(L_x \pm iL_y\right)\Theta\Phi = \left(m \pm 1\right)\left(L_x \pm iL_y\right)\Theta\Phi. \tag{9.68}$$

This result tells us that the L_- operator lowers the z-component of angular momentum by one unit, while the L_+ operator raises the z-component of angular momentum by one unit. From this, it may be concluded that if $m = l$, operating on the wave function with the raising operator L_+ must give a zero result, since m cannot be larger than l. Hence,

$$\left(L_x + iL_y\right)\Theta\Phi_l = 0. \tag{9.69}$$

If the lowering operator is now imposed upon this result, it is found that

$$\begin{aligned}\left(L_x - iL_y\right)\left(L_x + iL_y\right)\Theta\Phi_l &= \left(L_x - iL_y\right)\left(L_x + iL_y\right)0 \\ &= \left(L^2 - L_z^2 - L_z\right)0\end{aligned}. \tag{9.70}$$

Introducing the eigenvalues for the z-component of angular momentum, with $m = l$ in these results, one arrives at

$$L^2 = \hbar^2\left(\ell^2 + \ell\right) = \hbar^2 l\left(l+1\right). \tag{9.71}$$

This provides a value to support that used in (9.39).

9.2.4 Atomic Energy Levels

In treating the hydrogen atom, the central potential was taken to be the Coulomb potential that exists between the atomic core and the single electron. It would be desirable to continue to do this for the general atom, but there are problems. Certainly, the central-field approximation can still be used. However, the simple Coulomb potential of (9.38) is only valid for the first electron. Then, each additional electron sees that the core is shielded by the electron(s) already bound to the core, which reduces the amplitude of the potential outside the orbit of the first electron(s). In addition, for the innermost level, the charge is Ze, rather than e, where Z is the atomic number of the atom, or the number of protons in the core, and this charge is gradually reduced as more electrons are added. This means that the potential is not simply a $1/r$ potential but rather one whose amplitude and shape varies with the electron under consideration. The simple energy levels obtained in (9.50) are no longer valid, in that the angular momentum states are no longer degenerate with the states with no angular momentum—in essence, the energy levels now depend upon both n and ℓ. This change can be found by treating the difference between the actual potential (for a given electron) and the Coulomb potential as a perturbation and calculating the shift in energy for various values of n and ℓ. In fact, this is not particularly accurate, because the inner-shell electrons also interact with the outer electrons (and, in fact, each electron interacts with every other electron) through an additional Coulomb (repulsive) potential. Many schemes have been proposed for calculating the exact eigenvalues through various approximations, but usually some form of variational approach yields the best method. These approaches are, however, well beyond the level that is considered here. The most important result is that, for a given level index n, the states of lowest ℓ lie at a lower energy (are more tightly bound to the nucleus). This is sufficient information to begin to construct the periodic table of the elements.

The lowest, and only exact, energy level is that for hydrogen, where there are a single electron and the atomic core of unit charge. The lowest energy level from (9.50) is the one for $n = 1$. The allowed values of radial and angular momentum indices must satisfy $n' + \ell = 0$ from (9.51). Thus, the only occupied state has no angular momentum, and the radial wave function varies (in the asymptotic limit of a simple Coulomb potential) simply as $e^{-\rho/2}$, in reduced units. Because of electron spin, this energy level can hold two electrons, so this behavior actually holds also for helium (although the energy level is shifted to lower energy due to the non-Coulomb nature of the potential in helium). Because these wave functions are spherically symmetric, they have come to be termed the 1*s* levels. The 1*s* ($n = 1$, $\ell = 0$) level can hold two electrons, which then fill the shell for $n = 1$. This *shell* comprises the first row of the periodic table, H and He.

The next energy level, according to (9.50), has $n = 2$. For this level, (9.49) would tell us that $n' + \ell = 1$, so the lowest angular momentum state has $(n', \ell) = (1,0)$. This lowest state gives rise to the pair of electrons in the 2*s* state. For this state, the wave function is again spherically symmetric and has an asymptotic variation (for a simple Coulomb potential) proportional to $(\rho - 2)e^{-\rho/2}$. This wave function has two nodes (in the amplitude-squared value), and this behavior continues as the order of the *s* levels increases. The two elements that are added via these two energy states are Li and Be. For more electrons, the angular momentum states begin to be filled. These levels arise for $(n',\ell) = (0,1)$. There are $2\ell + 1 = 3$ values of m for these levels, and each will hold two electrons because of the electron spin angular momentum. This means that there can be an additional six electrons accommodated by these levels. From the discussion of Section 9.2.2, the angular part of the wave function for $m = 0$ varies as $\cos\theta$ only. This gives amplitude peaks along the z-axis, so this state is the P_z state. Similarly, the $m = \pm 1$ states give rise to the $L\pm$ states, which are combinations of the P_x and P_y states. These states have an angular variation as $(\sin\theta)e^{\pm i\phi}$. In Figure 9.2, the orientation of these three angular momentum wave functions is illustrated and has been oriented along the three principal axes in accordance with their notation given just above. The filling of these six states, which are termed the 2*p* states (the symbol *p* signifies the $\ell = 1$ states), is carried out by progressing through the elements B, C, N, O, F, and Ne ($Z = 5 - 10$). Thus, the $n = 2$ level can hold eight electrons, and this shell is completely filled by the element of atomic number ten (Ne). The elements Li through to Ne comprise row two of the periodic table.

The states for $n = 3$ arise in a similar manner. Here, $n' + \ell = 2$, so ℓ can take the values 0, 1, and 2. These give rise to the two 3*s* states and the six 3*p* states, which do not differ in principle from those described above. However, the ten possible states for $\ell = 2$ raise additional complications and are described by quite complicated wave functions. These are the *d* levels, as the symbol *d* describes these $\ell = 2$ levels. First, however, the third row of the periodic table is formed by filling the two 3*s* states and the six 3*p* states, in that order, by the elements Na, Mg, Al, Si, P, S, Cl, and Group A (elements number 11 through to 18). The 3*d* levels lie rather high in energy and actually may *lie above* the 4*s* levels! Thus, the next two elements, K and Ca, actually have their electrons in the 4*s* levels,

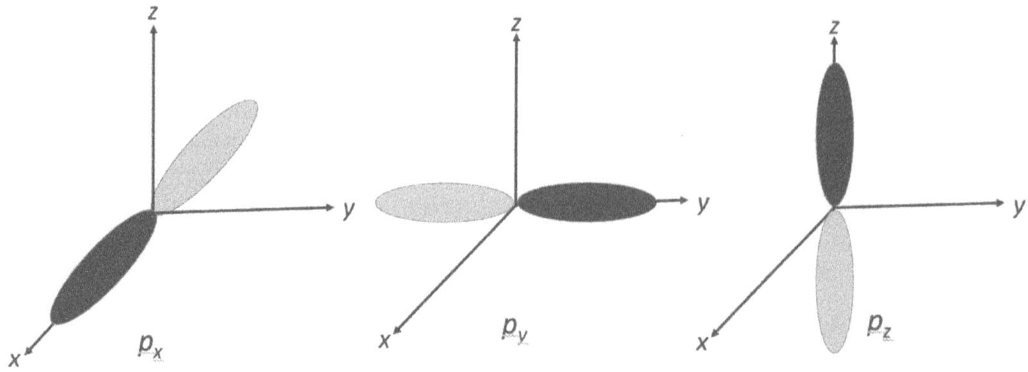

FIGURE 9.2 The orientation of the 2*p* orbitals is illustrated here. The black represents the positive part of the wave function; the gray represents the negative parts.

with the 3*d* levels completely empty. This is where the complications begin, as the 3*d* levels are now filled by the next elements, and the ten elements required to fill these levels are the first set of *transition metals.* In the periodic table, these are the elements that fill the Group B columns, as the original series is designated the Group A columns. The transition metals formed from the 3*d* levels are known as the Pb series, those formed by the 4*d* levels are known as the Pd series, and so on.

The remaining rows of the periodic table are filled in a very similar manner. The next angular momentum levels are denoted as the f levels (for $l = 3$). Many atoms now give rise to interstitial sets of elements: the rare-earth series and the actinide series, but the complications of these rows are somewhat beyond the simple assignments above. In Table 9.1, the energy levels of the *s* and *p* levels are given for a variety of atoms that find use in semiconductor technology (Herman and Skillman, 1963).

9.3 THE COVALENT BOND IN SEMICONDUCTORS

The basic structure of the energy bands in a semiconductor can be inferred from a knowledge of the atomic lattice and its periodicity. Indeed, the properties of periodic potentials were studied in one dimension in Chapter 4, and some properties of materials such as silicon were introduced there. In this section, it is desired to dig a little deeper into the nature of the covalent bond that exists in nearly all semiconductors. There are basically two steps in this process, both of which were introduced in Chapter 4. The first step after creating the periodic crystal is the manner in which the covalent bond

TABLE 9.1
Outermost Atomic Energy Levels for Selected Atoms (eV)

Atom	n	$-E_s$	$-E_p$	Atom	n	$-Es$	$-E_p$
H	1	13.6		Ga	4	11.37	4.9
B	2	12.54	6.64	Ge	4	14.38	6.36
C	2	17.52	8.97	As	4	17.33	7.91
N	2	23.04	11.47	Se	4	20.32	9.53
O	2	29.14	14.13	Cd	5	7.7	3.38
Al	3	10.11	4.86	In	5	10.12	4.69
Si	3	13.55	6.52	Sn	5	12.5	5.94
P	3	17.1	8.32	Sb	5	14.8	7.24
S	3	20.8	10.27	Te	5	17.11	8.59
Zn	4	8.4	3.38	Hg	6	7.68	3.48

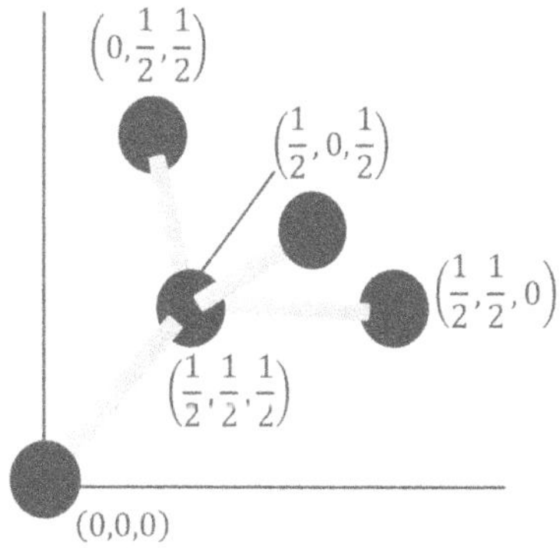

FIGURE 9.3 The tetrahedral coordination and bonding configuration of the diamond and zinc-blende lattices common to most semiconductors. The primary unit cell is composed of the two atoms at (0, 0, 0) and (0.5, 0.5, 0.5). The zinc-blende lattice has the same coordination as the diamond lattice except the two atoms of the unit cell are different. The inner atom of the two on the basis located at (0, 0, 0) is bonded to its four nearest neighbors (on the adjacent faces of the cube) by highly directional *orbitals*. These sp^3 orbitals give highly directional bonds.

is formed in these materials. Second is extending the covalent atomic bond into a set of bands that arise due to the periodicity of the lattice. In Chapter 4, the interaction between the atoms was simply expressed as a coupling energy. That energy arises from the covalent bond in semiconductors and indeed arises in nearly all organic compounds as these are typically covalent-bonded compounds. Now, it is the covalent bond that needs to be discussed.

The semiconductors in which we are interested are *tetrahedrally* coordinated, by which is meant that there are four electrons (on average) from each of two atoms in the basis of the diamond structure. (This is the face-centered cubic structure with a basis of two atoms per lattice site. See Figure 9.3.) These four electrons that create the bond are in the outer shell composed of s and p levels. For example, the Si bonds are composed of $3s$ and $3p$ levels, while GaAs and Ge have bonds composed of $4s$ and $4p$ levels. In each case, the inner shells are not expected to contribute anything at all to the bonding of the solid. This is not strictly true in levels 4 and higher, as the inner shell d levels often lie quite close to the outer s and p levels when the former are occupied. This would imply that there is some modification of the energy levels in GaAs and Ge due to the filled $3d$ levels of the inner shell. This correction is small and will not be considered further. The p wave functions are quite directional in nature (see Figure 9.2), and this leads to a very directional nature of the bonding electrons. Thus, the electrons are not diffusely spread, as in a metal, but are quite localized into a set of hybrids that join nearest-neighbor atoms together. This hybrid orbital bonding is depicted by the lines between the atoms in Figure 9.3.

The bonds between each pair of atoms are composed of *hybrids* that are formed by the composition of the various possible arrangements of the s and p wave functions and are known as sp^3 hybrids. There are, of course, four hybrids for the tetrahedrally coordinated semiconductors. These four hybrids may be written as

$$\begin{aligned}
|h_1\rangle &= \frac{1}{2}\left(|s\rangle + |p_x\rangle + |p_y\rangle + |p_z\rangle\right), \\
|h_2\rangle &= \frac{1}{2}\left(|s\rangle + |p_x\rangle - |p_y\rangle - |p_z\rangle\right), \\
|h_3\rangle &= \frac{1}{2}\left(|s\rangle - |p_x\rangle + |p_y\rangle - |p_z\rangle\right), \\
|h_4 &= \frac{1}{2}\left(|s\rangle - |p_x\rangle - |p_y\rangle + |p_z\rangle\right).
\end{aligned} \tag{9.72}$$

The first of these hybrids points in the (1 1 1) direction, while the other three point in the $(1\,\overline{1}\,\overline{1})$, $(\overline{1}\,1\,\overline{1}$, and $(\overline{1}\,\overline{1}\,1)$ directions, respectively (the bar over the top indicates a negative coefficient, and the three numbers are the multipliers for the unit vectors in the k_x, k_y, k_z directions, respectively). The factor of $\frac{1}{2}$ is included to normalize the hybrids properly so that $\langle h_i \mid h_j \rangle = \delta_{ij}$. These hybrids are now directional and point in the proper directions for the tetrahedral bonding coordination of these semiconductors. Thus, the bonds are directed at the four nearest neighbors of a given atom in the lattice. For Ga in GaAs, the four hybrids point directly at the four As neighbors, which lie at the points of the tetrahedron. This atomic configuration is the same as the methane molecule, where the C atom sits in the center and the four H atoms lie at the tips of the tetrahedron.

The energy levels of the four sp^3 hybrid orbitals are degenerate, as they all have the same energy eigenvalue. Each of the atomic levels of the individual atoms possesses a distinct energy level, and these were tabulated in Table 9.1. Thus, the s levels possess an energy given by E_s, and the p levels have the energy E_p. In general, these levels are properties of the atoms, so that the levels are different in the heteropolar compounds like GaAs. The levels will be marked with a superscript A or B, corresponding to the A–B compound that forms the basis of the lattice. As a general rule, the A atom is the *cation* (Ga), and the B atom is the *anion* (As) in chemical terms. In the following, the compound semiconductors will be treated, as they form a more general case, and the single component semiconductors, such as Si or Ge, are a special case that is easily obtained in a limiting process of setting A = B. In general, the separation between the two atomic energy levels has been given the name of a *metallic* energy (Harrison, 1980), so that

$$
\begin{aligned}
4V_1^A &= E_p^A - E_s^A, \\
4V_1^B &= E_p^B - E_s^B.
\end{aligned}
\tag{9.73}
$$

The hybrids themselves possess an energy that arises from the nature of the way in which they are formed. Thus, the *hybrid* energy is

$$
E_h = \frac{\langle h_i | H | h_i \rangle}{\langle h_i \mid h_i \rangle} = \frac{1}{4}\left(E_s + 3E_p\right). \tag{9.74}
$$

As remarked above, the four hybrid energy levels are degenerate so that (9.74) holds for all values of the index i (which runs from 1 to 4), where H is the Hamiltonian operator for the Schrödinger equation representing the crystal but neglecting any interaction between the atoms (without these interaction terms, the cross terms that would arise in (9.74) will vanish by orthonormality). However, the result (9.74) is different for the A atoms and the B atoms, because their atomic energies are different. Thus, in a zinc-blende lattice, there will be two sets of hybrids arising from the two dissimilar atoms. This is shown in Figure 9.4, where the hybrid energy is shown to form from the atomic energies in a compound semiconductor. The two hybrid energies are separated by the *hybrid polar energy* or the *heteropolar energy*, depending on whose definitions one wants to use. The notation C has been used here for this energy. This heteropolar energy arises from the ionic transfer of charge

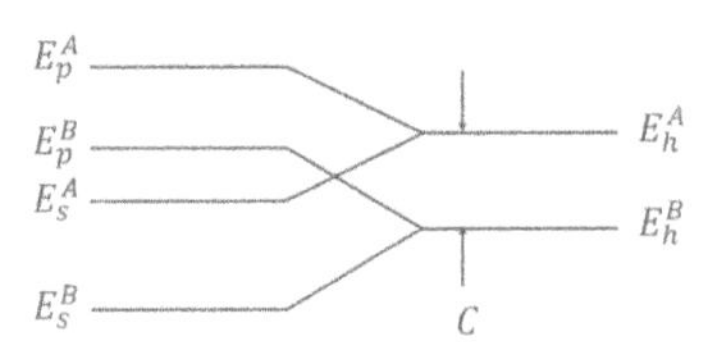

FIGURE 9.4 The atomic energies form hybrids, as indicated here, for each of the two atoms in the basis. These hybrids are separated by the *heteropolar* energy C.

that occurs in the compound semiconductor (since Ga has only three electrons and As has five electrons, there is a charge transfer in order to give the average four electrons of the tetrahedral bond). Of course, this energy vanishes in a pure single compound such as Si or Ge, which are referred to as homopolar materials. That is, they are composed of a homogeneous set of atoms, while the general compound semiconductor is heterogeneous in that it contains two (or more in the alloys) types of atoms. The heteropolar energy C may be easily evaluated using (9.74) as follows:

$$C = E_h^A - E_h^B = \frac{1}{4}\left[\left(E_s^A - E_s^B\right) + 3\left(E_p^A - E_p^B\right)\right]. \tag{9.75}$$

It is important to note that the hybrids are not eigenstates of either the isolated atom or of the crystal. Rather, they are constructed under the premise that they are the natural wave function for the tetrahedral bonds. However, they have been created to be what seems to be a natural form (at least to us). In principle, they will be stable in the crystal once the interactions between the various atoms are included. However, as pointed out above, they are not orthogonal under the action of the Hamiltonian, since the hybrid energies are the same for each atom, as defined in (9.74).

The interaction between hybrids localized on neighboring atoms begins to form the energy bands, as in Chapter 4. Only standing-wave interactions will be considered here ($k = 0$), and the propagating wave function–dependent changes are calculated by the methods of Chapter 4. The interaction energy between two atoms on different sites can arise from, for example, one atom's hybrid pointed in the (1 1 1) direction and the nearest-neighbor atom, displaced $\left(\frac{a}{4}, \frac{a}{4}, \frac{a}{4}\right)$, whose hybrid points in the opposite direction (there is a complete flip of the hybrid directions of the atoms as one moves along the body diagonal direction). The interaction energy between these nearest-neighbor hybrids is defined, using (9.74), to be

$$-V_2 = \frac{1}{4}\langle s^A|H|s^B\rangle + \frac{\sqrt{3}}{4}\left[\langle p^A|H|s^B\rangle + \langle s^A|H|p^B\rangle\right] + \langle\frac{3}{4}p^A|H|p^B\rangle. \tag{9.76}$$

Harrison (1980) has argued that these energies should depend only on the interatomic spacing $d\ (= \sqrt{3}a/4)$, where a is the edge of the face-centered cube, and that they should have the general form

$$V_2 \sim 4.37\frac{\hbar^2}{m_0 d^2}. \tag{9.77}$$

Phillips (1973) also argues that V_2 should be a function of the interatomic spacing, but that it should also satisfy another scaling rule. Since the atomic radii are quite similar for each row of the periodic table, the value of V_2 should be the same in AlP as in Si, the same in GaAs and ZnTe as in Ge, and so on. In other words, the value for this quantity is set by the distance between the atoms, and this really does not change much as one moves across a row of the table. Thus, this value is the same in heteropolar compounds as in homopolar compounds, and Phillips has termed V_2 the homopolar energy, $E_{h0} = 2V_2$ (this should not be confused with the hybrid energy E_h, which differs for the two atoms).

The *average energy gap* between bonding and antibonding orbitals is thus composed of a contribution from the homopolar energy E_{h0} and a contribution from the heteropolar energy C. It is often believed that this average energy gap between the conduction band and the valence band, which arises once the entire band structure has been formed, will scale with these various energies. This suggests that the two hybrids that arise from the two atoms at each end of a bond (Figure 9.4) create a bonding orbital that can be written as a linear combination of the two hybrids at each atom. These bonding orbitals then interact to form the bands which develop around the average energies of these bonding orbitals and the corresponding antibonding orbitals, which should be separated by approximately

$$\overline{E_G} = \frac{E_h^A + E_h^B}{2} \pm \sqrt{(2V_2)^2 + C^2} = \frac{E_h^A + E_h^B}{2} \pm \frac{\Delta}{2}. \tag{9.78}$$

It is important to note that the energy gap in (9.78) is *not* the normal bandgap between valence and conduction bands, but the difference in the average energy of these two bands. The bonding (lower sign) and the antibonding (upper sign) energy levels are symmetrically spaced about the average hybrid energy of the two atoms. The significance of the average energy is unexpected in heteropolar materials but is important for getting the positions of the average bonding and antibonding energies correct. The actual bands spread around these two average energies, separated by Δ, according to the techniques of Chapter 4.

It is clear that once the atomic energy levels are known, the hybrid energies can be immediately determined, and the heteropolar and homopolar energies will result. Using V_2 as the constant value given in (9.77), the average energies are now known. However, Phillips (1973) treats the homopolar energy and the heteropolar energy C as adjustable parameters to get better scaling for the average energy gap between the conduction and valence bands, building his results from a need to get the dielectric functions correct. His values are close to the Harrison (1980) values obtained by straight-forward application of the atomic energies, but there are differences. In the heteropolar materials, the differences are even greater, and the average gap can differ by more than a volt between the two approaches. To be sure, it is not at all clear that these two methods are comparable or that the two authors are talking about exactly the same quantities, even though it appears to be so. That the numbers are close is perhaps remarkable and points out the basic correctness of the overall picture of the composition of the energy bands in semiconductors. These average energies have been dealt with in this section (Ferry, 1991), because this is where the quantum mechanics enters the calculation, although dealing with the electrons as waves is also quantum mechanical.

9.4 HYDROGENIC IMPURITIES IN SEMICONDUCTORS

In Chapter 4, it was pointed out that in materials such as Si, the valence band was completely full of electrons, while the conduction band was completely empty. That is, there were just enough electrons with the two atoms in the primitive unit cell to completely occupy all the available states in the valence band, with none left over. Thus, Si was a blank canvas with which to introduce impurities to modify the properties and create many different devices. This process is achieved in semiconductors by placing impurity atoms into the host lattice at relatively precise locations. The idea is that the impurity atoms may have an additional electron or be short of one electron needed for the tetrahedral bond.

When the impurity atom has an additional electron over those (four) required for tetrahedral bonding, this extra electron can be ionized rather easily (which then puts the electron into the conduction band). In this case, the impurity atom has a single positive charge, while the ionized electron has a single negative charge. In many ways this interaction is quite like that in the hydrogen atom.

On the other hand, the impurity atom that is short of one electron allows for electrons to be excited from other atoms to this one (essentially exciting an electron from the bonding valence band to the impurity) in order to complete the tetrahedral bonding requirements. In this process, it is inferred that a "hole" moves from the impurity to the valence band. This hole (the absence of an electron in the valence band) can be ionized from the impurity, leaving a negatively charged impurity atom and a positively charged quasi-particle, the hole, in the valence band.

The energy required to ionize either the electron or the hole is generally much smaller than that required to lift an electron out of the valence band into the conduction band—the latter of which requires the band gap energy. Thus, these impurities introduce defect levels within the bandgap region of the semiconductor. The atom with an excess electron becomes a *donor*, with its energy level a few meV below the conduction band. The atom that lacks a fourth electron becomes an *acceptor*, with its energy level a few meV above the valence band. Because of the Coulombic nature of the interaction between the ionized particle (electron or hole) and the central cell of the ionized impurity, the perturbing potential is simply a screened Coulomb potential:

$$V(r) = -\frac{e^2}{4\pi\epsilon r}, \tag{9.79}$$

where ϵ is the permittivity of the semiconductor, not free space. It may generally be assumed that the carriers in the semiconductor are characterized by a scalar *effective mass m**, which is different for the electrons and the holes, as they respond to different momentum dependence of the conduction and valence bands, respectively (see Section 4.4). Thus, we are interested in either electrons near the minimum of the conduction band or holes near the maximum of the valence band. With the potential (9.79), and the effective mass, Schrödinger's equation for this reduced system becomes

$$-\frac{\hbar^2}{2m^*}\nabla^2\psi - \frac{e^2}{4\pi\epsilon r}\psi = E\psi. \tag{9.80}$$

This is the same equation as for the simple hydrogen atom (with the appropriate changes in the permittivity and the quasi-particle mass). This means that the allowed excitation energies for the donors or acceptors are simply given by the hydrogen energy levels (9.50) suitably adjusted. For example, consider the case of GaAs, where the effective electron mass is $0.067m_0$ and the low-frequency relative permittivity is 13.1. These parameters lower the ionization energy from 13.6 eV (Table 9.1) to about 70 meV. What is not accounted for in this hydrogenic model is variation according to just which atomic species is providing the impurity atom and just which semiconductor is being used as the host material. In the simple model, the results should be the same for all impurities, but this is not the case. For each such impurity atom, the Coulomb potential is modified by the electrons in the interior shells, as discussed in the previous section. However, the hydrogenic model is a good first approximation. Actual ionization energies vary from a few meV to a significant part of an eV in deep levels. The wave function for the first ionization state varies just like that for the hydrogen atom, but with an adjusted radius. In fact, we may write the wave function as

$$\psi(r) \sim e^{-r/a}, \tag{9.81}$$

where

$$a \sim \frac{4\pi\epsilon\hbar^2}{e^2 m^*} = a_0 \frac{\epsilon}{\epsilon_0}\frac{m}{m^*}, \tag{9.82}$$

and a_0 is the Bohr radius. For a hypothetical semiconductor, this is of the order of ten lattice constants. Consequently, the orbit of the electron, when it is captured by the impurity, actually samples a great many unit cells of the crystal, which implies that the effective mass approximation is a fairly good approximation for the hydrogenic impurity. Nevertheless, the potential probably deviates in reality from a Coulombic one, and this difference is likely to account for the variation in ionization energy seen for different impurities in the same semiconductor.

REFERENCES

Cohen-Tannoudji, C., Diu, B. and Laloë, F.1977. *Quantum Mechanics*, vol. 1. (New York: Wiley).
Darwin, C. G. 1931. The diamagnetism of the free electron. *Proc. Cambr. Phil. Soc.* **27** 86–90.
Ferry, D. K. 1991. *Semiconductors.* (New York: Macmillan).

Fock, V. 1928. Bemerkung zur Quantelung des harmonischen Oszillators im Magnetfeld. *Z. Phys.* **47** 446–448.
Harrison, W. A. 1980. *Electronic Structure and the Properties of Solids.* (San Francisco, CA.: Freeman).
Herman, F. and Skillman, S. 1963. *Atomic Structure Calculations.* (Englewood Cliffs, NJ.: Prentice-Hall).
McEuen, P. L., Foxman, E. B., Meirav, U., Kastner, M. A., Meir, Y., Wingreen, N. S. and Wind, S. J. 1991. Transport spectroscopy of a Coulomb island in the quantum Hall regime. *Phys. Rev. Lett.* **66** 1926–1929.
Phillips, J. C. 1973. *Bonds and Bands in Semiconductors.* (New York, NY: Academic).
Rössler, U. 1991. Electronic Structure of Small Systems, in *Quantum Coherence in Mesoscopic Systems*, Ed. Kramer, B. (New York: Plenum Press) 45–62.
Sikorski, C. and Merkt, U. 1989. Spectroscopy of electronic states in InSb quantum dots. *Phys. Rev. Lett.* **62** 2164–2166.
Tarucha, S., Austing, D. G., Honda, T., van der Haage, R. J. and Kouwenhoven, L. P. 1996. Shell filling and spin effects in a few electron quantum dot. *Phys. Rev. Lett.* **77** 3613–3616.
Zeeman, P. 1897a. On the influence of magnetism on the light emitted by a substance. *Phil. Mag., Ser. 5.* **43** 226–239.
Zeeman, P. 1897b. Doublets and triplets in the spectrum produced by external magnetic forces. *Phil. Mag., Ser. 5.* **44** 55–60.
Zeeman, P. 1897c. The effect of magnetization on the nature of light emitted by a substance. *Nature.* **55** 347.

PROBLEMS

1. Find the wave functions for the $n = 2$, $m = 0$ level of the two-dimensional harmonic oscillator. Use the angular momentum operators and show that this level is cylindrically symmetric.
2. Suppose that the x-axis harmonic oscillator is characterized by a frequency ω_1, which is slightly larger than that of the y-axis ($\omega_1 \simeq \omega + \delta\omega$). Compute the exact energy levels of the two-dimensional harmonic oscillator. Then, using the energy levels and wave functions for the symmetric two-dimensional harmonic oscillator, calculate the perturbation shift of these energy levels with the perturbation $m(\delta\omega)^2 x^2/2$.
3. Let us consider the application of an electric field $-E\mathbf{a}_x$ to the two-dimensional harmonic oscillator in the presence of a magnetic field. Determine the change in the energy levels introduced by this electric field.
4. What are the energy values of the lowest three energy levels in the hydrogen atom? (That of the lowest energy level, the ionization energy of the hydrogen atom, is termed one Rydberg.)
5. Verify that (9.40) is the correct equation for the angular variation of the hydrogen atom.
6. Develop the full wave functions (the form, neglect normalization) for the 3p states utilizing the angular momentum operators.
7. Show that L^2 in (9.64) commutes with each component of the angular momentum.
8. Using the known lattice constants and atomic energy levels, compute the values of the parameters C, V_2, and E_g for Si, Ge, GaAs, AlAs, InAs, InSb, and InP.
9. From the known lattice constant and density of Si, compute the number of atoms per cubic meter. Then, use this value of N, and the known dielectric constant of Si, to compute the valence plasma frequency from $\omega_P^2 = Ne^2/m_0\varepsilon_s$. What is the average energy gap required to satisfy the Penn dielectric function

$$\varepsilon_s = \varepsilon_0 \left[1 + \left(\frac{\hbar\omega_P}{E_G}\right)^2\right]?$$

10 Spin Angular Momentum

It has been pointed out several times in the previous chapters that each of our quantum states found in the various systems which have been studied will actually hold two electrons. These two electrons must have opposite spin. While only lip service has been paid to the fact that each electron possesses a *spin angular momentum* about its own axis (in one of the two possible directions—up or down), this became important at several points in the book, such as when we discussed superconductivity and Cooper pairs at the end of the last chapter. When one treats a single isolated electron, the various descriptions of the effect of spin do not cause very much trouble. However, most atoms and solids are densely populated with electrons, and the neglected properties can introduce new effects that need to be considered. Some of these effects may be quite small, while others can create macroscopic effects that are easily measurable. In any case, an understanding of them is necessary, if for no other reason than to be able to ascertain when they may properly be ignored.

In classical mechanics, it is quite easy to follow the individual trajectories of each and every electron, at least in principle, if not in the actual laboratory. However, the Heisenberg uncertainty principle prevents this from being possible in quantum mechanics. If we completely understand the position of a given particle, we can say little about its momentum and most other dynamical variables. The principle of the *indistinguishability* of identical particles (one cannot tell just which electron is being treated, as they all behave the same) leads to other more complex forms of the wave functions, especially as these electrons interact with each other through a Coulomb force. This interaction leads to effects in quantum mechanics that have no analog in classical mechanics. We use the word *identical* to indicate particles that can be freely interchanged with one another with no change in the physical system. While these particles may be distinguished from one another in situations in which their individual wave functions do not overlap, the more usual case is where they have overlapping wave functions because of the high particle density in the system. In this latter case, it is necessary to invoke a *many-particle wave function* to account for the interactions among the particles. To describe properly the properties of the many-particle wave function, it is necessary to understand the properties of the interactions that occur among and between these single particles. If the one-electron problems were our only interest, this more powerful development that is used for the interacting system would be worthless to us, for its complexity is not worth the extra effort. However, when multi-electron problems arise, the power of the approach becomes apparent. In this chapter, the spin properties of these general wave functions for the case of electrons, with their spin angular momentum, will be treated and the interaction among the electrons introduced.

Consider first just two electrons. Because these are indistinguishable particles, the physical state that is obtained by merely interchanging the positions of these two particles must be completely equivalent to the original one. This puts certain constraints upon the wave function. Let us consider the two-electron wave function $\psi(x_1,x_2)$. Here, the parameter x_1 refers to the position of the first particle and the spin orientation σ_1 of that electron (which will be further explained below). Thus, we can refer to this as a four-vector in space—three components for the position and the fourth for the spin orientation. Similar notation is used for the second particle. Now, under exchange of the positions and spins of the two particles, a phase shift may occur as

$$\psi\left(x_1,x_2\right)=e^{i\vartheta}\psi\left(x_2,x_1\right). \tag{10.1}$$

The phase ϑ is some real constant. By repeating the interchange (10.1), leads to

$$\psi\left(x_1,x_2\right)=e^{i\vartheta}\psi\left(x_2,x_1\right)=e^{2i\vartheta}\psi\left(x_1,x_2\right). \tag{10.2}$$

DOI: 10.1201/9781003717195-10

But this requires $e^{2i\vartheta} = 1$, or $e^{i\vartheta} = \pm 1$. Thus, the possible form for the wave function under exchange of the two particles is

$$\psi(x_1, x_2) = \pm\psi(x_2, x_1). \tag{10.3}$$

That is, the wave function that is found upon exchange of the two particles is going to be either symmetrical or anti-symmetrical. By symmetrical, we mean that it is unchanged, while anti-symmetrical implies a change of sign. Which of these two should be chosen?

The choice of either a symmetric or an anti-symmetric wave function lies in the imposition of the Pauli exclusion principle (Pauli, 1925). As previously stated, each eigenstate determined by solving the Schrödinger equation should hold two electrons, provided that they had opposite spin. In each situation, the positional wave functions for the two electrons are the same, but the spin wave functions must yield a spin eigenvalue σ that is oppositely *directed.* Now, this spin eigenvalue is one of the vectors included above in the description of the two-particle wave function. Thus, the electrons, which obey the Pauli exclusion principle, must have an *anti-symmetric wave function.* Particles that do not obey the exclusion principle (phonons, photons, etc.) are *bosons* and are found to obey Bose–Einstein statistics. These bosons have symmetric wave functions under interchange of the particles. On the other hand, particles that obey the exclusion principle (electrons and some others, which will not be of concern) are *fermions* and are found to obey Fermi–Dirac statistics. Fermions must have anti-symmetric wave functions under the exchange of particles. The use of the anti-symmetric wave function actually ensures that the Pauli exclusion principle is obeyed.

What is really meant by bosons and fermions? In the previous paragraph, it was stated that bosons do not obey the Pauli exclusion principle but do obey the Bose–Einstein distribution. On the other hand, fermions obey the Pauli exclusion principle—no more than two fermions, and these are of opposite spin can be accommodated in any quantum state. Bosons are particles with integer spin, such as phonons (zero spin) and photons (integer spin given by ± 1, corresponding to right- and left-circularly polarized plane waves, for example). Fermions are particles with half-integer spin, such as electrons. These two distributions are given by

$$f_{BE}(E) = \frac{1}{e^{E/k_BT} - 1} \tag{10.4}$$

for the Bose–Einstein distribution, and

$$f_{FD}(E) = \frac{1}{1 + e^{(E-E_F)/k_BT}} \tag{10.5}$$

for the Fermi–Dirac distribution, where E_F is the Fermi energy level, where the distribution function has the value 0.5. There is a significant difference in the behavior of these two distributions at low energy. The Bose–Einstein distribution diverges at $E \to 0$, as there is no limit to the number of bosons that can occupy the lowest energy state. On the other hand, the Fermi–Dirac distribution approaches unity in this limit, as the state is certainly occupied if it lies well below the Fermi energy E_F. As pointed out, one way of achieving the Pauli exclusion principle is to ensure that the wave function for a fermion is anti-symmetric, so that it satisfies (10.3).

If the many-particle wave function is inserted into the Schrödinger equation, it is found that, under the circumstances discussed above, it is possible to write this equation as

$$\left[H(x_1, x_2) - E\right]\psi(x_1, x_2) = 0. \tag{10.6}$$

The Hamiltonian itself is invariant under the exchange of the particles, as the spin does not appear in the common form of the Hamiltonian, and the equivalence of the two physical states implies that the energy has the same invariance. The imposition of the exclusion principle appears only in the wave function (unless some special spin-dependent interaction, such as the spin–orbit interaction—discussed below—is introduced into the Hamiltonian to distinguish one spin state from another). Also, the presence of a magnetic field creates a spin-dependent energy contribution.

Whether the wave function is symmetric or anti-symmetric has no impact upon (10.6) normally. The importance of this latter result is that, for simple Hamiltonians with no interaction among the electrons, it is usually possible to separate (10.6) into two equations, one for each particle, with the total energy being the sum of the single-particle energies. This separation is carried out in exactly the same manner as that in which separation of coordinates is done in a many-dimensional partial differential equation. As a result, it is possible to write the two-particle wave function as a product of the one-particle wave functions $\psi_1(x_1)$ and $\psi_2(x_2)$, where the two wave functions may be different. The subscripts on the wave functions themselves (as opposed to those on the variables) refer to the "number" of the particle. Carrying out his separation produces

$$\psi(x_1,x_2) \sim \psi_1(x_1)\psi_2(x_2). \tag{10.7}$$

However, this wave function does not possess the proper symmetry for electrons; for example, interchanging the positions of particles "1" and "2" does not produce the necessary anti-symmetry. This can be achieved with a somewhat cleverer summation, already introduced in (6.46). That is, the two-particle wave function is described as

$$\psi(x_1,x_2) = \frac{1}{\sqrt{2}}\left[\psi_1(x_1)\psi_2(x_2) - \psi_2(x_1)\psi_1(x_2)\right]. \tag{10.8}$$

This wave function has the desired anti-symmetry under the interchange of the two electrons.

This method of forming a properly anti-symmetric many-electron wave function from the single-electron wave functions has been extended to an arbitrarily large number of electrons by Slater (1929). The resulting wave function for N electrons is given by the *Slater determinant*

$$\psi(x_1,x_2,\ldots,x_N) = \frac{1}{\sqrt{N}}\begin{vmatrix} \psi_1(x_1) & \psi_1(x_2) & \cdots & \psi_1(x_N) \\ \psi_2(x_1) & \psi_2(x_2) & \cdots & \psi_2(x_N) \\ \vdots & \vdots & \vdots & \vdots \\ \psi_N(x_1 & \psi_N(x_2) & \cdots & \psi_N(x_N) \end{vmatrix}. \tag{10.9}$$

Equation (10.9) has the added usefulness that it has the property that it vanishes if two of the ψ_i are the same. This implies that no two electrons can have the same state (recall that spin angular momentum is included explicitly in the variables), and thus the Pauli exclusion principle is automatically satisfied. Thus, when the Hamiltonian can be separated into single-particle parts, and the wave function separated accordingly, the anti-symmetrized product of these wave functions in a Slater determinant ensures that the Pauli exclusion principle is satisfied.

10.1 SPIN ANGULAR MOMENTUM

As has been pointed out previously, electrons can have two spin orientations, which are normally taken as "up" and "down." But this is for an arbitrary orientation of the electron's own magnetic moment. Recognition of the two spin states arises from the introduction of the Pauli exclusion principle whenever a fully quantized state is considered. These two spin states must have opposite polarization, hence the designations as up and down. In fact, the spin can be oriented into any desired direction by the application of external forces, such as a magnetic field. The most common orientation, arising in the Zeeman (1897) effect, is along the z-axis which is achieved via an external magnetic field oriented in this direction. But other directions are useful in a variety of applications such as the use of spin orientation in a qubit (quantum bit) for quantum computation, which will be discussed in Chapter 11. Here, the values and orientations of the spin and its representation via the Pauli spin matrices will be examined.

General angular momentum for electrons around an atom was introduced in the previous chapter. For sure, the angular momentum of an electron orbiting a centrally symmetric potential, such as the Coulomb potential around an atom, possesses a quantized value for the angular momentum. In this atomic case, both the total angular momentum L^2 and the z-directed angular momentum L_z had eigenvalues and commuted. The fact that they could both be made to commute with the Hamiltonian tells us that they can both be diagonalized along with the Hamiltonian and therefore both could be measured at the same time as the total energy. In the present case, the only angular momentum to be considered is the spin angular momentum, which may be taken as oriented in the z-direction. Just as in the atomic case, the total spin S^2 will continue to commute with the z-component of spin S_z.

It may be recalled that in any quantum confinement problem such as the atom or a quantum well, the total wave function is a sum over a set of eigenfunctions $\varphi_i(x)$. For each of these functions, which are defined by a set of quantized values due to the confinement, one has not yet considered the spin. If it is now desired to also include the spin, then an additional part of each eigenstate wave function must be defined. Typically, this is a multiplicative term describing the spin state. Traditionally, spin is described by a two-component wave function called a *spinor*. From the Zeeman effect, one typically denotes the extra energy for the spin-up state by the value 1/2, and the value of the spin-down state by the value − 1/2 (and these are multiplied by $\hbar$ to get the angular momentum). This same notation is used to denote the two possible states and their spinors as

$$\begin{aligned}\varphi\left(\frac{1}{2}\right)&=\begin{bmatrix}1\\0\end{bmatrix},\\ \varphi\left(-\frac{1}{2}\right)&=\begin{bmatrix}0\\1\end{bmatrix}.\end{aligned} \tag{10.10}$$

The first row of each matrix refers to the *up* state and the second row refers to the *down* state. Thus, the eigenvalues correspond to those adopted in the Zeeman effect, as mentioned above.

Because the spin angular momentum has been taken to be oriented along the z-axis, the spin matrix for the z-component of angular momentum must be diagonal for two reasons. First, it must commute with the total spin and with the Hamiltonian and, second, it must produce the eigenvalues found from the Zeeman effect. This spin value is represented by a 2×2 matrix in order to act on one of the spinors and produce another spinor. Hence, this matrix for S_z may be written as

$$S_z=\frac{\hbar}{2}\begin{bmatrix}1&0\\0&-1\end{bmatrix}. \tag{10.11}$$

Operating on the spinors of (10.10) produces

$$\begin{aligned}S_z\varphi\left(\frac{1}{2}\right)&=\frac{\hbar}{2}\begin{bmatrix}1&0\\0&-1\end{bmatrix}\begin{bmatrix}1\\0\end{bmatrix}=\frac{\hbar}{2}\varphi\left(\frac{1}{2}\right),\\ S_z\varphi\left(-\frac{1}{2}\right)&=\frac{\hbar}{2}\begin{bmatrix}1&0\\0&-1\end{bmatrix}\begin{bmatrix}0\\1\end{bmatrix}=-\frac{\hbar}{2}\varphi\left(-\frac{1}{2}\right).\end{aligned} \tag{10.12}$$

From the study of normal angular momentum in the previous chapter, it is known that the value of L^2 is given as $\ell(\ell+1)\hbar^2$. Thus, a similar result, $S^2=s(s+1)\hbar^2$, is expected for the spin angular momentum, and using $s = 1/2$, we get $S^2 = = 3\hbar^2/4$. More strictly, this value is for the square of the magnitude of the total spin angular momentum. Again, the matrix representation of this total spin angular momentum must be diagonal, and this is given by

$$|S|^2 = \frac{3\hbar^2}{4}\begin{bmatrix} 1 & 0 \\ 0 & 1 \end{bmatrix}. \tag{10.13}$$

To find the other components of the spin angular momentum, and the other spin matrices, rotating coordinates are introduced, as done in the previous chapter, and

$$S_\pm = S_x \pm iS_y, \tag{10.14}$$

so that the squared magnitude of the total spin angular momentum becomes

$$S^2 = S_x^2 + S_y^2 + S_z^2 = S_+S_- + S_z^2 + \hbar S_z, \tag{10.15}$$

where the commutators

$$\begin{aligned} \left[S_x, S_y\right] &= i\hbar S_z, \\ \left[S_y, S_z\right] &= i\hbar S_x, \\ \left[S_z, S_x\right] &= i\hbar S_y, \end{aligned} \tag{10.16}$$

have been used. If the two rotating terms are reversed, then (10.15) becomes

$$S^2 = S_x^2 + S_y^2 + S_z^2 = S_-S_+ + S_z^2 - \hbar S_z. \tag{10.17}$$

Equation (10.15) and (10.17) can be combined to give

$$\frac{1}{2}\left(S_+S_- + S_-S_+\right) = \hbar^2/2, \tag{10.18}$$

and

$$\frac{1}{2}\left(S_+S_- - S_-S_+\right) = 2\hbar S_z. \tag{10.19}$$

The operators S_+ and S_- act as creation and annihilation operators for the spin angular momentum. That is, operating on a spinor with the first of these operators will raise the angular momentum, which can only occur if it acts on the spin-down state and produces the spin-up state, or

$$S_+\varphi\left(-\frac{1}{2}\right) = \hbar\varphi\left(\frac{1}{2}\right) \quad \to \quad S_+ = \hbar\begin{bmatrix} 0 & 1 \\ 0 & 0 \end{bmatrix}. \tag{10.20}$$

Similarly, acting with the operator S_- removes a quantum of angular momentum and lowers the spin angular. This can occur only if the operator acts upon the spin-up state and produces the spin-down state, or

$$S_-\varphi\left(\frac{1}{2}\right) = \hbar\varphi\left(-\frac{1}{2}\right) \quad \to \quad S_- = \hbar\begin{bmatrix} 0 & 0 \\ 1 & 0 \end{bmatrix}. \tag{10.21}$$

These last two equations can be used to find the individual components of S as

$$S_x = \frac{1}{2}(S_+ + S_-) = \frac{\hbar}{2}\begin{bmatrix} 0 & 1 \\ 1 & 0 \end{bmatrix},$$
$$S_y = \frac{1}{2i}(S_+ - S_-) = \frac{\hbar}{2}\begin{bmatrix} 0 & -i \\ i & 0 \end{bmatrix}. \tag{10.22}$$

The various square 2×2 matrices that appear in the above equations are known as Pauli spin matrices. They are often represented by a reduced notation as

$$\sigma_x = \begin{bmatrix} 0 & 1 \\ 1 & 0 \end{bmatrix},$$
$$\sigma_y = \begin{bmatrix} 0 & -i \\ i & 0 \end{bmatrix}, \tag{10.23}$$
$$\sigma_z = \begin{bmatrix} 1 & 0 \\ 0 & -1 \end{bmatrix}.$$

The fourth one, which appears in (10.13), is just the two-dimensional unit matrix. Such spinors are used in many cases beyond just spin to represent the behavior of two-level systems that have analogous behavior to spin systems (as will be discussed in Chapter 11). A typical two-level system is examined in the next section, and it is shown to have useful properties with which to describe spin angular momentum.

10.2 THE TWO-LEVEL SYSTEM

As discussed in the previous section, the spin of the electron is characterized by its eigenstate, for which the spin wave function is a two-component spinor. It makes natural sense then to spend a little more time with the idea of two-level systems (Merzbacher, 1970). To simplify the notation slightly, the generalized spinors (10.10)

$$\alpha = \varphi(\uparrow) = \begin{bmatrix} 1 \\ 0 \end{bmatrix},$$
$$\beta = \varphi(\downarrow) = \begin{bmatrix} 0 \\ 1 \end{bmatrix}, \tag{10.24}$$

may be introduced. The equation of motion for these spinors is, as usual, given by the Schrödinger equation in which the Hamiltonian is a 2×2 matrix. If there are no external forces acting upon our simple two-level system, then the Hamiltonian is independent of time and the wave functions evolve as

$$\varphi(t) = e^{-iHt/\hbar}\varphi(0), \tag{10.25}$$

where φ corresponds to either of the spinors α or β. In general, the Hamiltonian is diagonalized to provide the eigenvalues, which may be taken to be E_α and E_β. The total wave function can be written as a sum of the two spinors as

$$\psi = c_1\alpha + c_2\beta,$$
$$|c_1|^2 + |c_2|^2 = 1. \tag{10.26}$$

The above properties of the Hamiltonian and the spinors can be used to determine a number of intuitive properties. For example, to determine how β evolves into α, only the total Hamiltonian is required, as

$$\begin{aligned}\beta^{\dagger}e^{-\frac{iHt}{\hbar}}\alpha &= \begin{bmatrix}0 & 1\end{bmatrix}e^{-\frac{iHt}{\hbar}}\begin{bmatrix}1\\0\end{bmatrix}\\ &= \frac{H_{21}}{F_{\beta}-F_{\alpha}}\left(e^{-iE_{\alpha}t/\hbar}-e^{-iE_{\beta}t/\hbar}\right)\end{aligned}, \tag{10.27}$$

where

$$H_{21} = \langle\beta|H|\alpha\rangle = \begin{bmatrix}0 & 1\end{bmatrix}H\begin{bmatrix}1\\0\end{bmatrix} \tag{10.28}$$

is the appropriate matrix element. If the magnitude squared value of (10.27) is considered, it will be discovered that it oscillates due to the exponential terms, and that this oscillation causes the occupation to move from one state to the other with a frequency given by the difference in the eigen-energies of these two states. This oscillation is suggestive, as it points out that the total state given by (10.27) is itself a single spinor whose components are the complex coefficients given in this equation. What this means is that the two spinors α and β define a two-dimensional space for our wave function. Thus, a spinor whose coefficients are the coefficients in (10.27) is a vector in this two-component space and can be written as

$$\psi = \begin{bmatrix}c_1\\c_2\end{bmatrix} = \begin{bmatrix}|c_1|e^{i\vartheta_1}\\|c_2|e^{i\vartheta_2}\end{bmatrix}. \tag{10.29}$$

This description is not unique since the squares of the magnitude must sum to unity. Thus, the two phases can be rather arbitrary, and only the relative phase $\vartheta_1 - \vartheta_2$ can be specified in detail. So, if one writes $\vartheta_2 = \vartheta_1 + \varphi$, the difference between the states ψ and $e^{i\vartheta_1}\psi$, so that one has to somehow believe that these are the same state. In a sense, this all arise from the periodicities of the angles and that is a property that will be used below.

A useful quantity for dealing with two-level systems is the corresponding density matrix (see Section 6.7), defined via

$$\rho = \psi\psi^{\dagger} = \begin{bmatrix}c_1\\c_2\end{bmatrix}\begin{bmatrix}c_1^* & c_2^*\end{bmatrix} = \begin{bmatrix}|c_1|^2 & c_1c_2^*\\c_2c_1^* & |c_2|^2\end{bmatrix}. \tag{10.30}$$

An important property of this matrix is that its trace is unity, as required by Equation (10.26). Now, this is an important point. A fundamental property of any 2×2 matrix, whose trace is unity, is that it can be written in terms of the Pauli spin matrices and a quantity known as the *polarization* as

$$\rho = \frac{1}{2}(1 + \boldsymbol{P}\cdot\sigma), \tag{10.31}$$

where σ is a vector of the Pauli spinors (10.23). Comparing this result with (10.30), the individual components can be connected as

$$\begin{aligned}|c_1|^2 &= \frac{1}{2}(1+P_z),\\ |c_2|^2 &= \frac{1}{2}(1-P_z),\\ c_1c_2^* &= (c_2c_1^*)^* = \frac{1}{2}(P_x - iP_y).\end{aligned} \tag{10.32}$$

These can now be inverted to define the components of the polarization as

$$\begin{aligned} P_x &= 2Re\{c_1c_2^*\}, \\ P_y &= -2Im\{c_1c_2^*\}, \\ P_z &= |c_1|^2 - |c_2|^2. \end{aligned} \tag{10.33}$$

The trace of the density matrix is required to be the sum of the eigenvalues of the two wave function components, as well as summing to unity. Since the eigenvalues are 0 and 1, we also satisfy this requirement. This now implies that one of the eigenvectors of this density matrix is ψ itself. The other eigenvector must be orthogonal to ψ so that their inner product yields 0, but this other eigenvector is still arbitrary at this point.

An important property arises from the expectation values of the various spinors. We examine this with the x-spinor as

$$\begin{aligned} \langle \sigma_x \rangle &= Tr\{\rho\sigma_x\} = \frac{1}{2}\left[Tr\{I\sigma_x\} + \boldsymbol{P}\cdot Tr\{\sigma_x\}\right] \\ &= \frac{1}{2}P_x Tr\{\sigma_x^2\} = P_x \end{aligned}. \tag{10.34}$$

This can be repeated for each spinor, so that the important result

$$\boldsymbol{P} = \sigma = Tr\{\rho\sigma\} = Tr\{\sigma\rho\} \tag{10.35}$$

is obtained. That is, the polarization of the two-level system is defined by the expectation value of the spinor vector. The direction of the spin is uniquely connected to the polarization of the system. Another important point is that since the wave function ψ is an eigenfunction of the density matrix, so that it is also an eigenfunction of the last term in the density matrix (10.31), as

$$\boldsymbol{P}\cdot\psi = \psi. \tag{10.36}$$

To examine the nature of the polarization itself, some angular definitions of the various components of the wave function (10.29) can be introduced as

$$\begin{aligned} c_1 &= e^{i\gamma}cos(\vartheta/2), \\ c_2 &= e^{i(\gamma+\varphi)}sin(\vartheta/2), \end{aligned} \tag{10.37}$$

so that

$$\begin{aligned} P_x &= cos(\varphi)sin(\vartheta) \\ P_y &= sin(\varphi)sin(\vartheta). \\ P_z &= cos(\vartheta) \end{aligned} \tag{10.38}$$

These angles may be recognized as the angles in a spherical coordinate system which relates to the normal rectangular coordinates. The angle ϑ is the polar angle and the angle φ is the azimuthal angle. Hence, moving around within the state ψ means that moving around on the surface of a sphere by varying the two angles. This sphere is called the *Bloch sphere* and is pictured in Figure 10.1. From the definition of our spinors and the above coefficients, we recognize that our states may be defined as

$$\begin{aligned} \beta &= |\downarrow\rangle, \quad \vartheta = 0, \quad P = a_z, \\ \alpha &= |\uparrow\rangle, \quad \vartheta = \pi, \quad P = -a_z. \end{aligned} \tag{10.39}$$

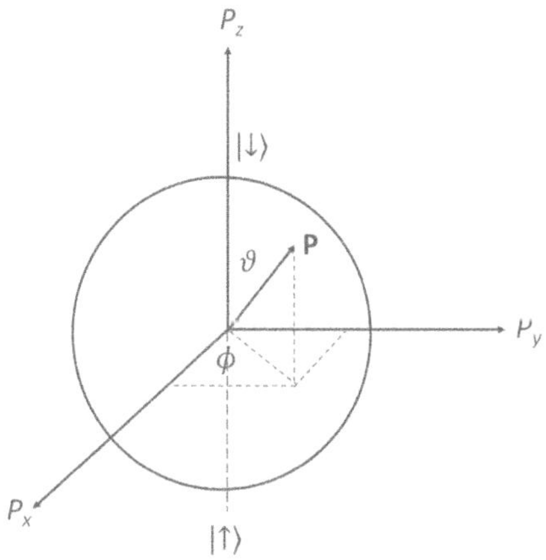

FIGURE 10.1 The Bloch sphere. All spin rotations are constrained to lie on the surface of this sphere, since the wave function is normalized to unity. The sphere has unit radius and the polarization vector moves on the surface of the sphere.

One of the properties of the spin that is familiar is the fact that it precesses under a variety of different forces. This can be investigated by introducing a simple form for the Hamiltonian, that is suggested by the density matrix itself. This form is written as

$$H = \frac{1}{2}\left(Q_0 I + \boldsymbol{Q}\cdot\sigma\right). \tag{10.40}$$

The time rate of change of the polarization is then given by the well-known relation that has been developed as

$$\begin{aligned} \frac{d\boldsymbol{P}}{dt} &= \frac{d\langle\sigma\rangle}{dt} = \frac{1}{i\hbar}\langle\sigma H - H\sigma\rangle \\ &= \frac{1}{2i\hbar}\left\langle\sigma\left(\boldsymbol{Q}\cdot\sigma\right) - \left(\boldsymbol{Q}\cdot\sigma\right)\sigma\right\rangle \\ &= \frac{1}{2i\hbar}\left\langle\boldsymbol{Q}\times\left(\sigma\times\sigma\right)\right\rangle = \frac{1}{\hbar}\boldsymbol{Q}\times\langle\sigma\rangle \end{aligned} \tag{10.41}$$

If σ was a simple vector, the last term in the parentheses of the third line would vanish, but this is not a simple vector—it is a vector of tensors, so it behaves rather differently. In fact, the result is easy to obtain, as

$$\left(\sigma\times\sigma\right) = \begin{vmatrix} a_x & a_y & a_z \\ \sigma_x & \sigma_y & \sigma_z \\ \sigma_x & \sigma_y & \sigma_z \end{vmatrix} = 2i\sigma. \tag{10.42}$$

This now leads us to the result

$$\frac{d\boldsymbol{P}}{dt} = \frac{1}{\hbar}\boldsymbol{Q}\times\boldsymbol{P}. \tag{10.43}$$

So, the precession of the spin polarization arises from any vector term in the Hamiltonian that is not parallel to the polarization itself. It is easy to show that this motion does not change the amplitude of the polarization, which is of course required for the spin itself. If the vector $\boldsymbol{Q}$ is a constant amplitude vector, then it defines a precession energy corresponding to a precession frequency $\omega_Q = Q/\hbar$.

10.3 SYSTEMS OF IDENTICAL PARTICLES

At the beginning of this chapter, the multi-particle wave function was introduced. In particular, this wave function was shown to be expressable with products of single-particle wave functions through the Slater determinant that ensured the anti-symmetry properties of the multi-particle wave function for fermions. Here, it is desired to expand this topic, and to explore some of the properties of the multi-electron wave functions in further detail.

Quite generally, we may introduce the N-particle wave function $\psi_N(x_1, x_2, \ldots, x_N)$, which represents the probability (when the magnitude squared is computed) of finding particles at positions and spins x_1, x_2, … , x_N. This wave function must satisfy the condition

$$\langle \psi_N | \psi_N \rangle = \int dx_1 dx_2 \ldots dx_N \left| \psi_N(x_1, x_2, \ldots, x_N) \right|^2 = 1. \tag{10.44}$$

As has been mentioned previously in Chapter 6, the coordinate space, often called a Hilbert space, for the N-particle system is simply the Nth tensor product of the single-particle Hilbert spaces with their corresponding spin spaces. The wave function of the N fermions is properly anti-symmetric under the exchange of any two particles, and therefore for a large number of permutations must satisfy

$$\psi_N(x_1', x_2', \ldots, x_N') = (-1)^P \psi_N(x_1, x_2, \ldots, x_N). \tag{10.45}$$

The P in this last equation is the total number of permutations of positions that is required to match the configuration of the wave function on the left with that on the right of (10.45). Thus, if we have for example, $\psi_4(x_1, x_2, x_3, x_4)$, then to get to $\psi_4(x_4, x_3, x_2, x_1)$ requires a total of six permutations (three to get x_4 moved to the beginning, then two to bring x_3 adjacent to it, and finally one more to interchange x_2 and x_1). Hence, this new example yields a symmetric product after the six interchanges. The general permuted wave function (10.45) then has the inner product with the original of

$$\langle \psi_N(x_1', x_2', \ldots, x_N') | \psi_N(x_1, x_2, \ldots, x_N) \rangle = (-1)^P. \tag{10.46}$$

It is important to note that the normalization of the many-electron wave function (10.44) arises from the normalization of the single-electron wave functions that go into the Slater determinant (10.9). The factor of $(-1)^P$ arises from the need to get these into the "right" product order for taking the associated inner products of the single-electron wave functions. The normalization factor in (10.9) then goes to cancel the multiplicity of terms that arise from the determinantal form. Let us consider just the two-electron case of (10.8) as an example (remember that the adjoint operator on the left of the inner product reverses the order of terms)

$$\begin{aligned}
\langle \psi_2 | \psi_2 \rangle &= \frac{1}{2} \left\langle \left[\psi_1(x_1)\psi_2(x_2) - \psi_2(x_1)\psi_1(x_2) \right] \left[\psi_1(x_1)\psi_2(x_2) - \psi_2(x_1)\psi_1(x_2) \right] \right\rangle \\
&= \frac{1}{2} \Big[\left\langle \psi_1(x_1)\psi_2(x_2) | \psi_1(x_1)\psi_2(x_2) \right\rangle + \left\langle \psi_2(x_1)\psi_1(x_2) | \psi_2(x_1)\psi_1(x_2) \right\rangle \\
&\quad - \left\langle \psi_1(x_1)\psi_2(x_2) | \psi_2(x_1)\psi_1(x_2) \right\rangle - \left\langle \psi_2(x_1)\psi_1(x_2) | \psi_1(x_1)\psi_2(x_2) \right\rangle \Big] \\
&= \frac{1}{2} \left[(-1)^0 + (-1)^0 - (-1)^1 0 - (-1)^1 0 \right] = 1
\end{aligned} \tag{10.47}$$

The zeroes arise because the wave functions at the same variables are different and orthogonal to one another. Thus, one needs to be careful that the one-electron wave functions are orthonormal and that the permutations are properly computed.

Finally, it is usually the case that the coordinates are the same for all of the single-electron wave functions; only the spin coordinates differ. Thus, the difference between the single-particle states is the particular member of the Hilbert space and the particular spin variable that is excited for each electron. No two combinations can be occupied by more than one electron. With this realization, it is possible to think of a simple set of parameters to characterize each single-electron wave function—its index in the Hilbert space and its spin angular momentum. Thus, we may write the single-particle wave functions as $\psi_{i\sigma}(x)$. Here, the index i signifies the particular member of the basis set, and σ denotes the spin state. These may be combined for simplicity into the index $\lambda = (i, \sigma)$. This will be done in the following discussion. However, this notational change is a major change of paradigm. In the case treated above, it was assumed that the coordinates (including spin) of a particle defined a particular localized wave function in the position representation, and the index of the wave function described the type of wave function at that position. Thus, at the point $\xi = (x, \sigma)$, there may be many types of wave function available, and the $\psi_{i\sigma}$ correspond to these types. In the latter description, however, it is assumed that there is only a single type of wave function at each site, or that there is no localization and that there are a number of wave function types. An example of the first of these options is a Gaussian wave packet localized at x. This is characteristic of, for example, localized delta function wave packets. The latter formulation is characteristic of, for example, the wave functions of an electron in a quantum well, where the "space" is the region that lies within the well, and the various wave functions are those that arise from the different energy levels. A second example of the latter, and one that will be heavily used, is that of momentum eigenfunctions, which are plane waves extending over all space. The indices are then the momentum values and the spin indices. It is important in evaluating the anti-symmetrized wave functions to understand fully just which interpretation is being placed on the individual single-electron wave functions.

In the treatment of the harmonic oscillator, it was found to be useful to introduce a set of non-commuting operators which described the creation or annihilation of one unit of energy in the harmonic oscillator. This also changed the wave functions accordingly. Can such a similar description be used to enhance the understanding of the many-electron picture? The answer is obviously yes, but it must be carefully examined as the rules will be changed by the requirement of anti-symmetry and the corresponding Pauli exclusion principle. New views that arise will be based upon the fact that the energy of the harmonic oscillator could be continuously raised by pumping bosons into the system; the introduction of n bosons could be achieved by using the operator $(a^\dagger)^n$. Here, however, operation with a single creation operator (the notation c for will be used for fermions) $c_\lambda^\dagger$ creates one fermion in state λ. This state may be a momentum eigenstate, a state in a quantum well, or any other state in a system in which the wave functions span the entire allowed variable space. If an attempt is made to create a second fermion in this state, the result must be forced to vanish because of the exclusion principle; for example,

$$\left(c_\lambda^\dagger\right)^2 |0\rangle = 0. \tag{10.48}$$

In this last expression, $|0\rangle$ is the so-called vacuum state in which there are no fermions present. In fact, however, Equation (10.48) must hold for any wave function in which the state λ is empty or even filled. This carries over to

$$\left(c_\lambda\right)^2 |\ldots\rangle = 0 \tag{10.49}$$

whatever the initial state happens to be. If the state is full, with an electron in it, it cannot be emptied twice (remember, these are spin states, so only a single electron is allowed in the state). If it is empty, it cannot be emptied again.

This suggests that a different combination of products must be used for these fermion operators. Consider, for example, the product

$$c_\lambda^\dagger c_\lambda + c_\lambda c_\lambda^\dagger. \tag{10.50}$$

If the state $|\lambda\rangle$ is empty, the first term immediately gives zero. The second term creates a fermion in the state, then destroys it, so the result is (0 + 1) = 1. Similarly, if the state $|\lambda\rangle$ is occupied, the second term gives zero (another fermion cannot be created), while the first term is the number operator and yields 1. Thus, the result in either case is the *anti-commutator* product

$$\left\{c_\lambda^\dagger, c_\lambda\right\} = c_\lambda^\dagger c_\lambda + c_\lambda c_\lambda^\dagger = 1. \tag{10.51}$$

The curly brackets are used here to indicate that the positive sign is used in the anti-commutator, as opposed to the negative sign used in the commutator relation. This may be extended to operators in other states as

$$\left\{c_\lambda^\dagger, c_\mu\right\} = \delta_{\lambda\mu}, \quad \left\{c_\lambda^\dagger, c_\mu^\dagger\right\} = \left\{c_\lambda, c_\mu\right\} = 0. \tag{10.52}$$

An important point in this discussion is that the number operator is idempotent: that is, $n_\lambda = n_\lambda^2 = (c_\lambda^\dagger c_\lambda)^2 = c_\lambda^\dagger c_\lambda (1 - c_\lambda c_\lambda^\dagger) = c_\lambda^\dagger c_\lambda$.

The operators $c_\lambda^\dagger$ and c_λ used here are fermion operators. The relationship (10.48) ensures that no more than a single fermion can exist in the given state. The attempt to put a second particle in this state must yield zero, as given by (10.48). Similarly, only a single fermion can be removed from an occupied state, as indicated by (10.49). These statistics differ markedly from the operators used in previous chapters, which were for bosons. While the wave function differed for each state, we referred to a particular harmonic oscillator mode as being described by its occupation. In increasing the occupation, we raised the energy of the mode (given by the number of bosons in that particular harmonic oscillator mode). Thus, there could be a great many bosons occupying the mode, even though the distinct mode wave function would change. In the case of fermions, however, the range of the occupation factor n_i is limited to only 0 or 1. This is imposed by using the same generation properties of the creation and annihilation operators, but with the limitations for fermions that are given by (10.48) and (10.49).

The general many-electron wave function is created by operating on the empty state, or vacuum state, with the operators for positioning the electrons where desired. For example, a three-electron state may be created as

$$\left|\mu\nu\lambda\right\rangle = c_\lambda^\dagger c_\nu^\dagger c_\mu^\dagger \left|0\right\rangle. \tag{10.53}$$

It should be noted that the order of creation of the particles is important, since changing the order results in a permutation of the indices, with its accompanying minus sign(s). Notice that the right-most creation operator creates the first electron state, while the second operator adds the second electron, and so on. It is possible to recognize that the creation and annihilation operators do not operate in a simple Hilbert space. In general, the creation operator operates in a space of n electrons and moves to a space with $n + 1$ electrons. Similarly, the annihilation operator moves to a space with $n - 1$ electrons. The general space may then be a product space of Hilbert spaces, and elements may be combined with a variety of partially occupied wave functions. This complicated structure is called a *Fock space*, but the details of this structure are beyond the simple treatment that we desire here.

10.4 SPIN EFFECTS IN SEMICONDUCTORS

The spin of an electron can be manipulated in a large number of ways. Now, it may fairly be asked why would one want to do such a thing. Of course, spin is a quantum property, with two states: up and down. In the new world of quantum computing, this is an ideal system with which to replace the classical logic bit, which is either zero or one. And, indeed, it will be seen in Chapter 11 that a majority of current effort in quantum computing is focused on the spin qubit (for quantum bit).

In order to take advantage of current semiconductor processing technology, it would be preferable to find a purely electrical means of achieving control of the spin and its direction. For this reason, a great deal of early attention has centered on the spin Hall effect in semiconductors. The idea was apparently first suggested by Dyakonov and Perel (1971) and later, and independently, by Hirsch (1999). The basic idea was that, in the presence of scattering centers, it was possible for spins of one orientation to be scattered in a different direction than spins of the opposite orientation. This would lead different spins to accumulate on opposite sides of the sample, a result of the presence of anisotropic scattering in the presence of the spin–orbit interaction. Thus, a transverse spin current arises in response to a longitudinal charge current, without the need for magnetic materials or externally applied magnetic fields. The spin–orbit interaction is a common part of the energy bands of a semiconductor, where it splits the otherwise triply degenerate top of the valence band at the Γ point. But there are other forms of the spin–orbit interaction that are of interest in situations in which symmetries are broken in the semiconductor device. In the spin Hall effect, edge states are created, as in the quantum Hall effect (Section 5.4.3), but here these edge states are spin polarized. The impurity-driven spin separation is known as the extrinsic spin Hall effect, and there can be an intrinsic spin Hall effect directly from the spin–orbit interaction when asymmetries exist in the device.

10.4.1 The Spin–Orbit Interaction

The quantum structure of atoms can lead to the angular momentum of the electrons mixing with the spin angular momentum of these particles. Since the energy bands are composed of both the *s*- and *p*-orbitals of the individual atoms in many of the semiconductors (Section 9.3), it has been found that the spin–orbit interaction affects band structure calculations. The spin–orbit interaction is a relativistic effect in which the angular motion of the electron interacts with the gradient of the confining potential of the atom to produce an effective magnetic field. This field couples to the spin in a manner similar to the Zeeman effect. Early papers, which used first-principles calculations of the band structure, clearly demonstrated that the spin–orbit interaction was important for the detailed properties of the bands. Not the least of these effects is the splitting of the threefold bands at the top of the valence band, producing the so-called split-off valence band (Kane, 1957). This latter band lies from a few meV to a significant fraction of an eV below the top of the valence band in various semiconductors.

The spin-orbit interaction modifies the transport by introducing an additional term to the Hamiltonian for the electrons, even in a periodic potential $V(\mathbf{r})$, as

$$H(p,r) = \frac{p^2}{2m_0} + V(r) + \frac{\hbar}{4m_0^2c^2}(\sigma \times \nabla V)\cdot \boldsymbol{p}, \tag{10.54}$$

where $\boldsymbol{p}$ is the normal momentum operator and the last term represents the spin–orbit interaction. The quantity σ is the normal Pauli spin vector, whose individual elements are the spinors of (10.23). It is clear from these matrices that the wave function is now more complicated as it includes a spin component, defined in (10.10). The direct effect of the spin–orbit interaction may be qualitatively understood as an extra energy cost for the alignment of the intrinsic magnetic moment of the electron with the magnetic field that arises from its own orbital motion. As a result, this term leads to a modification of the momentum of the carriers, through a gauge transformation of the wave function much like the Peierls' modification from the presence of the magnetic field. The most dominant effect is the separation of one of the three degenerate levels of the valence band at $k = 0$ to a lower energy. This lower energy is separated from the valance band maximum by what is known as the spin-orbit energy, which is about 44 meV in Si (Ferry, 2020).

10.4.2 Bulk Inversion Asymmetry

Bulk inversion asymmetry arises in crystals which lack an inversion symmetry, such as the zinc-blende materials. In these crystals, the basis pair at each lattice site is composed of two dissimilar atoms, such as In and P, or Ga and Sb. Because of this, the crystal has lower symmetry than, for example, the diamond lattice, where the basis pair is two Si atoms. Without this inversion symmetry, one still can have symmetry of the energy bands $E(\mathbf{k}) = E(-\mathbf{k})$, but the periodic part of the Bloch functions no longer satisfies $u_k(\mathbf{r}) = u_k(-\mathbf{r})$. As a result of this, the normal twofold spin degeneracy is no longer required throughout the Brillouin zone (Dresselhaus, 1955), as the spin–orbit interaction can split the energies of these two electrons. The importance of this interaction is already recognized by people who study the electronic band structure. The inclusion of this term via the spin–orbit interaction leads to the warped surface of the valence bands (Ferry, 2020). For the conduction band, the perturbing term that is added to the Hamiltonian arises from the added term in (10.54) and can be written in expanded form as

$$H_{BIA} = \eta\left[\left\{k_x, k_y^2 - k_z^2\right\}\sigma_x + \left\{k_y, k_z^2 - k_x^2\right\}\sigma_y + \left\{k_z, k_x^2 - k_y^2\right\}\sigma_z\right], \tag{10.55}$$

where k_x, k_y, and k_z are aligned along the [100], [010], and [001] axes, respectively. The terms in curly brackets are anti-commutation relations given by (10.51). The parameter η comes from the spin–orbit interaction and is given by Sakarai (1967) as

$$\eta = \frac{4i}{3}PP'Q\left[\frac{1}{(E_G + \Delta)(\Gamma_0 - \Delta_c)} - \frac{1}{E_G\Gamma_0}\right], \tag{10.56}$$

and the various terms defined by the band structure and wave functions of the material of interest. Here, E_G and Δ are the energy gap and the spin-orbit splitting of the valence band in a direct-gap material such as GaAs (1.4 eV and 0.34 eV, respectively). Γ_0 is the splitting of the two lowest conduction bands at the zone center (~3.1 eV in GaAs) and Δ_c is the spin-orbit splitting of the lowest conduction band at the zone center (usually only a few meV or less). The quantities P, P', and Q are matrix elements in the spin–orbit interaction (Ferry, 2020) and have the general form

$$P = \frac{i\hbar}{m_0}\langle S|p_z|Z\rangle, \tag{10.57}$$

where P and P' refer to the two atoms of the unit cell, and Q an interaction within the valence band. A simpler form of (10.55) is used by Kane (1957) in which

$$\eta = \frac{4i}{3}P\left[\frac{1}{(E_G + \Delta)} - \frac{1}{E_G}\right], \tag{10.58}$$

where P is now an average over the two atoms. The interaction is stronger in materials with small band gaps, as may be inferred from Equation (10.55). Note that (10.55) is cubic in the magnitude of the wave vector and is often referred to as the k^3 term.

While the above expressions apply to bulk semiconductors, much of the interest in recent years has been directed at quasi-two-dimensional systems in which the carriers are confined in a quantum well such as exists at the interface between AlGaAs and GaAs (see Section 2.4). Often, this structure is then patterned to create a quantum wire. For example, a common configuration is with growth along the [001] axis, so that there is no net momentum in the z-direction, and $\langle k_z\rangle = 0$, while

$\langle k_z^2 \rangle \neq 0$ is a representation of the quantization energy in the z-direction. Then, Equation (10.55) can be written as

$$H_{BIA} = \eta\left[\langle k_z^2 \rangle\left(k_y\sigma_y - k_x\sigma_x\right) + k_x k_y\left(k_y\sigma_x - k_x\sigma_y\right)\right]. \tag{10.59}$$

This is an important result. The prefactor of the first term in the square brackets is constant and depends upon the material and the details of the quantum well. This average over the z-momentum corresponds to the different sub-bands in the quantum well, so that only a single value will result when the carriers are only in the lowest sub-band. But, this structure has now split Equation (10.55) into a k-linear term and a k^3 term.

To explore (10.59) a little closer, a set of spinors will be chosen to represent the spin-up and -down states as

$$\begin{aligned} |+\rangle = |\uparrow\rangle &= \begin{bmatrix} 1 \\ 0 \end{bmatrix}, \\ |-\rangle = |\downarrow\rangle &= \begin{bmatrix} 0 \\ 1 \end{bmatrix}, \end{aligned} \tag{10.60}$$

as in (10.10). Then, the linear first term in (10.59) gives rise to an energy splitting according to

$$\Delta E_1 \sim -\langle \eta k_z^2 \rangle\left(k_x \pm i k_y\right) \tag{10.61}$$

in the rotating coordinates discussed in Section 10.1. Now, the spin "up" state rotates around the z-axis in a right-hand sense (with the thumb in the z-direction) with the spin polarization tangential to the constant energy circle in two dimensions. On the other hand, the spin "down" state rotates in the opposite direction, but with the spin polarization still tangential to the energy circle (this rotation will be illustrated below).

If the cubic terms in momentum in the last term of (10.59) are ignored for the moment the Hamiltonian for the two spin states becomes

$$H = \begin{bmatrix} \dfrac{\hbar^2 k^2}{2m^*} & -\langle \eta k_z^2 \rangle\left(k_x + i k_y\right) \\ -\langle \eta k_z^2 \rangle\left(k_x - i k_y\right) & \dfrac{\hbar^2 k^2}{2m^*} \end{bmatrix}, \tag{10.62}$$

where it has been assumed that the normal energy bands are parabolic for convenience. Then, the energy levels of the two states are given as

$$E_{\pm} = \frac{\hbar^2 k^2}{2m^*} \pm \eta\langle k_z^2 \rangle k. \tag{10.63}$$

Thus, not only is the energy splitting linear in k, but it is also isotropic in the two-dimensional momentum space. The resulting energy bands for the two states are composed of two interpenetrating paraboloids, and a constant energy surface is composed of two concentric circles (see Figure 10.2). The inner circle represents the positive sign in the above equation, while the outer circle corresponds to the negative sign. The eigenfunctions are no longer pure spin states but are an admixture given by

$$\varphi_z^{(\pm)} = \frac{1}{\sqrt{2}}\begin{bmatrix} 1 \\ e^{\pm i\vartheta} \end{bmatrix}, \tag{10.64}$$

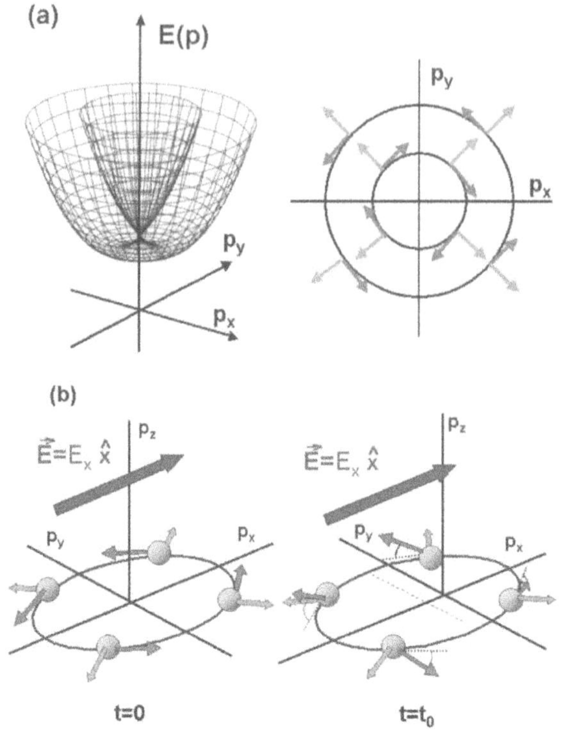

FIGURE 10.2 (a) Energy structure for the spin-orbit coupled system. The momentum is shown by light arrows and the spin polarization is shown by the dark arrows. (b) When an electric field is applied for some time, the spin is rotated out of the plane according to the direction of the transverse momentum. (Reprinted with permission from Sinova *et al.* (2004). Copyright 2004, the American Physical Society.)

where ϑ is the angle that k makes with the [1 0 0] axis of the underlying crystal in the heterostructure quantum well. This angle is the angle defined by the polar coordinates in the two-dimensional momentum space. Hence, the root with the upper sign in (10.63), which we think of as being mostly spin "up," has the spin polarization tangential to the inner circle. Correspondingly, the root with the lower sign in (10.63), which we think of as mostly spin "down," has the spin polarization tangential to the outer circle. If the cubic terms are now added in from (10.59), the energy levels of (10.63) are modified to be

$$E_{\pm} = \frac{\hbar^2 k^2}{2m^*} \pm \eta \langle k_z^2 k \rangle \left[1 + \left(\frac{k^4}{\langle k_z^2 \rangle^2} - 4\frac{k^2}{\langle k_z^2 \rangle} \right) sin^2\vartheta cos^2\vartheta \right]^{1/2} . \tag{10.65}$$

This is a much more complicated momentum and angle dependence but suggests that the constant energy circles are now warped in the same manner as the valence band constant energy surfaces. The transport is no longer isotropic in the transport plane. Similarly, the phase on the down spin contribution to the eigenfunction (10.64) is no longer simply the angle but acquires a "wobble" as it rotates around the circle.

10.4.3 Structural Inversion Asymmetry

The spin Hall effect most commonly originates from the Rashba form of spin–orbit coupling (Bychov and Rashba, 1984), which is present in a two-dimensional electron gas formed in an asymmetric semiconductor quantum well. Such an asymmetric quantum well is the quasi-triangular well found at the interface between AlGaAs and GaAs or at the interface of a Si MOSFET shown in Section 2.4, although the spin–orbit interaction is small in Si (but still very important). This is known as *structural inversion asymmetry.* In either of the quantum wells mentioned above, the structure is asymmetric around the heterojunction (or oxide) interface. As a result, there is a relatively strong electric field in the quantum well, and motion normal to this can induce an effective magnetic field. This is the structural inversion asymmetry known as the Rashba effect. Just as in the bulk inversion asymmetry, the electric field in the quantum well can lead to spin splitting without any applied magnetic field, due to the spin–orbit interaction. If we take the z-axis as normal to the heterojunction interface, then the spin–orbit interaction Equation (10.55) can be written as

$$H_{SIA} = \xi\sigma\cdot(\mathbf{k}\times\nabla V)\rightarrow\left[\sigma\cdot(\sigma\times\mathbf{k})\right]_z, \tag{10.66}$$

where

$$\xi = \frac{P^2}{3}\left[\frac{1}{E_G^2}-\frac{1}{(E_G+\Delta)^2}\right]+\frac{P'^2}{3}\left[\frac{1}{\Gamma_0^2}-\frac{1}{(\Gamma_0+\Delta_c)^2}\right] \tag{10.67}$$

arises from the spin–orbit interaction in the creation of the band structure, and the symbols have the same meanings as in (10.56). Again, the simpler version of Kane (1957) would neglect the last term in (10.67). The parameter ξ is composed of the constants and the electric field, and for the configuration discussed, we have

$$H_{SIA} = \alpha_z\left(k_y\sigma_x - k_x\sigma_y\right), \tag{10.68}$$

where

$$\alpha_z = \frac{1}{4}\left(\frac{\hbar}{m_0c}\right)^2\frac{\partial V}{\partial z} \tag{10.69}$$

is the normalized electric field. If we continue to use the spinors defined in (10.60), then the Rashba contribution to the energy is simply

$$E_R = \mp i\alpha_z\left(k_x \pm ik_y\right), \tag{10.70}$$

where the signs correspond to those in (10.60).

While the spin states are separated in energy, this does not simply add to the bulk inversion asymmetry. First, the two spin states are orthogonal to each other, and then they are phase shifted (with opposite phase shift) relative to the previous results. It is easier to understand the effect of this Rashba term if we diagonalize the Hamiltonian for the two spin states with the Rashba contribution. We can write the Hamiltonian, for parabolic bands using (10.70) as

$$H = \begin{bmatrix} \dfrac{\hbar^2k^2}{2m^*} & \alpha_z\left(k_x+ik_y\right) \\ \alpha_z\left(k_x-ik_y\right) & \dfrac{\hbar^2k^2}{2m^*} \end{bmatrix}. \tag{10.71}$$

When this Hamiltonian is compared to (10.62), the two off-diagonal terms are shifted not only by the minus sign in front but also by a phase factor of $-\pi/2$ within the term in parentheses in the upper right term and $\pi/2$ in the lower left term. Nevertheless, the Hamiltonian can still be diagonalized to give the new energies

$$E = \frac{\hbar^2 k^2}{2m^*} \pm \alpha_z k, \tag{10.72}$$

where k lies in the two-dimensional layer normal to the z-axis. As in the case of the linear term in the bulk inversion asymmetry, the energy splitting is linear in k, and also isotropic with respect to the direction of **k**. Thus, the energy bands also are composed of two interpenetrating paraboloids, and a constant energy surface is composed of two concentric circles (see Figure 10.2). The inner circle represents the positive sign in (10.72), while the outer circle corresponds to the negative sign. The two eigenfunctions are given by

$$\varphi_z^{(\pm)} = \frac{1}{\sqrt{2}} \begin{bmatrix} 1 \\ e^{i(\vartheta \mp \pi/2)} \end{bmatrix}, \tag{10.73}$$

where ϑ is the angle that k makes with the [1 0 0] axis of the underlying crystal. This direction lies within the heterostructure quantum well described in the previous section. The form of (10.73) clearly shows the phase shift relative to the bulk inversion asymmetry wave function. The spin direction remains tangential to the two circles but pointed in the negative angular direction for the inner circle and in the positive angular direction for the outer circle. When both spin processes are present, the spin behavior becomes quite anisotropic in the transport plane. However, the Dresselhaus bulk inversion asymmetry is generally believed to be much weaker than the Rashba terms discussed here. The strength of the Rashba effect can be modified by an electrostatic gate applied to the heterostructure, as it modifies the potential gradient term in (10.69), which allows for a gate-controlled spin splitting, something especially useful in studies of quantum qubits and their interactions.

REFERENCES

Bychov, Y. A., and Rashba, E. I. 1984. Oscillatory effects and the magnetic susceptibility of carriers in inversion layers. *J. Phys. C: Sol. State Phys.* **17** 6039–6046.

Dresselhaus, G. 1955. Spin-orbit coupling effects in zinc-blende semiconductors. *Phys. Rev.* **149** 580–586.

Dyakonov, M. I. and Perel, V. I. 1971. Possibility of orienting electron spins with current. *JETP Lett.* **13** 467–469.

Ferry, D. K. 2020. *Semiconductors: Bonds and Bands, 2nd Ed.* (Bristol: IOP Publishing).

Hirsch, J. E. 1999. Spin Hall effect. *Phys. Rev. Lett.* **83** 1834–1837.

Kane, E. O. 1957. Band structure of Indium antimonide. *J. Phys. Chem. Sol.* **1** 249–261.

Merzbacher, E. 1970. *Quantum Mechanics*, 2nd ed. (New York: Wiley) Chapter 13.

Pauli, W. 1925. Über der Zussamenhang des Abschlusses der Eleckronengruppen in Atom mit der Komplexstructur der Spektren. *Z. Phys.* **31** 765–783.

Sakarai, J. J. 1967. *Advanced Quantum Mechanics.* (Reading, MA: Addison-Wesley).

Sinova, J., Culcer, D., Niu, Q., Sinitsyn, N. A., Jungwirth, T., and MacDonald, A. H. 2004. Universal intrinsic spin Hall effect. *Phys. Rev. Lett.* **92** 126603.

Slater, J. C. 1929. The theory of complex spectra. *Phys. Rev.* **34** 1293–1322.

Zeeman, P. 1897. On the influence of magnetism on the nature of light emitted by a substance. *Phil. Mag., Ser.* 5, **43** 226–239.

PROBLEMS

1. Consider two electrons in a state in which the radius (around some orbit center) is set normalized to unity, and the angular wave function is $\phi(\theta) = f\,[\cos(\theta)]$. If these two electrons have opposite spins and are located at $f\,[\cos(\theta)]$ and $f\,[\cos(\theta + \pi)]$, discuss the anti-symmetric properties of these electrons. Can you infer the nature of the function describing the angular variations?

2. Show that the number operator

$$\langle n\rangle = \sum_i \langle c_i^\dagger c_i\rangle$$

commutes with the Hamiltonian.

3. Consider a perturbation that scatters an electron from state k to state k' while creating or annihilating one unit of lattice vibration. In Section 5.3, it was shown how the vibrating lattice could be represented in Fourier space as a summation of a set of harmonic oscillators. Devise a perturbing potential that accounts for the scattering of the above electron by the lattice (it will take three operators, two for the electrons and one for the boson).

11 An Introduction to Quantum Computing and Information

The idea of using quantum mechanics to provide more power than that available from classical computers seems to have begun around 1981–82 (Benioff, 1982), but Deutsch (1985) put the concept on a sounder footing a short time later when he reformulated the Turing (1937) thesis with the assertion that every finite realizable physical system can be perfectly simulated by a universal model computing machine operating by finite means. However, he asserted that the output of a quantum machine, although fully determined by the input state, is not an observable itself. Hence, the user must provide some cleverness if this state is to be determined. Benioff pointed out the need for the quantum state to experience interference, some indeterminism, and was what we would today call an entangled state. He also realized that non-dissipative evolution was unphysical, so immediately it became clear that a great deal of effort would be required to minimize the decoherence of the physical states. Deutsch (1989) then provided some input as to what type of states or systems might be usable in this process. Jozsa (1991) mathematically showed that one could associate a quantum state to each function that one desired to evaluate, and that this should lead to a rapid solution of problems via the quantum computer (Deutsch and Jozsa, 1992). Since these introductions, the world has progressed rapidly into a determined search for applications of quantum mechanics throughout the information processing world.

Progress beyond mathematical approaches came later when Lloyd (1993) suggested that arrays of weakly coupled quantum systems might be able to serve as the quantum computer if they were subjected to a sequence of controlled electromagnetic pulses, which could place the quantum bits into the desired superpositions. This shifted the problem to arranging for the superpositions, and the need for suitable *algorithms*. Perhaps the first such algorithm came from Peter Shor (1994), who introduced the idea of the quantum Fourier transform, a key element in breaking many encryption techniques that utilized large numbers generated by the product of prime numbers. Another algorithm, for fast data base searches, was developed by Grover (1996). Now, it turns out that, some three decades later, very little progress in creating and using quantum computers has been made, particularly with a lack of new algorithms.

Our purpose here is not to cover the world of quantum computers. There are many other books that do this just fine (Ferry, 2025; Marinescu and Marinescu, 2004; Nielsen and Chuang, 2000). However, the quantum mechanics of these processes is a natural evolution of the quantum mechanics discussed in the previous ten chapters. What we want to deal with is the quantum mechanical ideas of what creates the heart of the quantum computer, and how it functions with our normal approaches discussed in the previous chapters. We will begin with the idea of the quantum mechanical replacement for the computer bit—the *qubit*, short for quantum bit. Then we will discuss some current implementations of circuits and materials that create one or more qubits.

Before proceeding, however, it has to be realized that a fair amount of thought on quantum computing is really not correct. As pointed out above, Deutsch (1985) extended the ideas of Turing (1937). Hence, it is important to point out that the Turing machine is irreversible for a great many reasons, so that it is most likely that a realizable quantum computer will also likely be irreversible as pointed out by Benioff (1982). In describing his machine, Turing formulated basic principles of operation. These required that the sequences of operations must be "circle-free" and must have definite start and stop states. He used a blank space on the memory tape to signal that the stop state. This requirement meant that the computation had to proceed to a numerical answer and could not wind up in an endless loop. Turing pointed out that a force is required so that the machine proceeded toward the stop state. This force, by itself, meant that the machine was not symmetric in time and

DOI: 10.1201/9781003717195-11

therefore was irreversible. Suggestions that a quantum computer could be reversible, just as in text-book quantum mechanics, are misguided.

11.1 BITS AND QUBITS

The quest for quantum computers begins with the qubits. Normal (classical) computers work on the use of logical bits, which have values of 0 and 1. These are digital bits, and their use has expanded to the new quantum computers. However, there is a difference, as quantum mechanics lives in the analog world. Hence, qubits are analog—they extend beyond the digital world. This creates some advantages and some disadvantages. The differences will be expanded in the first subsection below, and the following paragraphs will describe the interactions and entanglement in qubits.

11.1.1 Bits

While it may not appear to be the case, due to the similarity of name between bit and qubit, these two quantities are quite different. To begin with, the normal bit, the basis of our binary computation in everyday classical computers, has only two values, 0 and 1, which form the binary base. While the qubit also has a 0 and a 1, it is analog in that we may define arbitrary complex values in any manner in which the magnitude of the qubit remains 1. But the earliest computers relied on base 10 computation because that is the normal way in which the world does arithmetic. Nevertheless, it was realized that the electronic devices within the computer had mainly "on" and "off" states, so even the base 10 numbers were implemented with binary numbers.

It was John von Neumann (1945) who suggested that the base 2 binary system should be used for simplicity in building computers. He also suggested the stored program approach, which along with binary arithmetic, formed the idea of the von Neumann architecture. Essentially, all computers from this time forward have used the binary system, as it is eminently suitable to two-state bits. In Figure 11.1, the idea of the bit/qubit is presented as two orthogonal axes in a configuration space (for reasons that will become apparent later). In this space, we define the 2 unit values for the bit as the states 0 and 1. Then, using quantum notation, we can define this space, and the bit, as the function

$$\psi = a|0\rangle + b|1\rangle, \quad a,b \subset (0,1) \ni a+b=1. \tag{11.1}$$

The notation on the right says that a and b are either 0 or 1, but both cannot be the same. Either $a=1, b=0$ or $a=0, b=1$. Hence, the bit value is merely 0 or 1, and not anything in between.

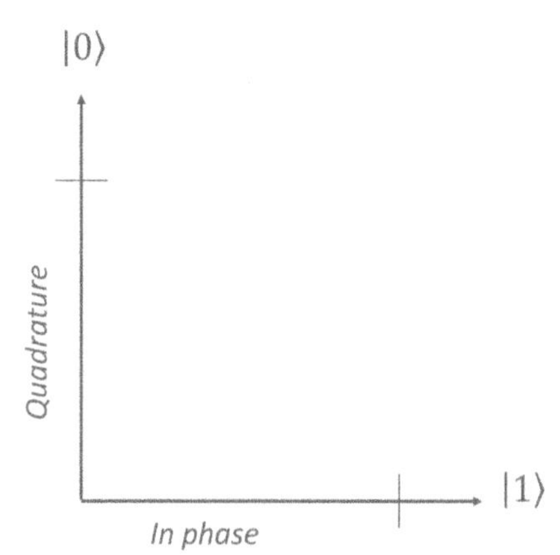

FIGURE 11.1 The coordinate set for an arbitrary bit, showing how the axes correspond to the values 0 and 1. The bit must be oriented along one of these two axes, as the value must be either 0 or 1.

Typically, in a computer, these levels are defined by either the system ground ($V = 0$), or a proper bias voltage usually denoted as V_{DD}. The secret of modern microchips is that the core ingredient, a transistor, is quite good at resetting these voltage values at each bit operation, minimizing any errors or noise (Mead and Conway, 1979). The idea of the binary system was not new to von Neumann, or to the logic world. The concept had originated almost a century earlier when George Boole (1854) proposed using it to evaluate logic values in philosophical discussions as a way to evaluate different arguments. It seemed a natural to use this system with computers created with what seemed to be devices with binary output that helped reduce noise by resetting the logic values at each stage.

In the computer itself, a particular state is defined by the values of the bits used to describe the state. Then, one can define a state transition diagram, in which each possible state is a node, and transitions between these nodes indicate the action taken when a control signal is applied. Various outputs arise from the set of transitions and the state of the machine after the transition. Hence, an algorithm is a set of instructions that guide the system through a set of transitions, changing the various bits under the guidance of the control signals, to a desired result. Good algorithms lead to the machine halting at the end of the algorithm and yield a desired output (and with bad algorithms, the machine sometimes never stops). In fact, in the Turing (1937) model, it was pointed out that a valid computer must have a "stop" state and must not go into interminable "loops, or rings." This same operational model will exist also in the quantum computer. That is, the qubits will define a state, while the state transitions will depend upon control signals.

11.1.2 Qubits

When we move from the bit to the qubit, the change is rather subtle, although exceedingly important. The same diagram of Figure 11.1 for the bit can be adapted to the qubit, but new properties, as shown in Figure 11.2, arise from the analog property of the qubit. There can still be the two coordinate axes representing the 0 and 1. But now the wave function is analog and can take any value on a unit circle that exists in the plane defined by the two coordinate axes. Hence, (11.1) is still valid, but now the coefficients are going to be different, as they are also analog and subject to the new constraints given by

$$\psi = a|0\rangle + b|1\rangle, \quad |a|^2 + |b|^2 = 1. \tag{11.2}$$

The constraint on the right of (11.2) means that the value of the wave function moves on the unit circle as a and b change their values (just as the state moved on the Bloch sphere in Chapter 10).

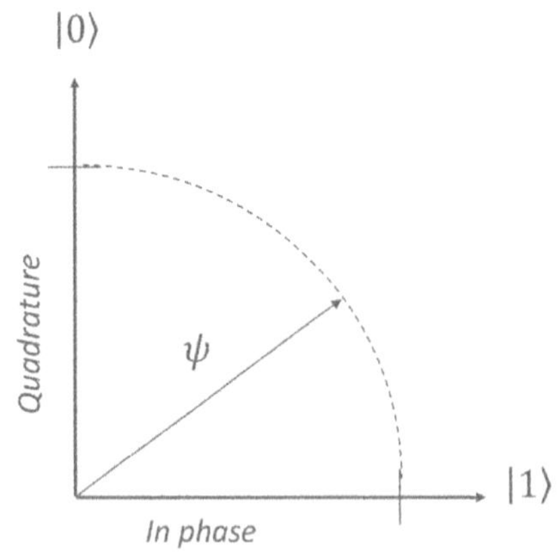

FIGURE 11.2 The value of the qubit now is analog and the magnitude of the wave function is unity, but the phases are arbitrary, but controllable.

Moreover, the two coefficients may be complex numbers, and (11.2) tells us that the net wave function must have a magnitude of 1, as required by any quantum wave function. The problem that arises is that this formulation is not completely descriptive for the actual wave function for the quantum qubit, as it says nothing about the phases of the two coefficients.

The two states $|0\rangle$ and $|1\rangle$ can be thought of as representing real energy levels of a two-level system, as discussed in Section 10.2. Equally well, they can be thought of as the two possible spin states of a fermion—spin up and spin down. Normally, these are aligned along the z-axis in spherical coordinates. From the latter concept, one may write the Hamiltonian representing this two-level system as

$$H = \frac{1}{2}\Delta\sigma_z, \tag{11.3}$$

where Δ is the energy spacing between the two levels and σ_z is a Dirac spinor, given by (10.23). Hence, we have mapped the (0,1) states into the states (−1,1), in terms of $\Delta/2$. The upper level is taken to be the 1 state and the lower level the −1 state, in keeping with the normal spin wave functions, defined in (10.10). In terms of these two states, the wave function can be written as

$$\psi = \begin{bmatrix} a \\ b \end{bmatrix} = \begin{bmatrix} |a| e^{i\phi_1} \\ |b| e^{i\phi_2} \end{bmatrix}. \tag{11.4}$$

Here, each coefficient has been given its own phase, although it is only the relative phase that becomes important, as will be seen.

It is convenient to write the wave function in a slightly different form that still characterizes the system but makes the two phases less arbitrary. For this, the two-level system is written in terms of the density matrix as

$$\rho = \begin{bmatrix} |a|^2 & ab^* \\ a^*b & |b|^2 \end{bmatrix}, \tag{11.5}$$

which may be recognized as the density matrix for the two-level system of Section 10.2, and given as (10.30), albeit with a slightly different notation. Hence, all the properties developed in this earlier section can be used to describe the qubit, particularly the polarization vector representing the direction of the qubit on the unit surface of the Bloch sphere shown in Figure 10.1. As previously, the two angles are defined as the polar angle and the azimuthal angle, where the former is the angle away from the $+z$-axis, and the latter lies in the (x,y) plane and is measured from the x-axis, as is normal for spherical coordinates. Thus, many qubit operations are easily expressible as rotations around the three axes shown in the figure. However, a slightly different notation is used with the qubits. With spin, the value can go from 1 to −1 as the cosine as the polar angle ϑ varies from 0 to π. Here, however, the value of the variables go from 0 to 1, so that the z-component of polarization (or the relevant qubit state) varies as $cos(\vartheta/2)$ to achieve this. Notice in Figure 11.3 that the state $|0\rangle$ is at $cos(\vartheta/2) = 1$, at the top of the Bloch sphere, while the $|1\rangle$ is at the bottom where $cos(\vartheta/2) = 0$. This is a reversal from the earlier notation but is common in the qubit world.

In treating the qubit as a two-level system, this could be two specific levels of a real atom, or in an artificial atom such as a quantum dot. The important point is that the phrase "two-level atom" is often a "code word" for the qubit system. This is because the energy levels in an atom are nonlinearly spaced; that is, they vary as $1/n^2$. Hence, the two desired levels are separated by an energy that is not the same as the separation of any other two levels. If the levels were linearly spaced, as in a normal harmonic oscillator, one has a problem. For example, if the lower two levels are those of interest, they are separated by exactly the same energy as any other two adjacent levels. This

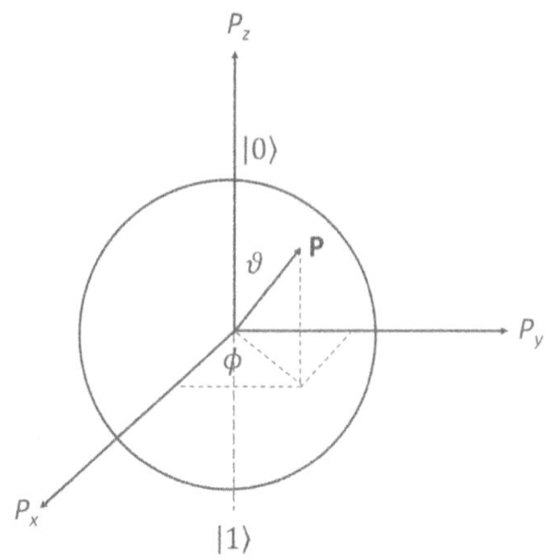

FIGURE 11.3 The Bloch sphere allows the polarization to rotate in three directions. Then, many qubit operations are expressible as rotations around one of the three axes. This is the same Bloch sphere from Figure 10.1, but the spin directions are replaced with the qubit values.

means that excitation of the lower two levels can also be absorbed by any other two levels, and this constitutes a sizable source of decoherence in the qubit. Hence, one desires the energy levels to be nonlinearly spaced, so that the two levels of the qubit do not couple to other unused energy levels. Since the atom has this property, the phrase "two-level atom" is often applied to any qubit system to indicate it has the desired properties.

11.1.3 Entanglement

Although entanglement has been discussed in previous chapters, it is good to remind ourselves of the key features that entanglement brings to a problem. Erwin Schrödinger (1935) called entanglement the most important aspect of quantum mechanics. When two particles interact, they are no longer independent of one another. Rather, they must now be described by a single entangled wave function incorporating both particles. It is this entanglement that differentiates the quantum system from the classical one, although correlation is well-known in classical systems. However, in classical systems the two correlated particles/waves remain separate in their individual descriptions. In quantum mechanics, this is no longer the case. The typical example is when a photon is absorbed and causes two electrons to be emitted. In order to conserve spin, these two electrons must have opposite spin, as well as moving in opposite directions with opposite momentum. Since the spin on either of the two electrons is unknown, it can only be assumed that either the first electron is spin up and the second is spin down, or the first electron is spin down and the second is spin up, with these two possibilities denoted by the wave function

$$\psi(1,2) = \frac{1}{\sqrt{2}}\left[|1,\uparrow\rangle|2,\uparrow\rangle - |1,\uparrow\rangle|2,\uparrow\rangle\right]. \tag{11.6}$$

In the strictest sense, this wave function describes a lack of knowledge of the spin states. However, if the spin of one of the two electrons is measured, then that measurement proscribes the opposite spin for the other particle, merely from the form of (11.6). That is, a measurement of one particle choses which of the two possibilities is correct.

Needless to say, it is entanglement that has become the crucial ingredient in quantum computing. It is this property that gives rise to the possibility of major speedup in any computation. For example, if we take the Fourier transform, the determination of the transform has to be computed once for each possible frequency component. If there are N frequencies, then we repeat the process N times, although this can be done on parallel processors to help speed it up. On the other hand, if qubits are used with the quantum Fourier transform, then in principle all N frequencies can be entangled into the wave function of each of the qubits and the transform done once. This, of course, may be the wrong answer since the wave function is a probability distribution, so we need to redo the transform many times to arrive at the most likely answer. But the theory goes that this will be much less than N times, and so using the quantum Fourier transform will speed up the computation dramatically, especially if N is a large number. This, of course, is a facetious argument. The speedup is usually compared with the discrete Fourier transform for classical systems. However, it has been known for more than half a century that a fast Fourier transform (FFT) method exists to speed up the process if the frequencies are chosen to be powers of 2 (Cooley and Tukey, 1965), a method that dramatically speeds up the computation. Interestingly enough, the quantum Fourier transform basically has the same level of difficulty as the FFT as the speedup grows from this factor and can offer gains only through the use of entanglement.

In the quantum computer, it is desired to arrange for qubits to interact with each other under controlled circumstances. When two qubits interact, they become entangled, just as the two particles discussed above. They remain entangled until some mechanism gives rise to decoherence, which breaks up the entanglement. As a result, one needs long decoherence times in the qubits. In general, the quantum computer works on controlled gates, as these gates provide the interaction between the qubits. As an example, consider the so-called CNOT gate, or controlled NOT. In the classical system, any transistor (or CMOS gate) is naturally a NOT gate. This is because the transistor inverts its input. If a high voltage is taken as a logical 1 and a low voltage as a logical 0, then a 1 input to the gate causes the transistor to turn on, which lowers the output voltage, and this gives a 0 output. Similarly, when the input voltage is low, the transistor is in the off state and the output voltage is high. Thus, the input is inverted, which means that the logical output is NOT(input). The CNOT differs from this slightly, in that the action of the transistor gate is controlled in a manner that makes the classical XOR gate (which is the "exclusive OR" gate). The XOR gate has two inputs, which are called a and b. If either a or b is high (a 1), then the output is also high. But, if both inputs are either high or low, the output of the gate is low. In other words, the output of the XOR gate is 1 if either a or b is high, but not when both are high.

In the quantum CNOT gate, one of the two inputs is taken to be the control signal, which then operates upon the other input, as shown in Figure 11.4. Here, the control input (our a, for example) is labeled c. The input wave function (our b, for example), from a previous qubit, is denoted by x. The symbol on the bottom line indicates binary arithmetic modulo 2. So, if the control bit is high, the current state of the signal is inverted (the carry is ignored). On the other hand, if the control is 0, the current state of the signal x is passed unchanged to the output. This then gives the XOR function. This is usually illustrated with a "truth table" as

$$\begin{array}{ccc} x & c & y \\ 0 & 0 & 0 \\ 1 & 0 & 1 \\ 0 & 1 & 1 \\ 1 & 1 & 0 \end{array} \cdot \tag{11.7}$$

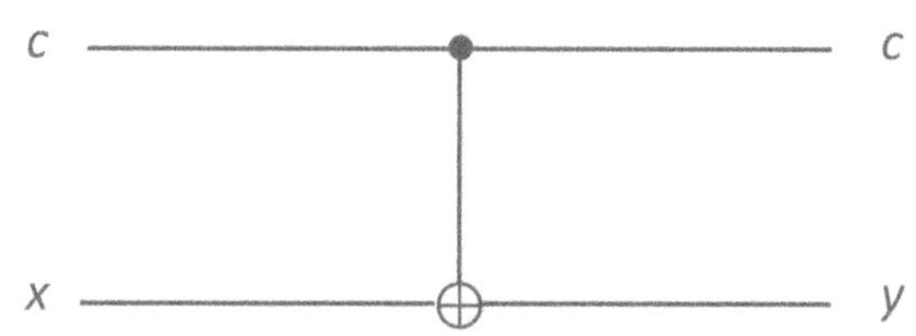

FIGURE 11.4 The CNOT gate, where c is the control input and x is the target input. The two are added modulo 2 to produce the output y.

Thus, the output y is high when either of the two inputs is high, but not when both are the same. The output can be written as a wave function, in which the state of the control bit is the first signal and the state of x is the second variable exactly with (11.6) (the sign between the terms can actually either positive or negative in the qubit case).

If we take the control bit as a two-state Hilbert space, and the input bit as a two-state Hilbert space, then the entangled output lies in a tensor product Hilbert space, denoted as $H^2 \otimes H^2$. As discussed previously, the wave function (11.6) is not a simple product, and it cannot be separated into parts that lie in only one of the two individual Hilbert spaces. This is what defines the entanglement. Hence, the wave function in (11.6) is an entangled wave function. The plus/minus sign gives two possibilities, which both give the same value of 1 for the square magnitude of the wave function (the factor of ½ provides proper normalization).

The introduction of the control signal is readily accomplished by the manner in which the interaction between qubits is accomplished and managed. To carry out an algorithm, one has to manipulate the control signals, which actually change the interactions in a specific manner so that entanglement is created, manipulated, or destroyed. This is no different than the classical computer where bit strings are pushed around, manipulated, and then stored or erased. Here, however, one has the entanglement which increases the power of individual qubits. But the qubit remains the heart of the quantum computer, and various forms of creating qubits will be discussed in the sections below. First, however, one must understand the interaction itself.

The discussion above appears to clearly indicate how the quantum CNOT gate operates in much the same manner as the classical XOR gate. But, is it all that clear? There is so much going on that it is not that obvious that a number of assumptions are being made without being discussed, and the difference between the Copenhagen interpretation and the EDS (for Einstein, de Broglie, and Schrödinger) lies at the heart of these assumptions. Consider the Copenhagen view in which nothing really exists in the gate until measurements are made. Then, the measurements of the output y and control signal c determine what the output of the gate actually is. Could these measurements actually override the inputs? Quantum mechanics is a probabilistic theory, and the CNOT gate is logically reversible—that is, if two such gates are connected in series, the output of the second gate is identically the input to the first gate (if it all works right). In short, there is only a likely chance, based upon the probability statistics of the quantum system that the output actually obeys the truth table of (11.7), and an even smaller likelihood that this occurs in the Copenhagen view. Indeed, since there is some feeling that the CNOT gate is physically reversible, the output that is measured could determine what the actual inputs turn out to be, and perhaps not what was intended at all. There is even a chance that the entire gate reduces to simple noise as it cannot determine what is the input and the output. While many still propose reversible logic, such gates require a unique next state and a unique previous state, a condition which leads to the state transition matrix describing a cyclic permutation group. This means that all the states lie on one or more rings (Ferry *et al.*, 1989), violating one of the Turing conditions for a computer.

On the other hand, quantum computers have been shown to work, albeit in the probabilistic sense. And, since the world continues to move forward, even though one finds quantum mechanics at the root of almost all modern systems, it is quite likely that reality is also found at the quantum level. Thus, there is reasonable cause to expect that the truth table of (11.7) will be found to hold, although perhaps only in the statistical sense. The word "cause" here is used explicitly to express the fact that the truth table actually suggests full deterministic behavior occurs and thus causality exists even in the quantum world. This is the EDS version of quantum mechanics. Such assumptions of causality and determinism tend to exist through any algorithm for computation regardless of whether it is classical or quantum mechanical.

So, if one has a truth table, such as (11.7), from where does the probability arise? There are many sources of uncertainty which lie at the heart of statistical theories, but one source is the gate itself. In creating an integrated circuit, the gate has a specific position; the input position and the output position are determined by the layout of the circuit on the chip. While the output position may not be known exactly (due, for example, to the length of an interconnect with the next gate(s)), it will still induce an uncertainty in the momentum of the particles defining the current or other logical state. Since it was shown in Chapter 1 that the momentum lies in the phase of the wave function, the uncertainty in the momentum affects both the time at which the output is formed and the entanglement that may be desired in the wave function. Some have suggested that uncertainty also exists at the initial conditions of the wave function, since these are analog and are themselves statistical. While correcting errors is relatively simple in classical digital logic, it is more problematic in analog quantum computation. How is an error actually recognized by error correcting qubits? Do these latter qubits also introduce further errors? These are only a few of the questions that are being asked in this world of quantum computing.

11.1.4 Entanglement of Two Harmonic Oscillators

A somewhat more detailed model of entanglement is provided when two distinct harmonic oscillators are coupled through an interaction. The quantized oscillators are the basis of Planck's (1900) original theory of black-body radiation and actually provided the quantization conditions. These harmonic oscillators were also used in quantized electromagnetic fields. The coupling of two harmonic oscillators is a classical problem in nonlinear dynamics, where the energy oscillates from one oscillator to the other and back again. The treatment of coupling quantized harmonic oscillators has been discussed previously by many, but here the approach of Kim and Iafrate (2004) will be followed. The Hamiltonian for the two coupled harmonic oscillators can be written as

$$H = \hbar\omega_1\left(a_1^\dagger a_1 + \frac{1}{2}\right) + \hbar\omega_2\left(a_2^\dagger a_2 + \frac{1}{2}\right) + \hbar\kappa\left(a_1^\dagger a_2 + a_1 a_2^\dagger\right), \tag{11.8}$$

where the operators have their normal meanings and the interaction describes the creation of a boson in one oscillator along with the annihilation of a boson in the other. Without loss of generality, it is fair to set $\omega_1 > \omega_2$. While two such oscillators appeared as the two-dimensional harmonic oscillator, these were not interacting, and the present case is more difficult. In this former case, there was only a single angular momentum, while here each oscillator may have its own angular momentum, in which the system angular momentum can be defined as

$$\begin{aligned} L_x &= \frac{\hbar}{2}\left(a_1^\dagger \otimes a_2 + a_1 \otimes a_2^\dagger\right), \\ L_y &= \frac{\hbar}{2}\left(a_1^\dagger \otimes a_2 - a_1 \otimes a_2^\dagger\right), \\ L_z &= \frac{\hbar}{2}\left(n_1 - n_2\right), \\ L_T &= \frac{\hbar}{2}\left(n_1 + n_2\right). \end{aligned} \tag{11.9}$$

The first two terms are defined in terms of a tensor product space as introduced in Chapter 10. Each creation or annihilation operator was defined in Chapter 5 as a combination of a position and a momentum operator, and this defines a direction of rotation. The angular momentum can then be understood in comparison with the two-dimensional harmonic oscillator.

In addition to the individual operators, each oscillator has its own number operator, used in the last two lines of (11.9). With these, the coupled two-oscillator system is defined completely in analogy with a two-spin system to be

$$\left|n_1 n_2\right\rangle = e^{-iL_y\gamma/\hbar}\left|n_1\right\rangle\left|n_2\right\rangle, \tag{11.10}$$

where γ is a parameter defined in terms of the coupling constant κ (whose dimension is that of a radian frequency) and the detuning $\Delta\omega$ as

$$tan(\gamma) = 2\kappa/\Delta\omega, \; \Delta\omega = \omega_1 - \omega_2. \tag{11.11}$$

The combined energy can be written as

$$E_{n_1 n_2} = \frac{\hbar}{2}(n_1 + n_2 + 1)(\omega_1 + \omega_2) + \frac{\hbar}{2}(n_1 - n_2)\Omega, \tag{11.12}$$

where an interacting hybrid frequency has been defined as

$$\Omega^2 = (\omega_1 - \omega_2)^2 + 4\kappa^2, \tag{11.13}$$

which is a combination of the detuning and the interaction parameter.

Now, consider any arbitrary function that is described in terms of the two oscillator states. Any such quantum function defined in such a manner will have the time variation determined by the Hamiltonian as

$$i\hbar \frac{df}{dt} = [H, f], \tag{11.14}$$

as was found in Chapter 6. When such a function is defined (it still is undetermined), it can be expressed with the Wigner transform version given from (6.89) as

$$f_W(\boldsymbol{x}, \boldsymbol{p}) = \frac{1}{h^3} \int d^3 s \left\langle \boldsymbol{x} - \frac{\boldsymbol{s}}{2} \middle| f \middle| \boldsymbol{x} + \frac{\boldsymbol{s}}{2} \right\rangle e^{i\boldsymbol{p}\cdot\boldsymbol{s}/\hbar}. \tag{11.15}$$

While the function *f* remains undefined, it may have the expectation value determined by the two-oscillator wave functions (11.10), which includes summations over the various number operators, as

$$\begin{aligned} f_W(\boldsymbol{x}, \boldsymbol{p}) = \frac{1}{h^3} \int d^3 s \sum_{n_1, n_2} \sum_{n_1', n_2'} \left\langle \boldsymbol{x} - \frac{\boldsymbol{s}}{2} \middle| n_1 n_2 \right\rangle \\ \times \langle n_1 n_2 | f | n_1' n_2' \rangle \langle n_1' n_2' \rangle \left| \boldsymbol{x} + \frac{\boldsymbol{s}}{2} e^{i\boldsymbol{p}\cdot\boldsymbol{s}/\hbar} \right. \end{aligned}. \tag{11.16}$$

To be a little clearer, the wave functions themselves are now introduced to replace the Dirac bracket representation of them, as

$$\begin{aligned} f_W(\boldsymbol{x}, \boldsymbol{p}) = \frac{1}{h^3} \int d^3 s \sum_{n_1, n_2} \sum_{n_1', n_2'} \psi^{\dagger}_{n_1 n_2} \left(\boldsymbol{x} - \frac{\boldsymbol{s}}{2} \right) \\ \times \langle n_1 n_2 | f | n_1' n_2' \rangle \psi_{n_1' n_2'} \left(\boldsymbol{x} + \frac{\boldsymbol{s}}{2} \right) e^{i\boldsymbol{p}\cdot\boldsymbol{s}/\hbar} \end{aligned}. \tag{11.17}$$

The central expectation value of the function *f* that remains in the second line is a *c*-number and can be removed from the integral, so that

$$\begin{aligned} f_W(\boldsymbol{x}, \boldsymbol{p}) = \frac{1}{h^3} \sum_{n_1, n_2} \sum_{n_1', n_2'} \langle n_1 n_2 | f | n_1' n_2' \rangle \\ \times \int d^3 s \, \psi^{\dagger}_{n_1 n_2} \left(\boldsymbol{x} - \frac{\boldsymbol{s}}{2} \right) \psi_{n_1' n_2'} \left(\boldsymbol{x} + \frac{\boldsymbol{s}}{2} \right) e^{i\boldsymbol{p}\cdot\boldsymbol{s}/\hbar} \end{aligned}. \tag{11.18}$$

The integral itself, on the last line of (11.18), may be defined as a function

$$T_{n_1n_2,n_1'n_2'}(\boldsymbol{x},\boldsymbol{p}) = \frac{1}{h^3}\int d^3s\,\psi^{\dagger}_{n_1n_2}\left(\boldsymbol{x}-\frac{\boldsymbol{s}}{2}\right)\psi_{n_1'n_2'}\left(\boldsymbol{x}+\frac{\boldsymbol{s}}{2}\right)e^{i\boldsymbol{p}\cdot\boldsymbol{s}/\hbar}. \tag{11.19}$$

Moyal (1949) has shown that these *T*-functions satisfy a form of orthogonality which may be described via

$$\int d^3\boldsymbol{x}\int d^3\boldsymbol{p}\,T_{n_1n_2,n_1'n_2'}(\boldsymbol{x},\boldsymbol{p})\,T_{n_1''',n_2''',n_1''n_2''}(\boldsymbol{x},\boldsymbol{p}) = \frac{1}{h^3}\delta_{n_1'n_1'''}\delta_{n_2'''n_2'}\delta_{n_1n_1''}\delta_{n_2n_2''}. \tag{10.20}$$

Thus, it is easy to determine the value of the matrix element leading the integral in (11.18) from the Wigner function, once it is determined, by integrating the later with the appropriate *T*-function.

The approach above may seem a bit obtuse, but it leads to the ability to define the entanglement of the two oscillators with a measure M of this entanglement by

$$M = 1-\left|f^T_{n_1n_1'}\right|^2,\ f^T_{n_1n_1'} = \sum_{n_2}\langle n_1n_2|f|n_1'n_2\rangle, \tag{11.21}$$

which is a trace over the quasi-diagonal matrix element. To be more direct, the matrix element (the second term in the last equation) may be written as

$$f^T_{n_1n_1'} = h^3\int d^3\boldsymbol{x}\int d^3\boldsymbol{p}\sum_{n_2}T_{n_1n_2,n_1'n_2'}(\boldsymbol{x},\boldsymbol{p})\Big|_{n_2'=n_2} f_W(\boldsymbol{x},\boldsymbol{p}). \tag{11.22}$$

Many approximations can be made to the wave functions in order to evaluate this measure of entanglement. Other measures are discussed in the next paragraphs.

11.1.5 Measures of Entanglement

In the above treatment of two coupled harmonic oscillators, a measure of entanglement was introduced without much discussion. This particular measure was formed by subtracting the population of the quasi-diagonal states from the total system state, and this difference supposedly measured the off-diagonal parts of the density matrix. These off-diagonal parts were identified in Chapter 6 as being related to the entanglement. In fact, this approach might be too simple to serve as a definition of the entanglement.

Schrödinger (1935) described entanglement as an identification of the intrinsic level of statistical relations between subsystems of a compound quantum system. Even this leaves some doubt about the understanding of entanglement.

Some newer approaches focus upon (11.6), which is thought to originally to be due to Bohm (1952), which yields a view as to what is meant by entanglement. While this equation is written in terms of the eigenstates of two spin-particles, it cannot be separated into a product of the wave functions of the two particles. Hence, it must be an entangled state as it cannot be decomposed into a product of states. It is this view that suggests that the quasi-diagonal states of the density matrix are separable states, and the off-diagonal parts left represent the non-separable part of the system—the so-called entangled part. But, it is a common approach. For example, the von Neumann (1935) entropy

$$S_e = -Tr\{\rho ln(\rho)\} \tag{11.23}$$

(the subscript is added to differentiate this entropy from the action S introduced in Chapter 1). is a projection of the actual density matrix onto the pure states that make up the density matrix. If one separates the density matrix into parts for the pure states p and the mixed states $(1-p)$

$$\rho = \begin{bmatrix} p & 0 \\ 0 & (1-p) \end{bmatrix}, \tag{11.24}$$

the result allows a computation of the von Neumann entropy as (Licata and Fiscaletti, 2014)

$$S_N = -pln(p) - (1-p)ln(1-p), \tag{11.25}$$

which implies that the entropy of the pure states is zero. More importantly, this separation basically is the same rule that was applied to (11.21).

If there is no further information to be obtained than that from the pure states, for which there is no uncertainty or fluctuations, one can understand a reason for the entropy being zero. Thus, if the entropy is related to the difference between the pure states and the mixed states (essentially the off-diagonal terms of the density matrix), it becomes very close to what is known as the Fisher entropy in normal information theory, which is normally considered the amount of information a statistical variable carries about an unknown parameter.

Nevertheless, the above is just one possible measure of entanglement. The rise of quantum information theory (and quantum computation) has spawned a plethora of various measures of entanglement (Horodecki *et al.*, 2009). Many of these are based upon ideas from statistical physics and thermodynamics, in particular the ideas of entropy. But this is not the entropy of thermodynamics, although the name is the same. In statistical physics and thermodynamics, systems which are not isolated are normally considered. These systems are not strictly in equilibrium either. These systems are described not only with average values of position and momentum, but also with concepts of energy, temperature, and entropy. Even here, however, the visual concept of entropy remains somewhat vague (at least to the normal person). The view commonly accepted is that entropy is a measure of disorder in the system, or as a measure of the *inability* to be able to convert available thermal energy into work.

Quantum information theory, which includes quantum computing, is a rapidly expanding field. Here, the quantification of entanglement/coherence is an essential component. It recently has been demonstrated that any nonzero amount of coherence in a system can be converted into an equal amount of entanglement between that system and another initially incoherent one. Talking about entanglement implies that the idea of quantum coherence is also important. In a non-equilibrium system, it is assumed to evolve to a steady state in which a local entropy appears much as that of a physical thermodynamic entropy. However, entropy in information systems is not equivalent to physical entropy. For bits, one does not have the equivalent of pressure as bits do not undergo random thermal motion (first, they don't move about, and second, if they did have such randomness, they would not be bits). This holds for qubits as well. Thus, information entropy has no relation to physical or thermal entropy (Kish and Ferry, 2018). Physical entropy is connected to the actual physical states of a system, while information entropy is connected to the bits of information. There are a significantly larger number of physical states than there are bits or qubits.

11.2 THE JAYNES–CUMMINGS MODEL

In general, in both the last chapter and the sections above, the qubit has been discussed in terms of a "two-level atom." Yet, it has not been mentioned how the various inputs are generated. Certainly, one input is the result of a previous qubit, just as in classical logic operations. The control signal, however, remains to be defined. Quite often, in the quantum computer, the qubit is adjusted often by an electromagnetic signal that itself is generated according to the algorithm. In a great many cases, the coupling is between a two-level atom and a resonant (quantized) electromagnetic cavity. It turns out that this is actually a very common situation in chemistry as well as in quantum computing, and a relatively standard model is given by the Jaynes and Cummings (1963) model. More recently, the process of coupling a coherent optical beam to such an atom has become of great interest in quantum optics, and, of course, in optical qubits. In this section, this model will be described in some detail, because it is a simple model for coupling electromagnetic radiation to a qubit, a process that is found in many implementations of quantum operations. Following this, some variations in the model will be discussed. A schematic diagram of the Jaynes–Cumming model is shown in Figure 11.5.

The principle of the model is analogous to the problem of two coupled classical pendulums that are connected to one another in order to produce an interaction. This interaction leads to a coupling of the two oscillating modes in a way in which all of the energy oscillates between the two pendulums. At a given instant of time, pendulum 1 may be static while all the energy is in the oscillation of pendulum 2, and at a later time the process is in the opposite state. In the quantum case, one pendulum is the harmonic oscillator that represents the quantization of a mode of the electromagnetic cavity (see Section 5.3.2), while the second is the atom, which is described as an oscillator by its Rabi frequency for oscillation of probability between the two states (energy levels). Hence, the field can interact with the atomic state and lead to Rabi oscillations of the atomic state population. As with the pendulums, the Rabi oscillation is not static in amplitude. Something different happens, in that the Rabi oscillations collapse and then reappear periodically, just as the energy moves from one pendulum to the other. The temporal time over which this collapse and reappearance occurs has been termed the *revival time* (Norozhny *et al.*, 1981). When the field in the cavity is in a coherent state, and the atom is prepared in an excited state, it is found that the atomic and field states become rapidly entangled and then subsequently disentangle at one-half the revival time (Gea-Banacloche, 1990). That is, the maximum entanglement of the atomic states occurs when the Rabi oscillations are strongest, and the minimum entanglement occurs when the Rabi oscillations are minimal. Thus, the model shows the full capability of providing entanglement under the control of an electromagnetic field.

The Hamiltonian can be described relatively simply in terms of the various components, as discussed above, and shown in Figure 11.5. First, the electromagnetic field, or the laser field, is quantized as usual in terms of a series of harmonic oscillators that describe each of the modes that

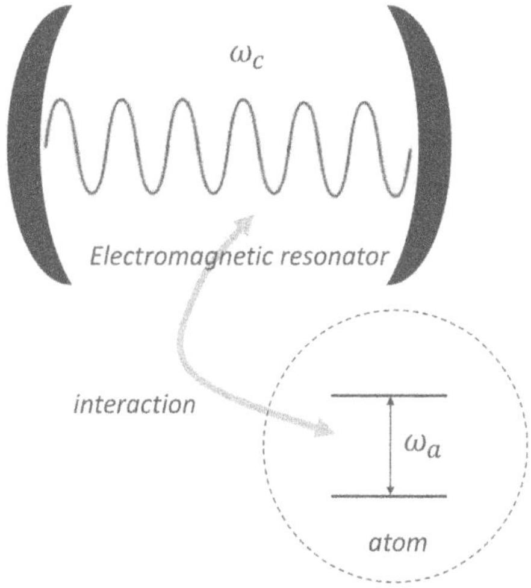

FIGURE 11.5 Schematic representation of the Jaynes and Cummings (1963) model in which a two-level atom (lower right) interacts with a mode of a resonant cavity (top). Details are in the text.

can exist in the cavity or wave, with each harmonic oscillator described as in Section 5.3.2. Since the interest is in a single mode, the wave energy may be written as

$$H_{EM} = \hbar\omega_c a^\dagger a, \tag{11.26}$$

where the operators are the normal creation and annihilation operators (5.12) for the harmonic oscillator representation of the single electromagnetic mode of the cavity, and ω_c is the frequency of this mode. The atomic levels are just the two levels of the atom, which we describe with a pseudo-spin index with the spin up state as the upper level and the spin down state as the lower level. The Hamiltonian for the two-level atom is then

$$H_a = \frac{1}{2}\hbar\omega_a\sigma_z. \tag{11.27}$$

Here, ω_a is the frequency determined from the difference between the two atomic energy levels and σ_z is the Pauli spinor from (10.23).

Now, the total electric flux density $\boldsymbol{D} = \varepsilon\boldsymbol{E} + \boldsymbol{P}$, where ε is the permittivity, arises from the applied electric field and the polarization of the atom constituents of the dielectric medium, which here is the atom. This polarization is given by the two states of the atom. This polarization is arbitrarily taken to be in the x-direction and use (10.22) to write this as

$$\boldsymbol{P} \to P_x = \sigma_x = \frac{1}{2}(\sigma_+ + \sigma_-). \tag{11.28}$$

The electric field $\boldsymbol{E}$ is arbitrarily oriented in the z-direction, so that the energy is $eE_z z$. If the charge is absorbed into a "coupling constant," the interaction energy can be written as

$$H_{int} = \hbar\Omega \cdot E_z z \cdot P = \frac{1}{2}\hbar\Omega E_z\left(a^\dagger + a\right)(\sigma_+ + \sigma_-), \tag{11.29}$$

where the first parenthesis is the operator representation of z, and the second parenthesis comes from (11.10). The parameter Ω incorporates all the various constants arising from the connection between the operators (from the harmonic oscillators). The creation and annihilation operators each have their time variation determined by their respective frequencies, so the four cross terms will have slow variations due to the differences of the two frequencies and fast variations due to the sum of the two frequencies.

Generally, the fast components are ignored and only the slow components are retained in what is called the *rotating wave approximation*. The Hamiltonian is then separated into two commuting parts, which are described by

$$\begin{aligned} H_1 &= \hbar\omega_c\left(a^\dagger a + \frac{\sigma_z}{2}\right), \\ H_2 &= \hbar\delta\frac{\sigma_z}{2} + \frac{1}{2}\hbar\Omega\left(a\sigma_+ + a^\dagger\sigma_-\right). \end{aligned} \tag{11.30}$$

In this form, the new parameter δ is

$$\delta = \omega_a - \omega_c \tag{11.31}$$

and is the so-called *detuning* of the frequency between the field and the atom. Equation (11.30) creates a proper 2 × 2 Hamiltonian, whose eigenvalues are given in terms of the number operator for the field as

$$E_{\pm} = \hbar\omega_c\left(n+\frac{1}{2}\right) \pm \frac{1}{2}\hbar\Omega_n, \tag{11.32}$$

where

$$\Omega_n = \sqrt{\delta^2 + \Omega^2(n+1)}. \tag{11.33}$$

This last expression defines the Rabi frequency for oscillation of density between the two atomic levels, in response to the interaction with the electromagnetic field. It should be noted that, as the mode occupation builds up, the Rabi frequency increases, and the amplitude of these oscillations increases. When the mode occupation is small, the Rabi frequency is basically just the detuning frequency of the system.

In this system, the electromagnetic wave, and its interaction with the two-level atom, provides the control over the state of the atom. By modulating, or changing, the interaction process, one changes the nature of the qubit itself. Here, the Jaynes–Cummings (JC) model provides a single example of how this process is accommodated. In the following sections, we will examine actual experimental techniques for the production of various types of qubits, hopefully explaining each approach sufficiently well for the reader to understand the process.

11.2.1 An Example in Optics

As discussed above, the field in the cavity and the atomic states rapidly become entangled when the field is in a coherent state and the atom is prepared in an excited state. As time advances, they subsequently disentangle at one-half the revival time (Gea-Banacloche, 1990). The maximum level of entanglement occurs when the Rabi oscillation is strongest. The lowest value of entanglement occurs at the minimum of the Rabi oscillation. Experimentally, it has been found that the photon parity operator exhibits similar behavior, but the amplitude of the parity operator actually occurs when the Rabi oscillations are at their minimum. That is, the oscillation of the parity operator seems to be shifted by half a period from the Rabi oscillations (Birrittella *et al.*, 2015). The above interaction parameter Ω_n and Rabi frequency Ω_n are related to the occupation number of photons in the cavity by (11.33) when the detuning is very small as ($n = a^\dagger a$, as usual)

$$\Omega_n = 2\Omega\sqrt{n+1}, \tag{11.34}$$

and the parity operator is related to the occupation as

$$\Pi_{par} = (-1)^n = e^{i\pi n}. \tag{11.35}$$

The parity operator develops a time variation in its commutation with the Hamiltonian above. This leads to

$$\langle\Pi_{par}\rangle = e^{-n}\sum_{j=0}^{\infty}(-1)^j\frac{n^j}{j!}cos\left(2\Omega\sqrt{j+1}\right). \tag{11.36}$$

This expectation operator now oscillates in time and appears almost as an optical pulse, with the maximum as indicated above. One can now develop the Wigner function in terms of this parity operator and the appropriate displacements of the wave function, as in (11.15)

$$f_W(\mathbf{x},\mathbf{p},\mathbf{a}) = \frac{1}{h^3}\int d^3\mathbf{a}\left\langle \mathbf{x}-\frac{\mathbf{a}}{2}\middle| f \middle| \mathbf{x}+\frac{\mathbf{a}}{2}\right\rangle e^{i\mathbf{p}\cdot\mathbf{a}/\hbar}, \tag{11.37}$$

where α is a complex number defining a point in phase space; e.g., the real and imaginary parts correspond to position and momentum (Cahill and Glauber, 1969). The expectation value involves the wave function as is normally described. As the parity operator oscillates, the Wigner function exhibits oscillations in the entanglement and appears very analogous to Figure 6.1 in that there are two peaks describing the two shifted wave functions and a central bit describing the entanglement (this is also apparent in Figure 11.7). It is this latter part which shows the oscillations in time which correspond to the variations in the entanglement, the Rabi oscillation, and the expectation of the parity operator. This will be more evident in the following paragraphs. Here, some results for the density matrix and the atomic inversion are plotted in Figure 11.6. The top panel is the trace of the square of the atomic density matrix, while gt is a reduced time (t is the actual time). The factor $g = d(\Omega_n/\hbar V\varepsilon_0)$, where d is the atomic dipole matrix element (related to Ω), V is the mode volume, and ε_0 is the permittivity of the medium.

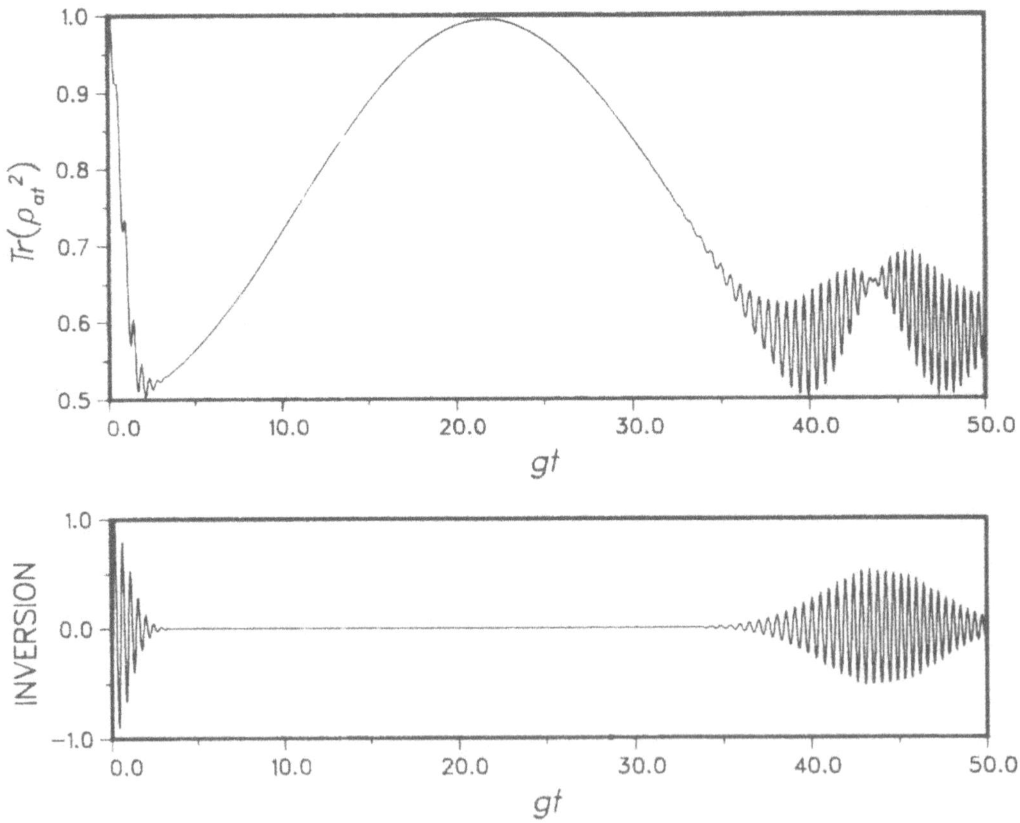

FIGURE 11.6 Top: The quantity $Tr\{\rho_{at}^2\}$ as a function of the reduced time for an atom prepared in the lower atomic state and the field initially in a coherent state with the average photon number $\bar{n} = 49$. Bottom: The average atomic inversion (probability to find the atom in the upper state minus the probability that it is in the lower state) as a function of time for the same initial conditions as above. (Reprinted with permission from Gea-Banacloche (1990), copyright 1990 by the American Physical Society.)

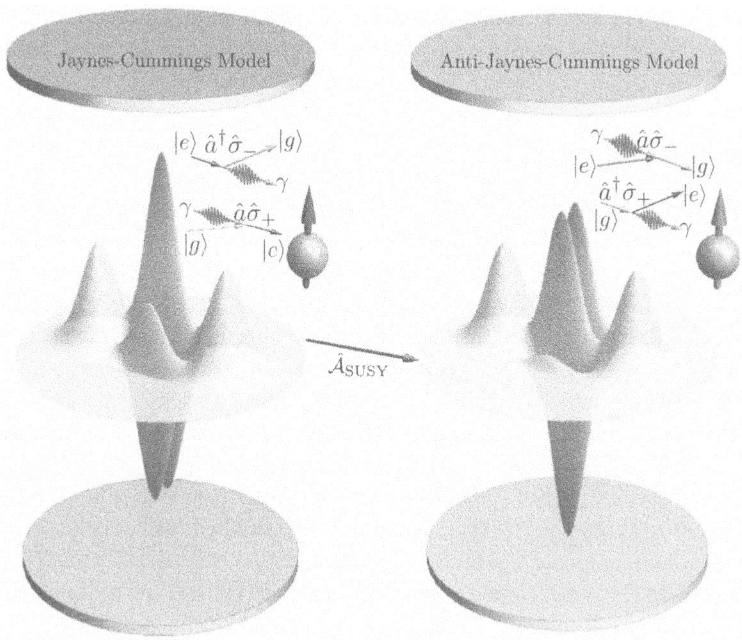

FIGURE 11.7 Schematic representation of the two models for optical coupling, shown via their Wigner functions. Experimental setups for studying the JC and AJC models traditionally involve confining a single electromagnetic mode between two mirrors (the optical cavity). This mode interacts with an atomic system processing two energy levels, here marked ground $|g\rangle$ and excited $|e\rangle$. In the JC model, excitations can be exchanged between photons (γ) and the atomic levels using the operators in (11.30). On the contrary, the AJC model of (11.38) allows for simultaneous creation or annihilation processes in both subsystems. (Reprinted from Bocanegra-Garay *et al.* (2024), under the creative commons attribution license 4.0.)

11.2.2 The Anti-Jaynes–Cummings Model

The JC model conserves the number of excitations in the system. Work has been going on over the past decade on a slightly different model, in which the number of excitations is not conserved. This modifies (11.30) ever so slightly by changing the positions of the creation and annihilation operators as (Bocanegra-Garay *et al.*, 2024).

$$\begin{aligned} H_1 &= \hbar\omega_c\left(a^\dagger a + \frac{\sigma_z}{2}\right), \\ H_2 &= \hbar\delta\frac{\sigma_z}{2} + \frac{1}{2}\hbar\Omega\left(a^\dagger\sigma_+ + a\sigma_-\right). \end{aligned} \tag{11.38}$$

This version is referred to as the Anti-JC (AJC) model. The difference between this and the JC model is explained slightly better in Figure 11.7. In the JC model, exciting the atom from the ground state $|g\rangle$ to the excited state $|e\rangle$ will absorb the photon, thus conserving the excitation. On the other hand, in the AJC model, the passing photon can change the number of photons and excite the atom through a process similar to stimulated emission. The figure illustrates the Wigner functions for the two models as well as the interactions. One can note that the central part of the Wigner functions, the entanglement part, appears inverted between the two models, and this changes the phase relationships between the sub-systems.

There is also a Rabi model, which in essence combines the JC and the AJC along with some other interaction terms (Anaya-Contreras *et al.*, 2025). This model gives more flexibility in defining the response of the system but almost becomes overkill. In addition, and more to the consideration of coupled qubits, the coupling of two such JC systems has also been considered (Castaños-Cervantes *et al.*, 2024).

11.3 SOME POSSIBLE QUBITS

There are a great number of systems that have been suggested as possible qubits. Many of these involve spin. However, the most common system remains the superconducting circuits, such as the superconducting quantum interference device (SQUID) introduced in Section 4.5.4. This will be treated in greater detail in Section 11.4. Optical qubits will be dealt with in Section 11.5. In the present section, several other qubits will be discussed, including that of using spin within a quantum dot. The spin–orbit interaction is relatively small in Si, but it is large enough to consider making spin-based qubits in this material. The idea for a qubit based upon Si effectively dates to a suggestion by Kane (1998) to use a ^{31}P dopant in isotopically pure ^{28}Si. In this approach, the half-integer spin of an electron bound to the ^{31}P dopant couples to the spin-1/2 nuclear spin of the dopant atom via the hyperfine interaction, with a characteristic spin-level spacing of 98 MHz, when a magnetic field is used to split the various spin states. This gives rise to a four-level system, which will be described further below.

The idea of using spin to create a qubit with quantum dots was apparently first proposed by Loss and DiVincenzo (1998). They developed a detailed scheme to achieve quantum computation with a pair of single-electron quantum dots (which create a potential similar to the two-dimensional harmonic oscillator discussed in Section 9.1). The qubit is realized with the spin of an excess electron in one of the quantum dots. Two-qubit quantum-gate operation is achieved by merely adjusting the barrier existing between the two dots. If the potential barrier is high, the two qubits, one in each dot, do not interact. If the potential is lowered, then the two qubits are allowed to interact, and the spins are affected by a coupling due to the spin–spin coupling energy. Many approaches burst upon the scene after this seminal paper.

11.3.1 ^{31}P Donors

As mentioned above, Kane (1998) suggested that the nuclear spin of the P atom could be used as a qubit. This approach is interesting because it has the promise of a long spin-relaxation time and is scalable (O'Brien *et al.*, 2001). The process of getting a P atom to a precise location turns out not to be so difficult. One method is to hydrogenate the dangling bonds on the Si surface, then to remove particular hydrogen atoms using a scanning tunneling microscope (in ultra-high vacuum) (Lyding *et al.*, 1994). Removing the hydrogen allows a phosphorus molecule to be chemically attach to the dangling Si bond at that particular site. In the case mentioned above, the molecules from phosphine gas are used to deposit the P atom at the desired position, and then the difficult task is to deposit isotopically pure Si while maintaining the P atom at the desired position. An alternative approach, more suited to mass production, is the implantation of a single atom (presumably P) at designated spots in the Si surface (Matsukawa *et al.*, 1997).

The operation of a single donor qubit under control of surface gates and measured with a single-electron transistor (SET) (described in Section 3.4.3) was evaluated by Dehollain *et al.* (2016). In Figure 11.8, the donor qubit formed on the P atom is illustrated and some of the operations are described. Here, a method termed gate set tomography (GST) is a tool for characterizing logic operations in the qubit. The qubit itself is formed by the spin states of an electron bound to a ^{31}P atom as mentioned above, and the atom is implanted into the isotopically pure silicon substrate. The spin states are split by an applied magnetic field, and switching is achieved via coupling of the electron spin to the nuclear spin, resulting in a two-spin, four-level system. Qubit preparation is handled by tunneling to/from a SET, which is also used for sensing the state. The aluminum gates are placed on top of a SiO_2 layer. Manipulation of the qubit state is done with an electromagnetic signal tuned to the electron spin resonance frequency that itself is set by the magnetic field. A magnetic field of

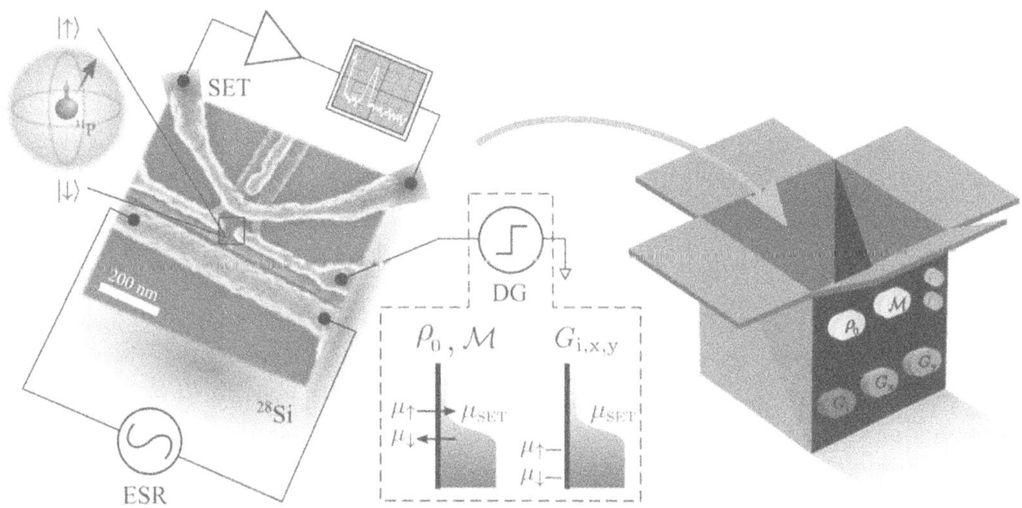

FIGURE 11.8 Diagram of a qubit device and the gate set tomography (GST) model of a qubit. SEM image of the on-chip gate structure is shown to the left. The aluminum gates have been false colored for clarity. Depicted in darkest grey are the source-drain n^+ regions, which connect the single-electron transistor (SET) to the current measurement electronics. For initialization and measurement, the donor gates (DG) are pulsed such that $\mu > \mu_{SET} > \mu^-$, inducing spin-dependent tunneling between the donor and the SET. When applying a gate sequence, the DG are pulsed to higher voltage to prevent the donor electron from tunneling to the SET. The inset diagram—(upper left) zoomed from the approximate donor location—represents the Bloch sphere of the qubit, consisting of the spin of an electron confined by an implanted ^{31}P donor, with its nuclear spin frozen in an eigenstate. The GST model treats the qubit as a black box with buttons (right), which allow one to initialize (ρ_0) and apply each Gate in the Gate set ($G_{i,x,y}$) and Measure (M) in the observable basis. (Reproduced under the Creative Commons Attribution 3.0 licence from J. P. Dehollain *et al.* (2016).)

1.55 T is applied to the system, and this splits the various spin states, with the electron spin coupled to the nuclear spin by the hyperfine interaction. The relaxation rate for the nuclear spin is orders of magnitude slower than that for the electron spin, so that the latter spin levels can be used as the two levels for the qubit. The qubit states $|1\rangle$ and $|0\rangle$ are the electron spin levels $|\uparrow\rangle$ and $|\downarrow\rangle$, respectively. Preparation of the initial state is done via single-electron tunneling to/from the donor atom, a process controlled by the aluminum gates. Then, during the state manipulation with the spin resonance signal, the pair of gates, just above the electromagnetic line connected to the electromagnetic source, is raised to a higher voltage to disconnect the donor spin from the SET. Finally, the SET can be used to sense the qubit state.

11.3.2 Double Quantum Dots

It was pointed out above that two coupled quantum dots could be used for a spin qubit (Loss and DiVincenzo, 1998). We can illustrate the double-quantum-dot approach with some relatively recent work (Brunner *et al.*, 2011). A structure fabricated on a GaAs/AlGaAs heterostructure with Schottky barrier metallic gates was used to define the active region, which may be glimpsed below the magnetic material in Figure 11.9(a). Along the bottom of this panel, one can identify the various gates (dashed lines connecting to the gray areas of the open region) as left barrier (L), plunger (PL), control (C), and right barrier (R). As above, this is a gate-defined double-quantum-dot (see Section 3.4.4), to which has been added a split Co micromagnet (the light regions in this figure), in order to provide a magnetic field. To the outside of the active double-dot, a quantum point contact (QPC) has been added (indicated in the figure by the current path arrow for I_{QPC}). When the dot on the right has an electron (or more) in it, the Coulomb interaction causes the opening for the QPC to narrow,

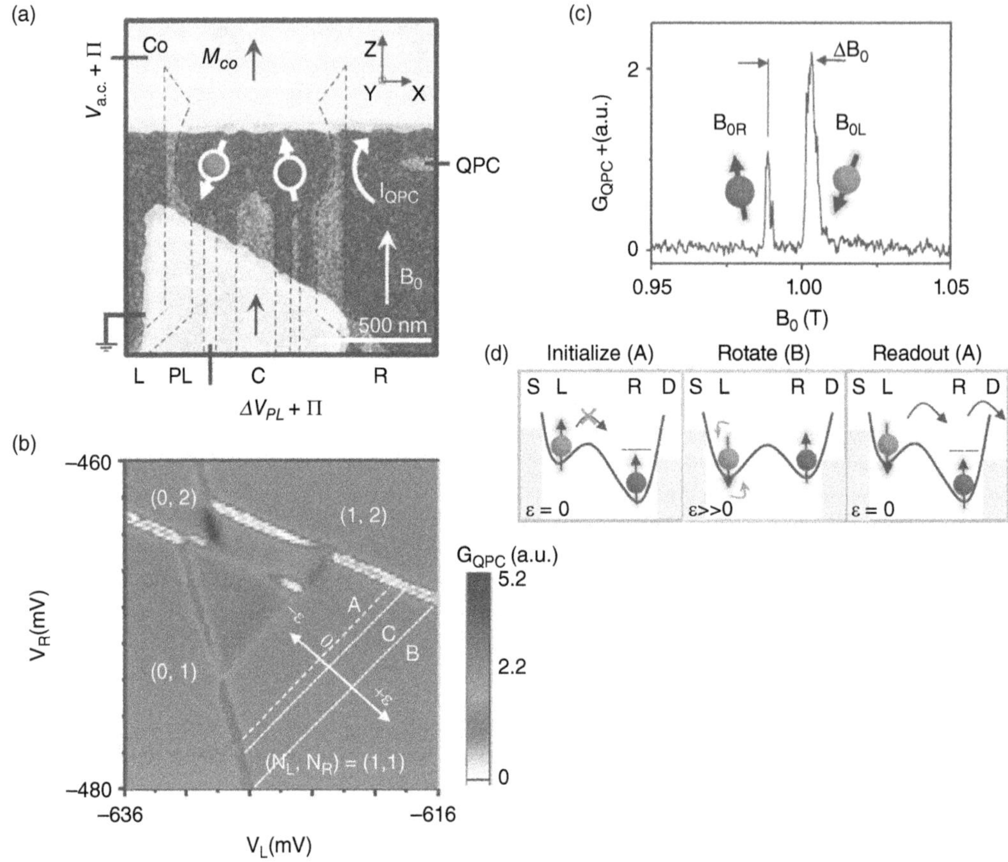

FIGURE 11.9 (a) A image of the device showing the two quantum dots and the electron spin. (b) The stability diagram around the area of interest. (c) Spin resonance signals for the left and right dots. (d) Measurement cycle for controlled spin rotations. (Reprinted with permission from Brunner *et al.* (2011), copyright 2011, the American Physical Society.)

and the resulting decrease in current can signal the charge state. This sensor can be used to map out the stability regime for the double-dot system, shown in Figure 11.9(b). Essentially, this is a plot of the conductance through the QPC as the potentials on the left (L) and right (R) barriers are varied. Here, N_L and N_R denote the charge in the left and right dots and are indicated by the numbers in parentheses. The single-spin rotations and qubit interactions are carried out in the (1,1) state (lower right of the stability diagram). To rotate each spin, electrically driven spin resonance is used, although the Co nanomagnets provide a spatially varying magnetic field due to the shape of these magnets. The spin resonance signal, for the two dots, is seen in panel (c) of this figure. Panel (d) shows the shape of the double-well potential, which is controlled by the various gate electrodes. During operation, a static magnetic field of 2 T in the left dot and 1.985 T in the right dot is applied, and the dot is excited with a microwave signal of 11.1 GHz. As indicated in the figure, various gate voltage pulses are also applied to sequence the interaction. For stage A in the figure, the two dots are set in the spin blockade condition, where the spins cannot interact, and both spins are either up or down. In stage B, the dots are isolated from one another and one of the spins is rotated by applying a pulsed microwave signal. Finally, at the second stage A, to the right, the spin can be read out with the QPC sensor in which the signal is proportional to the probability that the two spins are oppositely polarized. This demonstrates the ability to control the spin polarization. Finally, this work demonstrates that two-qubit gates may be used to demonstrate operations via the interdot spin exchange operation, as suggested by Loss and DiVincenzo (1998). If all of this seems to be a great deal of work to operate a pair of qubits, it is. Yet, this technology is still in its infancy, and experiments such as this demonstrate the feasibility of the scientific basis. In the experiments, the authors demonstrated that an all electrical two-qubit gate could be realized, and it could perform simple operations.

It is also possible to marry the previous double-dot qubit with a single-atom transistor. A double-dot qubit has been realized using two single phosphorus atoms recently (Hollenberg *et al.*, 2004). The double-well potential is created from the local potential well created by the individual P atoms. This qubit is a charge-based qubit in which the charge resides on one of the atoms, while the other is ionized, thus creating a P–P$^+$ charge system. The qubit is manipulated by a set of gates and a SET, operated at a high a.c. frequency, is used for the readout. The device is created in a Si–SiO_2 structure with the pair of P atoms implanted into the Si substrate. The control gates are used to move the charge from one P atom to the other, and this position can be sensed by the SET.

Much progress has been made in the Si-based quantum dot qubit in recent years, but most of this work remains in the academic world. That is, there is to date no working quantum computer that is based upon this technology, although several industries have asserted an interest in utilizing the approach as it fits with their classical Si technology.

11.3.3 NV Centers

The nitrogen-vacancy (NV) complex centers have been considered for possible qubits for the past few years (Jelezko and Wrachtrup, 2006). In some sense, this is like the donor qubit, as the complex produces an energy level in the band gap of wide-band-gap materials. The most common material in which this has been created is diamond, a wide-band-gap semiconductor (5.47 eV). These NV centers are one kind of defect often referred to as a "color center." There are a variety of actual N and V configurations within the crystal structure, each of which produces a different range of defect energy levels. While the center also has been found in GaAs and SiC, diamond remains the most common carrier. The various levels for these centers can be in the infrared or visible for use in these qubit applications. And, the nature of the qubit structure varies, as some approaches use the coupling of the electronic and nuclear spins, while others couple the electronic spin to a strain in the crystal. In the latter case, the divacancy has also been used.

Normally, the configuration used is with a substitutional nitrogen atom on the diamond lattice with an adjacent vacancy. This NV center is usually discussed in terms of a six-electron model. In this model for the negatively charged center, there is a ground state of an ^{3}A spin triplet (usually denoted as a ^{3}A energy level), with a zero-field splitting of 11.88 μV between the three angular momentum levels: $m_S = 0$ and the $m_S = \pm 1$. The degeneracy of the $m_S = \pm 1$ can be lifted with an external magnetic field, so that either one of these levels, together with the $m_S = 0$ level, forms a well-defined two-level system. This two-level system is the physical realization of the qubit. The spin states can be polarized by excitation of an electron to the upper energy level, denoted as the ^{3}E triplet, with a green laser, and then readout is performed by using the floresence of this excited level. The quantum state of this spin qubit can be manipulated with the application of a microwave signal, especially to affect the interaction between this two-level system and the nuclear spins of the with neighboring ^{13}C (in diamond) atoms via the hyperfine interaction. This role of the microwaves can be reversed in that the qubit can be used to sense the presence of the microwaves. However, there is currently no effort at the quantum computer level to utilize this approach.

11.3.4 Trapped Ions

Radio frequency electromagnetic potential traps have been around for quite some time (Neuhauser *et al.*, 1980). This has led to the development of qubits with either ionized atoms (Monroe *et al.*, 1995) or neutral atoms (Ebadi *et al.*, 2021). In the following, the methods of creating traps for the ion qubit will be discussed as it is further along in commercial realization. Creating ion traps is

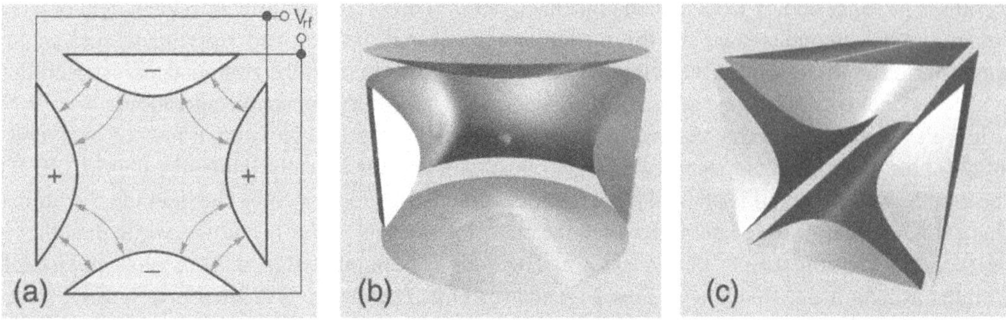

FIGURE 11.10 The most basic geometry. The use of hyperbolic shape of the electrodes can produce a perfect quadrupole field. Other configurations are usually just a variation on this theme. (b) The 3D structure, with rotationally symmetric geometry, can produce a confining potential in three dimensions. (c) The linear electrodes produce confinement in only two dimensions and must supplemented with an axial confinement potential. (Reprinted with permission from (Brownnutt *et al.*, 2015), copyright 2015 by the American Physical Society.)

achieved by a shaped potential in a region of space, in which the potential seen by the ion rises in all directions so that it sits in a three-dimensional potential well. While there are several types of traps, two distinct types of traps are primarily in use in the experimental world today, although the most common, and easiest to understand, is the quadrupole trap illustrated in Figure 11.10(a) (Brownnutt *et al.*, 2015). Here, a planar field configuration is shown with four electrodes that produce a potential that varies as one moves around the inner space. Generally, this potential can be an oscillating microwave potential that has the general form

$$V \sim V_0 + \frac{V_{ac}}{2r_0^2} cos\left(\omega_t t\right)\left(x^2 - y^2\right). \tag{11.39}$$

which is shown in panel (a) of the figure. The quantity V_0 is a constant bias voltage that often appears in these devices. V_{ac} is the electromagnetic potential at frequency ω_t that creates the confinement, while r_o is the physical dimension between the various electrodes. In the 3D quadrupole trap, (11.39) is extended to three dimensions, as shown in Figure 11.10(b). A linear version of the trap, in which the quadrupole fields (11.39) remain in the (x,y)-plane and two end electrodes are used to provide a linear, and axial, electric field to help confine such as shown in Figure 11.10(c). This latter is more usual as they can be segmented to provide a series of traps along the axial direction, in which the atom can be induced to move along the axis of the trap.

With trapped ions, it was recognized that various energy levels of the Yb^+ ion could be used for a qubit. This is usually achieved with the hyperfine levels, which are energy levels shifted by small interactions due to the electromagnetic multipole interaction between the nucleus and the electron cloud. In previous chapters, the spin-orbit interaction was discussed. This has several components. One is that discussed in Section 10.4.1, in which the orbital angular momentum interacts with a gradient in the radial potential to produce an effective magnetic field that affects both the spin and orbital angular momenta. The orbital angular momentum is governed by the phase angle φ, so that rotates in the positive angle direction, while $m \leq -1$ rotates in the negative angle direction. This vector varies as $sin(\vartheta)$ so that this produces a magnetic field with a multiple behavior. The second component arises because the charge rotation in the φ-direction produces a current loop. Current on its own produces a magnetic field and the magnetic dipole of the moving atomic charge generates a z-directed magnetic field that changes sign with the sign of m. The total magnetic field can interact with the nuclear magnetic dipole as well, and this contributes the shift in energy levels in the fine structure. Mostly this is dominated by the potential gradient fields.

Progress toward an actual processor arrived when it was shown that the ion traps could be created on a silicon surface using standard fabrication techniques (Allen *et al.*, 2017). In this approach, the linear ion traps use a combination of 3.07 and 0.27 MHz radio frequencies and d.c. voltage excitation, and the qubits are manipulated using laser excitation. Since then, the approach has been commercialized with some success.

11.3.5 Braiding Quasi-Particles

In another method of quantum computing, the various quantum qubits must be moved around each other's path, in a manner termed *braiding*. The concept of braiding is actually connected to the process in which strands of hair are interleaved to form braids, or the laces of a shoe are interleaved to actually achieve the *tie*. A mathematical theory for this is known as *knot theory* (Adams, 1994). Knot theory in mathematics and physics lies in the analysis of graphs, which describe how one system state connects to its successor by the application of a program. Such a graph leads to the state transition graphs in computer science. The idea of braiding is shown conceptually in Figure 11.11 for both a scientific calculation and as a computing approach. The two trajectories (black lines, both dashed and solid) represent the movements of two (presumably quantum) qubit states, and the labels on left and right denote terminology for science and computing. The actual computation gate lies in the center of the figure where the two trajectories are wrapped around one another (braided). The left path, as it moves through the center, passes under the right path and then passes over the right path, creating a "tie" (as in the shoe lace) or *braid*. It is worthwhile to mention that braiding of qubits has been achieved with trapped ions (Iqbal *et al.*, 2024a).

The idea of braiding quantum states has been around for some time, especially with regard to coupling qubits formed from sometimes weird particles. A normal semiconductor, or even normal physics, is three-dimensional. On the other hand, transport at the surface, or at the interface between two dissimilar materials, can be nearly two dimensional. This is the case in the MOSFET where the channel is formed between silicon dioxide and silicon. It was in fact a normal MOSFET at very low temperature, in which that the quantum Hall effect (Section 5.5.3) was discovered (Von Klitzing *et al.*, 1980). Some commercial research laboratories have made the assertion that using such quantum states, which are topologically stable (cannot be distorted, so should have fewer, if

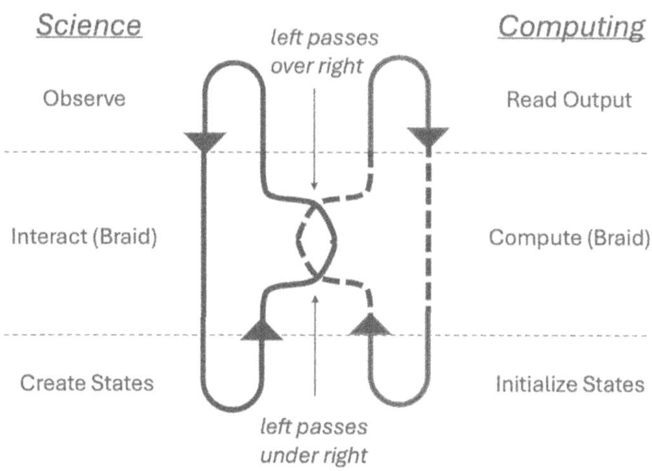

FIGURE 11.11 A conceptual view of braiding. The lines are qubit paths. The actual computation gate lies in the center of the figure where the two trajectories are wrapped around one another (braided). The left path, as it moves through the center, passes under the right path and then passes over the right path, creating a "tie" (as in the shoe lace) or *braid*.

not zero, errors), along with a weird particle, such as the Majorana fermion. This latter particle is a strange beast, which is its own anti-particle, and has zero kinetic energy, but is a relativistic fermion in Dirac's relativistic quantum mechanics identified by the Italian Majorana (1937). However, it is not at all clear that this magic particle has ever been seen in experiments. This is especially important since braiding has been seen in trapped ion approaches, which also have been shown to be topologically stable (Iqbal *et al.*, 2024b).

11.4 JOSEPHSON-BASED QUBITS

The Josephson junction and superconductivity have already been discussed in Sections 3.3.4 and 3.4.5. The Josephson junction is a tunnel junction in which the materials on either side of the insulator are superconductors and the tunneling particles are Cooper pairs (two electrons with opposite spin). However, single-electron tunneling can occur under the appropriate conditions. Normally at low temperature, no tunneling occurs until $eV_a > 2\Delta$, where the Cooper pairs are broken up by the applied voltage. However, if the barrier is relatively thin, say 1–2 nm, then the superconducting wave functions can extend through the barrier, so that they interact with those wave functions on the other side, which leads to coherent mixing of the wave functions on either side of the junction to produce a current at zero bias! This is termed the d.c. Josephson current.

In recent years, qubits have been made from superconducting circuits that include Josephson junctions. The step from a Josephson junction toward a qubit based upon this device was taken in 1999, with macroscopic qubits based on quantum levels representing collective degrees of freedom such as the persistent current in a superconducting ring (Mooij *et al.*, 1999) or excess charge on a superconducting island (Nakamura *et al.*, 1999). When superconducting components are used in electrical circuits, then these circuits become quantum objects, and the qubit has several modes of operation.

The dominant form of the qubit using Josephson junctions is the SQUID, described in some detail in Section 4.5.4. Most applications of the SQUID for quantum computing applications use the two (or more) junction version of the d.c. SQUID. In the drive to make the qubits small, one actually introduces the quantum dot, QD, shown in Figure 3.20. The short part of the ring, between the two Josephson junctions, forms the QD, while the two capacitors shown are actual Josephson junctions. This leads to two energies being important in the qubit. One arises from the Josephson current (3.85) given as

$$I_J(B) = I_{J0}cos\left(\frac{eBA}{h}\right), \tag{11.40}$$

which leads to a Josephson energy

$$E_J = \Phi_0 I_{J0}/2\pi. \tag{11.41}$$

The second energy is that of the single-electron box tunneling, given as

$$E_c = e/2C. \tag{11.42}$$

The operation of both the qubit and the SQUID depends upon the relative size of these two energies. There is a third energy that is important and that is the superconducting gap $E_G = 2\Delta$. There is still another consideration in which one needs to distinguish between two types of qubits: microscopic and macroscopic. The microscopic qubit is based upon an internal charge or spin, much like the Si qubits of the previous section, or upon two natural levels as the two-level atom in the JC model. Macroscopic qubits are based on quantum levels representing collective degrees of freedom (often called *macroscopic quantum states*) like the persistent current in a superconducting ring or

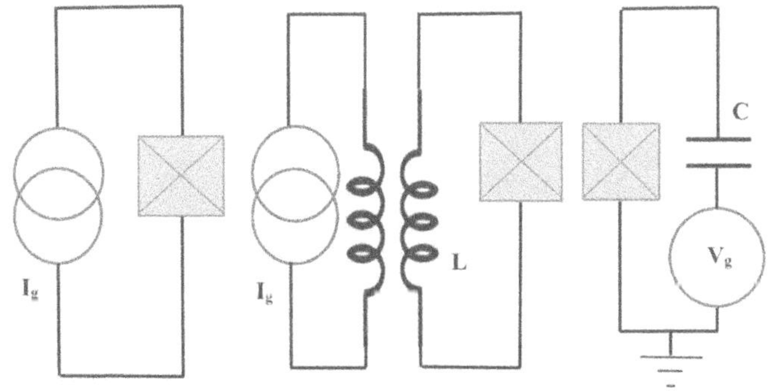

FIGURE 11.12 Basic superconducting qubits: phase qubit (left), flux qubit (center), and charge qubit (right). An X across a line, or in the box as here, represents a Josephson junction. (Reprinted from Sharifi (2023), under the creative commons attribution license 4.0.)

excess charge on a superconducting island, such as the quantum dot in Figure 3.20. These give different types of qubits, which will be discussed below. To compare these, we show three different types of SQUID-based qubits in Figure 11.12 for comparison (Sharifi, 2023). From left-to-right, these are the phase qubit that contains an a.c. current source and a Josephson junction within the loop, the flux qubit that couples the flux from the current into the loop via a transformer, and the charge qubit that replaces the current source with a voltage and a capacitor upon which to store the charge. These will be treated in the reverse order in the following paragraphs.

11.4.1 Charge Qubits

The charge qubit is composed primarily of the Cooper pair box, which itself is the region between the two junctions when this region is small such that there are a set of distinguishable energy levels. This box plays the role of our two-level atom, which dominates the behavior when $E_G > E_c$. The box is accessed through a pair of small Josephson junctions, so that single-electron (or single Cooper pair) tunneling governs the excitation of the box. A single Cooper pair box is a unique artificial solid-state system, analogous to the single-electron box in Section 3.4.4. Although there are quite a few electrons on the island (box), they all form Cooper pairs in the superconducting state and then condense into a single macroscopic quantum state. This state is separated from the normal electrons by the superconducting gap Δ, discussed in Chapter 4. The only low-energy excitations arise from Cooper pair tunneling when the gap is larger than the charging energy (11.42). Notice that this differs by a factor of 2 from Section 3.4.4, due to their being two electrons in each Cooper pair. On the other hand, if the charging energy is larger than the gap energy, and the thermal fluctuations are suppressed, the system can be considered to be a two-level system in which the lowest two energy states differ by one Cooper pair. The separation of the two levels of interest can be controlled by an additional gate voltage.

The ease of coupling charge qubits is illustrated in Figure 11.13 (You *et al.*, 2002). Here, three charge qubits are shown for a possible quantum computer application. Each of the three qubits has two SQUIDS (top and bottom) coupled to a charge box by a pair of Josephson junctions (the box is coupled to each SQUID at both left and right points). The charge on the box is varied by the applied voltage V_{Xi} $(i = 1,2,3)$ and the capacitor C_i. In each case, the two SQUIDs are assumed to be symmetric and all junctions to be the same, with the same Josephson energy E_{Ji}^0 and capacitance C_i. Each of the junctions is pierced with a flux, given by ϕ_{iA} at the top and ϕ_{iB} at the bottom. The SQUIDs have an internal flux given as Φ_{Xi} and a coupling energy

$$E_{Ji} = 2E_{Ji}^0 cos\left(2\pi\, \Phi_{Xi}/\Phi_0\right), \tag{11.43}$$

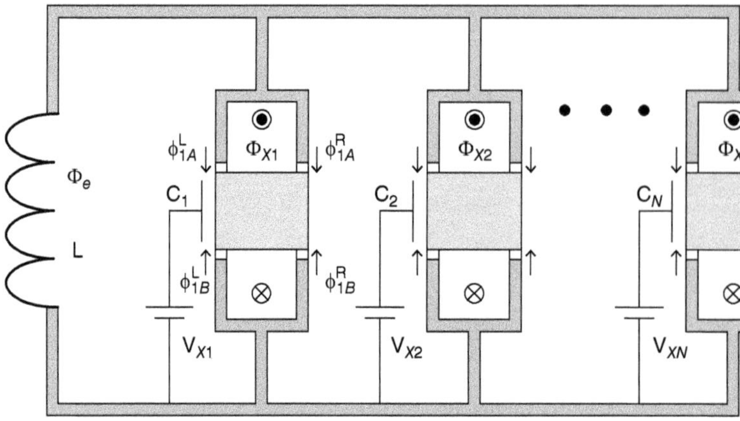

FIGURE 11.13 A schematic diagram of three qubits for a quantum computer. All of the Josephson charge qubits are coupled the common superconducting inductor. Each charge-qubit is operated in the regime where the charging energy is larger than the Josephson energy and smaller than the superconducting gap, so that quasi-particle tunneling is forbidden. (Reprinted with permission from You *et al.* (2002), under the creative commons attribution license 3.0.)

where $\Phi_0 = h/e$ is the quantum unit of flux (note that sometimes the 2 is shifted to the charge here that is appropriate for Cooper pais, so that ($\Phi_0 = h/2e$). The entire structure is shunted by a superconducting inductance which preserves a persistent current

$$I = \sum_{i=1}^{N} I_i, \tag{11.44}$$

where I_i is the net current through each qubit. This leads to a total Hamiltonian for the system of

$$H = \sum_{i=1}^{N} \left[E_{ci} \left(n_i - n_{Xi} \right)^2 - E_{Ji} \right] + \frac{1}{2} L I^2, \tag{11.45}$$

where E_{ci} is the charging energy of each box, n_i is the electron population on each box, and n_{Xi} is an offset charge induced through C_i and V_{Xi}. In this case, the net chargimg energy of the box is given by both the external capacitor and that of the junctions as

$$E_{ci} = 2e^2 / \left(C_i + 4C_{Ji} \right). \tag{11.46}$$

The superconducting inductor actually produces the coupling between the Cooper-pair boxes and thus moderates the interaction between the qubits. Of course, the interaction has to be "turned on" by switching on the SQUIDs in the chosen qubits.

Another approach moves the actual control outside the qubits, which themselves involve a ring with the Josephson junction and a normal capacitor, the latter of which acts as a charge qubit. This is illustrated in Figure 11.14 (Yu *et al.*, 2022). Here, there are two charge qubits, shown across the bottom of the figure. These are coupled through the two capacitors C_c and grounded through the top SQUID, which includes two Josephson junctions. Bias is applied to the two qubits (to adjust background charge, for example) through the capacitors $C_{gi}, i = 1,2$, and the two voltage sources. Here, $E_{Ji}, C_{Ji}, C_{gi}, V_{gi}, i = 1,2$ are the Josephson energy, junction capacitance, gate capacitance (to the voltages), and voltage sources. E_{Ji} and C_s are the equivalent quantities for the SQUID. The authors obtain full control of the operation of the qubits using magnetic flux, shown in the figure at the three circuit mesh points. This circuit allows both analog and/or digital operation.

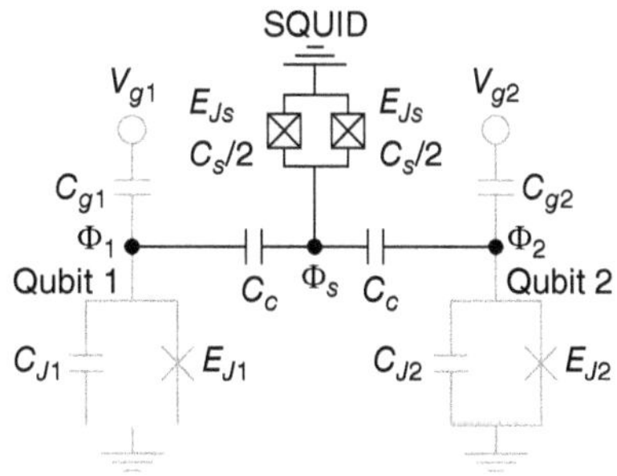

FIGURE 11.14 A schematic diagram of the two-qubit circuit, using charge qubits. The qubits are grounded through the top SQUID, and the two coupling capacitors C_c. The three fluxes indicated in the circuit are node fluxes that control the degrees of freedom of the circuit. (Reprinted from Yu *et al.* (2022), under the creative commons attribution international license 4.0.)

When the charge qubit is shunted by a capacitor, as a method of reducing current noise, the device has been called a *transmon* (Koch *et al.*, 2007). Transmon is an abbreviation of the *trans*mission line shunted plas*mon* oscillation qubit (note that this qubit does not employ a SQUID *per se* but uses electromagnetic striplines instead). The shunt capacitor reduces the charging energy and hence increases the size of the Josephson energy relative to the charging energy, so that we have $E_J \geq E_c$. This puts the ratio between that of the charge qubit and the flux/phase qubit (discussed below). Although this takes us away from a simple charge qubit, the transmon is often called a charge qubit because the circuit retains the Cooper pair box as a "quantum dot" charge site. Measurement and control of the box qubit are commonly done by means of microwave resonators. This design allows one to reverse the inequality above and make the Josephson energy much smaller than the single-electron tunneling energy, which is a characteristic of the transmon qubit. The advantage of the transmon is based upon the fact that the dispersion in the box charge is quadratic, but the various energy levels have an anharmonicity that is only linear in this charge. Thus, this allows a finer control on just which levels will be used for the qubit.

11.4.2 Flux Qubits

An interesting aspect of electrical circuits is that a resonant circuit can be actually be a quantum oscillator, corresponding to the harmonic oscillator as shown in Section 5.2, and in the JC model above. In the absence of dissipation, the Hamiltonian can be written as

$$H = \frac{q^2}{2C} + \frac{\Phi^2}{2L}, \tag{11.47}$$

where the charge q and flux Φ are the conjugate variables and satisfy a commutator relation with one another (see Section 5.2). Of course, to reach the quantum limit, we must work at low temperature, which naturally makes a connection with superconducting circuits. In the presence of dissipation, a good quality factor Q (= $\omega\tau \gg 1$) is needed. Hence, coupling a Josephson junction to a superconducting ring containing an inductor (either discrete or distributed) gives rise to the persistent-current flux qubit, sometimes called an RF SQUID (discussed above). The operation of this circuit needs $E_J \gg E_C$, so the charging energy cannot be too large. The eigenstates of the ring represent two counterrotating persistent currents, corresponding to a fixed number of flux quanta (h/e) in the loop. The inductance of the ring gives rise to a parabolic potential, like the harmonic oscillator, and adding the Josephson oscillating potential provides the needed nonlinearity to separate off the lowest states from the linear chain, as discussed earlier. For the two levels of the qubit, the Hamiltonian can be written as

$$H = 4E_C n^2 + E_L\left(\varphi - \varphi_e + \varphi_{int}\sigma_z\right)^2 - E_J cos(\varphi), \tag{11.48}$$

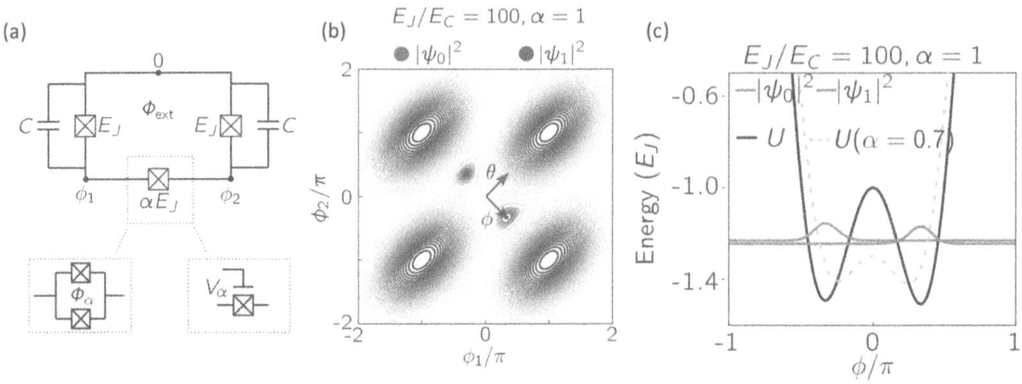

FIGURE 11.15 (a) Circuit layout for the double-shunted flux qubit with a variable junction by either a SQUID (lower left) or gate voltage tunable nanowire junction (lower right). (b) Potential landscape of the structure with the two lowest-energy eigenstates shown in color/shading with $E_J/E_C = 100, \alpha = 1, \Phi_{ext} = 0.997\pi$. (c) A one-dimensional cut of (b) along $\varphi = (\varphi_1 - \varphi_2)/2$ along with the wave functions showing their exponential separation for $\alpha = 1$. The potential at $\alpha = 0.7$ is shown as the grey dashed line. (Reprinted from Krøjer *et al.* (2024), under the creative commons attribution international license 4.0.)

where the interaction term (second on the right-hand side) arises from the effect of an external loop of electromagnetic flux that is coupled into the SQUID ring and affects the two persistent current states.

When $\varphi_e = \Phi_0/2$, the lower part of the potential is a symmetric double-well, creating nearly degenerate $|L\rangle$ and $|R\rangle$ states which lead to bonding and antibonding hybrids of these two states. These latter two states are created by the macroscopic tunneling through the Josephson junction which couples the two persistent current states, and these hybrid states provide the two levels of the qubit. This is shown in more detail in Figure 11.15 (Krøjer *et al.*, 2024). Panel (a) of the figure shows a three-junction flux qubit, in which two of the junctions are shunted with capacitors (a so-called double shunted flux qubit), while the third junction may be either the simple Josephson junction or a pair of parallel junctions (shown to the lower left) or even a junction with a capacitively coupled voltage for tuning. The potential profile is shown in panel (b) with the minimum in the potential indicated by the two colored/shaded regions. A cut of the potential from upper left to lower right is shown in panel (c), with the waves for the two states indicated in color/shading. The double well potential and its two minima are clearly recognized in this last panel.

11.4.3 Phase Qubits

The phase qubit was illustrated in Figure 11.12 (left panel) as composed of a Josephson junction and a current source (either a.c. or d.c.) in a superconducting loop. As the charging is supposed to be small, the ratio of the Josephson energy to the charging energy is huge, 10^3–10^4 or more. Nevertheless, it can be used either alone or in combination with other types of Josephson-based qubits. Such a combination is shown in Figure 11.16 (Fay *et al.*, 2008).There are a total of five Josephson junctions in this circuit. The right-loop including the current source I_b and the junction with energy E_{J2}^S forms the phase qubit, and this is attached to two voltage leads that connect to the differential amplifer at far right to measure the voltage between the top and bottom lines. The charge qubit is created by the asymmetric pair of Josephson junctions on the left end, which are biased through capacitor C_0 and V_g and have the charged quantum dot between them. The phase qubit is the SQUID consisting of the center two Josephson junctions and their loop. This latter is current biased by the source on the right and coupled to, and controled with, the fluxes Φ_T and Φ_S, which are created through the mutual inductances coupling to the d.c. magnetic field (top) and a.c. magnetic field (bottom). A image of the transistor, as fabricated, is shown on the left.

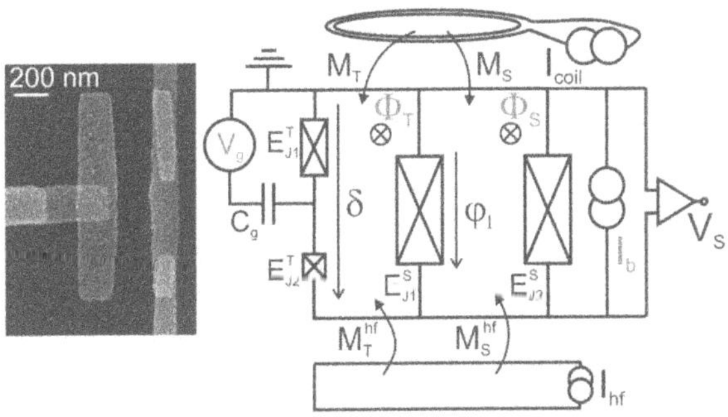

FIGURE 11.16 Electrical Schematic of the coupled circuit. The operating point is controlled by the d.c. voltage and a.c. current sources and by the fluxes coupled from the d.c. and a.c. magnetic fields at top and bottom. Details are discussed in the text. (Reprinted with permssion from Fay *et al.* (2008), copyright 2008 by the American Physical Society.)

11.5 OPTICAL QUBITS

It is not surprising that optical qubits have been developed as well, and these are usually closely based upon the JC model. The optical gates are not as far advanced as those for silicon and superconducting technologies. But the optical approach has the advantage that it can be foreseen to operate efficiently at room temperature, an advantage that should not be overlooked. In addition, optical qubits are natural for use in optical communications systems. Clearly, one can extend the JC model to the case of multiple (qubit) atoms and a resonator that yields the extension of (11.26–11.29) to

$$H(t) = \hbar\omega_C \hat{a}^\dagger \hat{a} + \frac{\hbar\omega_a}{2}\sum_{i=1}^{N}\sigma_z^i + \frac{\hbar\Omega}{2}E_z\sum_{i=1}^{N}\left(\hat{a}+\hat{a}^\dagger\right)\hat{\sigma}_x^i, \tag{11.49}$$

where it is assumed that there are N atoms in this extended system. The prefactor of the second term is now interpreted as the qubit/dimer splitting in each atom. This has been used to show that a pulsed source can be used to generate many-body correlations in light-matter systems and to determine the time evolution of an N qubit system coupled to the global resonator (Gómez-Ruiz *et al.*, 2018). This is just one advance, and most of the qubits described above have their analogs in optical systems.

For example, coupling an optical waveguide to a ring resonator can lead to entangled quantum pulses. The ring resonator is an optical waveguide in which the required periodicity around the ring leads to resonances that give quite small line widths, with the lines equally spaced in wavelength. Hence, a signal in the waveguide can couple into the ring through a tunneling process, and then the third-order nonlinearity in the polarization gives rise to a spontaneous four-wave mixing process. In this process, two incident (tunneled) photons are taken from the wave to produce two new photons, one at a higher frequency and one at a lower frequency (Vernon and Sipe, 2015). These two new photons are entangled and can exit back to the waveguide to be used as an entangled pair in a quantum optical communications system.

Another approach is to use so-called color centers, of which the NV center discussed above is one example. For optical applications, one approach is to use the spin of the color center to create an optically addressable qubit. Recently, it has been suggested that SiC, with its wide bandgap, provides a semiconductor substrate in which the neutral divacancy (denoted VV^0 in the literature), the silicon vacancy (denoted V_{Si}), or a substitutional transition metal ion that produces a state in the gap (such as Cr^{4+} or V^{4+}) (Crook *et al.*, 2020). In such a system, the electron spin on the defect acts as a qubit node, which is interconnected to other nodes by single photons acting as the carrier of optical communications. Entanglement is limited for the intrinsic emission into a zero-phonon state transition, which produces indistinguishable photons for interference between spatially separated spins. One can enhance a coupled defect's zero-phonon emission by the use of nano-cavities. The enhancement is called the Purcell factor, which actually is quantified by the enhancement of the excited state's lifetime reduction via (Crook *et al.*, 2020),

$$F \equiv \frac{\Gamma_{cavity}}{\Gamma_{bulk}} = F_1 F_2 \frac{3Q}{4\pi^2 V}\left(\frac{\lambda_{cavity}}{n}\right)^3 + 1, \tag{11.50}$$

where Γ_{cavity} and Γ_{bulk} are enhanced cavity and unenhanced emission rates, so that $F = 1$ means no enhancement. For the photonic cavity, Q is the quality factor, V is the mode volume, λ_{cavity} is the mode volume, and n is the index of refraction.

11.6 QUANTUM INFORMATION

As one may envisage, it is clear that the use of the properties of quantum mechanics, specifically qubits, to provide efficient information storage and transfer might provide new approaches to make these tasks somewhat more secure. Secret communications have been known and used for as long as people have communicated with one another. But, it should be remembered that methods of breaking into these secret messages have existed just as long. One of the most famous from the last century is the German's continued reliance upon their Enigma coding machines in World War II, even when the allies were reading their messages almost faster as the intended recipient. Today, the hope is that quantum techniques will provide methods to alleviate break-ins to private systems. It is already the case that a considerable amount of communication is classically encrypted, either routinely by a cell phone, or on purpose using a variety of efficient encryption schemes. Yet, there also are a variety of rather well-known techniques to break various encryption methods, and major entities (banks, governments, etc.) spend an enormous sum each year to avoid having their systems hacked.

There is a multitude of various communications channels in use throughout the world. Perhaps the most familiar is over-the-air communications such as that with cell phones and radio. Of course, there are cables, either microwave or optical fiber versions. Today, a significant amount of information is transmitted via satellites. Nevertheless, cables have been tapped since they were first laid, and over-the-air and satellite communications have been monitored almost as long. Cryptography is not just an interesting science; it is the backbone of modern communications. Quantum mechanics offers new hopes for more secure communications. There are a number of reasons for this but most revolve around the concept of entanglement.

Any attempt to read an entangled qubit will cause it to collapse into either a 1 or a 0, so such an attempt actually constitutes a measurement on the qubit. Thus, it becomes relatively easy to determine if anyone has actually compromised the information. On the other hand, if the recipient wants to know the information, the "key" necessary to decrypt the signal must be known. An example is the four-digit code on a bank card that is used by the bank in decrypting the signals from an ATM. Similarly, the recipient in communications needs to have a key that aids in the decryption of the message. So, in order to have secure quantum communications, one needs three important things: (1) the encryption method, or *protocol*, (2) a method of providing the recipient the key, and (3) a transmission method that preserves both the encryption and the entanglement.

11.6.1 Qubit Gates

The quantum CNOT gate was discussed above and illustrated in Figure 11.4. The signal and the control inputs come from two preceding qubits that create these signals. So, in fact, many gates are required all of which form the system. Consequently, there are a variety of different gate operations. As in classical systems, complicated algorithms in most cases can be expressed as a combination of single qubit operations (and perhaps some two qubit operations like the CNOT). To begin, some simple gate operations will be illustrated here. The first is a unity gate, which actually does nothing. It may be represented by the operation (I is the two-level unit matrix)

$$O = I = \sigma_0 = \begin{bmatrix} 1 & 0 \\ 0 & 1 \end{bmatrix}, \tag{11.51}$$

which means, if we write the wave function as a two level system with probabilities a and b for the two levels,

$$O\psi = \begin{bmatrix} 1 & 0 \\ 0 & 1 \end{bmatrix}\begin{bmatrix} a \\ b \end{bmatrix} = \begin{bmatrix} a \\ b \end{bmatrix}. \tag{11.52}$$

Of course, this "gate" does nothing, but the equation illustrates how to apply the gate to a "normal" two-level state (recall that most qubits have two states at least, such as the double-well potential that arises in some SQUIDs).

An X-gate provides an x-rotation and, logically, its matrix is the appropriate Pauli spin matrix, as

$$X\psi = \begin{bmatrix} 0 & 1 \\ 1 & 0 \end{bmatrix}\begin{bmatrix} a \\ b \end{bmatrix} = \begin{bmatrix} b \\ a \end{bmatrix}. \tag{11.53}$$

Hence, the X-gate is actually a NOT gate that merely inverts the two occupations in the two states. The Y-gate follows a similar pattern as

$$Y\psi = \begin{bmatrix} 0 & -i \\ i & 0 \end{bmatrix}\begin{bmatrix} a \\ b \end{bmatrix} = i\begin{bmatrix} -b \\ a \end{bmatrix}. \tag{11.54}$$

In general, a rotation around the y-axis involves the x- and z-coordinates but can generally be written in classical mechanics as

$$\begin{aligned} R_y(\beta) &= \begin{bmatrix} cos(\beta) & -sin(\beta) \\ sin(\beta) & cos(\beta) \end{bmatrix} \\ &= \begin{bmatrix} cos(\beta) & 0 \\ 0 & cos(\beta) \end{bmatrix} - i\begin{bmatrix} 0 & -isin(\beta) \\ isin(\beta) & 0 \end{bmatrix}. \\ &= cos(\beta)O - isin(\beta)Y \end{aligned} \tag{11.55}$$

Thus, an actual y-rotation involves both the Y gate and the O gate (or identity gate). Now consider the normalized state

$$\frac{1}{2}\left(\sqrt{3}|0\rangle + |1\rangle\right) = \frac{\sqrt{3}}{2}\begin{bmatrix} 1 \\ 0 \end{bmatrix} + \frac{1}{2}\begin{bmatrix} 0 \\ 1 \end{bmatrix}. \tag{11.56}$$

If this state is addressed with the Y-gate, the result is

$$-\frac{i\sqrt{3}}{2}\begin{bmatrix} 0 \\ 1 \end{bmatrix} + \frac{i}{2}\begin{bmatrix} 1 \\ 0 \end{bmatrix} = \frac{i}{2}\left(|0\rangle - \sqrt{3}|1\rangle\right), \tag{11.57}$$

which may be seen to be a rotation on the Bloch sphere of Figure 10.1.

There are many other gates beyond the few that have been mentioned here. Quantum circuits are created with a series of gates arranged according to a particular algorithm being implemented, just as with classical computer programming. It is easy to create the gate arrangement needed, and this circuit is likely to be independent of the technology chosen for the actual physical gates. Today, there are several approaches to quantum computing that have actually created hardware that is in use.

11.6.2 Communications

Information networks utilizing qubits for security are currently one of the biggest topics in communications. However, even with communications, the process of measurement remains a problem. Consider the polarization of a photon. As mentioned before, photons have an integer spin (and are thus bosons), usually denoted by ±1. The upper sign denotes right-circular polarization of the wave, and the lower sign denotes left-circular polarization of the wave. Normally, the polarization is taken to be linear, which is a combination of these two results in either the vertical or horizontal direction. Thus, this and a couple of other directions can be described in qubit notation as

$$\begin{aligned} |\uparrow\rangle &= |0\rangle \\ |\rightarrow\rangle &= |1\rangle \\ |\nearrow\rangle &= \frac{1}{2}\left(|0\rangle + |1\rangle\right)^{\cdot} \\ |\searrow\rangle &= \frac{1}{2}\left(|0\rangle - |1\rangle\right) \end{aligned} \tag{11.58}$$

The last two mixed states are useful in processing qubits, particularly in optical applications.

Communication of secure information between two distant points can be expressed by a common argument appearing almost universally in discussions of the topic. This begins with two people, Alice and Bob (the usual suspects in quantum communications). Bob decides to travel to a point that lies a significant distance away, while leaving Alice at home. For some reason, Alice feels it is necessary to tell Bob what the weather is at home (looking it up in the paper doesn't convey the quantum requisites). Before leaving, Bob and Alice thought that they would use two qubits to encode the weather in the following form

$$\begin{aligned} |00\rangle &\rightarrow freezing, clear \\ |01\rangle &\rightarrow freezing, cloudy \\ |10\rangle &\rightarrow non-freezing, clear \\ |11\rangle &\rightarrow non-freezing, cloudy \end{aligned} . \tag{11.59}$$

Now, using entangled qubits to send the information to Bob will permit Alice to use only a single qubit, a compression known as dense coding. It is generally felt that the use of quantum entanglement allows a denser form of this coding. If we use two qubits, then there are four possible states:

$$\begin{aligned} |0\rangle, |0\rangle &= |00\rangle, \\ |0\rangle, |1\rangle &= |01\rangle, \\ |1\rangle, |0\rangle &= |10\rangle, \\ |1\rangle, |1\rangle &= |11\rangle. \end{aligned} \tag{11.60}$$

These four states represent the weather through their relationship with (11.59). To proceed, Alice and Bob create two qubits which are entangled into the state

$$|\psi_0\rangle = \frac{1}{\sqrt{2}}\left(|00\rangle + |11\rangle\right) = \frac{1}{\sqrt{2}}\begin{bmatrix}1\\0\\0\\1\end{bmatrix}. \tag{11.61}$$

That is, the two qubits both have the same polarization, but this is not known. Bob will now take one of these two qubits with him on his trip, while Alice retains the other. To encode the weather Alice will prepare a qubit to be sent to Bob with one of the four *I*, *X*, *Y*, or *Z* transforms. She will choose one of these transforms depending upon the state of the weather. Because the weather is freezing and clear, she chooses the identity transform which leave the qubit (11.61) unchanged. For the second weather state, she uses the *Z* transform, which is applied only to the first value of each of the pairs in (11.61) (this multiplies the first zero by 1 and the second 1 by −1) thus yielding

$$|\psi\rangle_{01} = \frac{1}{\sqrt{2}}\left(|00\rangle - |11\rangle\right) = \frac{1}{\sqrt{2}}\begin{bmatrix}1\\0\\0\\-1\end{bmatrix}. \tag{11.62}$$

For the third weather state, she uses the *X* transform (11.53), again on the first numbers in each case, which provides the state

$$|\psi\rangle_{10} = \frac{1}{\sqrt{2}}\left(|10\rangle + |01\rangle\right) = \frac{1}{\sqrt{2}}\begin{bmatrix}0\\1\\1\\0\end{bmatrix}. \tag{11.63}$$

For the last weather state, she uses the *-iY* transform (11.54) to give

$$|\psi\rangle_{11} = \frac{1}{\sqrt{2}}\left(-|10\rangle + |01\rangle\right) = \frac{1}{\sqrt{2}}\begin{bmatrix}0\\1\\-1\\0\end{bmatrix}. \tag{11.64}$$

Once Bob receives the qubit from Alice, which is one of (11.61–11.64), he operates upon it with the CNOT gate, which provides

$$\begin{matrix} T_{CNOT}|\psi\rangle_{00} = \frac{1}{\sqrt{2}}\begin{bmatrix}1\\0\\1\\0\end{bmatrix} & T_{CNOT}|\psi\rangle_{01} = \frac{1}{\sqrt{2}}\begin{bmatrix}1\\0\\-1\\0\end{bmatrix} \\ T_{CNOT}|\psi\rangle_{10} = \frac{1}{\sqrt{2}}\begin{bmatrix}0\\1\\0\\1\end{bmatrix} & T_{CNOT}|\psi\rangle_{11} = \frac{1}{\sqrt{2}}\begin{bmatrix}0\\1\\0\\-1\end{bmatrix} \end{matrix}. \tag{11.65}$$

This operation has merely flipped the last two values of each part of the qubit. These new wave functions can be written as

$$|\psi'\rangle_{00} = \frac{1}{\sqrt{2}}\left(|00\rangle + |10\rangle\right), \quad |\psi'\rangle_{01} = \frac{1}{\sqrt{2}}\left(|00\rangle - |10\rangle\right),$$
$$|\psi'\rangle_{10} = \frac{1}{\sqrt{2}}\left(|01\rangle + |11\rangle\right), \quad |\psi'\rangle_{11} = \frac{1}{\sqrt{2}}\left(|01\rangle - |11\rangle\right). \tag{11.66}$$

The last digit, which is $|0\rangle$ in the case of the first row and $|1\rangle$ in the case of the second row, can be factored out. Measuring the second bit separates the results into two pairs of bits, and Bob can proceed with one of the two pairs, depending upon the result of his measurement of the second bit. This measurement does not affect the first bit in any way. Bob now applies the Hadamard transform to the four qubits. The Hadamard transform is given by

$$H = \frac{1}{2}\begin{bmatrix} 1 & 1 \\ 1 & -1 \end{bmatrix}, \tag{11.67}$$

and using this on the first bits of (11.66) produces the new results

$$|\psi''\rangle_{00} = \frac{1}{2}\begin{bmatrix} 1 & 1 \\ 1 & -1 \end{bmatrix}\begin{bmatrix} 0 \\ 1 \end{bmatrix} = \begin{bmatrix} 0 \\ -1 \end{bmatrix}, \quad |\psi''\rangle_{01} = \frac{1}{2}\begin{bmatrix} 1 & 1 \\ 1 & -1 \end{bmatrix}\begin{bmatrix} 0 \\ -1 \end{bmatrix} = \begin{bmatrix} 0 \\ 1 \end{bmatrix},$$
$$\psi''\rangle_{10} = \frac{1}{2}\begin{bmatrix} 1 & 1 \\ 1 & -1 \end{bmatrix}\begin{bmatrix} 0 \\ 1 \end{bmatrix} = \begin{bmatrix} 0 \\ -1 \end{bmatrix}, \quad |\psi''\rangle_{11} = \frac{1}{2}\begin{bmatrix} 1 & 1 \\ 1 & -1 \end{bmatrix}\begin{bmatrix} 0 \\ -1 \end{bmatrix} = \begin{bmatrix} 0 \\ 1 \end{bmatrix}. \tag{11.68}$$

The previous results were separated as either row 1 or row 2 by the measurement of the second bit. This last measurement separates the results as column 1 or column 2, and the result of the two measurements determines entirely which of the four states was sent.

Bob had the ability to decode Alice's signal because he possessed the "key," the set of transforms that he needed to apply to the qubit he received. In this case, Bob possessed the key because he knew the set of operations that would be applied by Alice. In the more general situation, the recipient of the message needs to know the key, and this must be distributed to him by the source of the message, and preferably in advance of the communication.

11.6.3 Quantum Sensing

At the core of sensing is the basic principal of measurement. Noise has always been a problem in measuring sensitive signals, and quantum sensing is no different. The belief is that using quantum mechanics in the sensing process will ultimately provide an improved signal-to-noise ratio, which is at the heart of evaluating how good a sensitive measurement actually is.

In the middle of the last century, a method of monitoring signals below the noise level was created with the lock-in amplifier, or synchronous detector. In this method, the incoming signal was modulated at a low frequency, and the output of the detector was then checked for this modulation. This approach is still used in most experimental laboratories. While the measurement of sensitive signals has progressed, one hopes that quantum sensors will provide new opportunities to achieve even more sensitive detection systems. The SQUID discussed above was first developed as a sensitive detector of magnetic fields and is perhaps the earliest detection method in modern magnetic resonance imaging.

The search for more sensitive detection mechanisms, including the use of quantum techniques, spans all of the sciences (Giovannetti *et al.*, 2004). In quantum sensing, the general approach is to achieve one of three effects: (1) use a quantum object to measure a physical quantity, (2) introduce quantum coherence as a method to enhance measurements of a physical quantity, or (3) improve the sensitivity of a measurement through the use of entanglement. The desire in this process is to go past classical limits in the measurement of the physical entity.

Measurements have always been limited in some way, and this limit is usually the noise that is always present with the measurement system. The problem is to find the desired signal in this background of noise, with a sufficient amplitude that the signal can be measured. There are normally two major noise sources in electronic systems. The first, and most common, is thermal noise, often called Johnson noise or Johnson–Nyquist noise (Johnson, 1928; Nyquist, 1928). This noise arises from thermal fluctuations and the power is generally restricted by the bandwidth of the amplifier, so that the noise power is

$$P = k_B T \cdot \Delta f, \tag{11.69}$$

where Δf is the frequency bandwidth of the amplifier/detector. The second type of noise is known as $1/f$ noise and this source has a spectrum that varies as $1/f^{\alpha}$, with the exponent α being $0 < \alpha < 2$. This noise is often the dominant noise in biological systems, but it is present in many, if not most, devices in the microwave and optical regions of the spectrum (Handel and Chung, 1993).

In more recent times, communications have switched from analog to digital signals. This has led to its own classical limit on communication channels. Generally, this limit is expressed with the power in the signal P, and the noise power N. The maximum bit rate C that can be used is (Shannon, 1948)

$$C = \Delta f \, log_2 \left(\frac{P+N}{N} \right). \tag{11.70}$$

This has led to further considerations that suggest that the minimum energy needed to measure, or erase, a bit is (Landauer, 1961)

$$E = k_B T ln(2), \tag{11.71}$$

although this has been somewhat argumentative. While most people believe that there is a minimum energy required to actually read the value of a bit (the actual measurement of this bit), Landauer posed (11.71) as the energy needed to erase a bit, and he also considered that this process would raise the entropy in the bit string.

A version of *entropy* has been discussed extensively in the digital world, but it does not seem to be understood well. Physical entropy and information entropy mean totally different things (Kish and Ferry, 2018; Shiffer, 1991). It is the "information entropy" that is important in communications, as this addresses the information-bearing states which can be the 2^N possible configuration of N bits or qubits. But, these are macro-states! While Boltzmann defined entropy as being a property of a macro-state, the physical entropy S is actually related to the number of *microscopic* configurations, W, that are compatible with that macro-state. This is not just the bit string, but the microscopic states of the entire physical system that supports any macro-state. But, this is a side argument to those wishing to use quantum states to enhance the ability to detect a weak signal in the presence of noise. There are as many suggestions for using quantum systems for this purpose as there are likely quantum systems in existence, but it is an important task.

REFERENCES

Adams, C. C. 1994. *The Knot Book: An Introduction to the Mathematical Theory of Knots.* (New York: W. H. Freeman).

Allen, S., Kim, J., Moehring, D. L., and Monroe, C. R. 2017. Reconfigurable and programmable ion trap quantum computer, in *Proc. IEEE Conf. Rebooting Comp.* (New York: IEEE Press) pp. 1–3.

Anaya-Contreras, J. A., Ramos-Rrieto, I., Zuñiga-Segundo, A. and Moya-Cessa, H. M., 2025. The anisotropic quantum rabi model with diamagnetic term. *Front. Phys.* **13** 1565407.

Benioff, P. A. 1982. Quantum mechanical Hamiltonian models of discrete processes that erase their own histories: Application to Turing machines. *Int. J. Theor. Phys.* **21** 177–201.

Birrittella, R., Chang, K., and Gerry, C. C. 2015. Photon-number parity oscillations in the resonant Jaynes-Cummings model. *Optics Commun.* **354** 286–290.

Bocanegra-Garay, I. A., Castillo-Celeita, M., Negro, J., Nieto, L. M., and Gomez-Ruiz, F. J. 2024. Exploring supersymmetry: Interchangeability between Jaynes-Cummings and Anti-Jaynes-Cummings models. *Phys. Rev. Res.* **6** 043218.

Bohm, D. 1952. A suggested interpretation of the quantum theory in terms of "hidden" variables. I. *Phys. Rev.* **85** 166–179.

Boole, G. 1854. *An Investigation of the Laws of Thought: The Mathematical Theories of Logic and Probability.* (London: Walton and Maberly).

Cahill, K. E. and Glauber, R. J. 1969. Density operators and quasiprobability distributions. *Phys. Rev.* **177** 1882–1902.

Brownnutt, M., Kumph, M., Rabi, P., and Blatt, R. 2015. Ion trap measurements of electric-field noise near surfaces. *Rev. Mod. Phys.* **87** 1419–1482.

Brunner, R., Shin, Y.-S., Obata, T., Pioro-Ladrière, M., Kubo, T., Yoshida, K., Taniyama, T., Tokura, Y., and Tarucha, S. 2011. Two-qubit gate of combined single-spin rotation and interdot spin exchange in a double quantum dot. *Phys. Rev. Lett.* **107** 146801.

Castaños-Cervantes, L. O., Procopio, L. M., and Enriquez, M. 2024. Coherent control of two Jaynes-Cummings cavities. *Sci. Rept.* **14** 3790.

Cooley, J. W. and Tukey, J. W. 1965. An algorithm for the machine calculation of complex Fourier series. *Math. Comp.* **19** 297–301.

Crook, A. L., Anderson, C. P., Miao, K. C., Bourassa, A., Lee, H., Bayliss, S. L., Bracher, D. O., Zhang, X., Abe, H., Ohshima, T., Hu, E. L., and Awschalom, D. D. 2020. Purcell enhancement of single silicon carbide color center with coherent spin control. *Nano Lett.* **20** 3427–3434.

Dehollain, J. P., Muhonen, J. T., Blume-Kohut, R., Rudinger, K. M., Gamble, J. K., Nielsen, E., Laucht, A., Simmons, S., Kalra, R., Dzurak, A. S., and Morello, A. 2016. Optimization of a solid-state electron spin qubit using gate tomography. *New J. Phys.* **18** 103018.

Deutsch, D. 1985. Quantum theory, the Church-Turing principle, and the universal quantum computer. *Proc. Roy. Soc. London A* **400** 97–117.

Deutsch, D. 1989. Quantum computational networks. *Proc. Roy. Soc. London A.* **425** 73–90.

Deutsch, D. and Jozsa, R. 1992. Rapid solution of problems by quantum computation. *Proc. Roy. Soc. London A* **439** 553–558.

Ebadi, S., Wang, T. T., Levine, H., Keesling, A., Semeghini, G. Omran, A., Bluvstein, D., Samajdar, R., Pichler, H., Ho, W. W., Choi, S., Sachdev, S., Geiner, M., Vuletic, V., and Likin, M. D. 2021. Quantum phases of matter on a 256-atom programmable quantum simulator. *Nature.* **595** 227–232.

Fay, A., Hoskinson, E., Lecocq, F., Levy, L. P., Hekking, F. W. J., Guichard, W., and Buisson, O. 2008. Strong tunable coupling between a superconducting charge qubit and phase qubit. *Phys. Rev. Lett.* **100** 187003.

Ferry, D. K. 2025. *Quantum Information in the Nanoelectronic World.* (Zurich: Springer Nature Switzerland).

Ferry, D. K., Grondin, R. O., and Akers, L. A. 1989. *Two-Dimensional Automata in VLSI. Submicron Integrated Circuits.* Watts, R. K., Ed. (New York: Wiley) pp. 377–412.

Gea-Banacloche, J. 1990. Collapse and revival of the state vector in the Jaynes-Cummings model: An example of state preparation by a quantum apparatus. *Phys. Rev. Lett.* **65** 3385–3388.

Giovannetti, V., Lloyd, S., and Maccone, L. 2004. Quantum enhanced measurements: Beating the standard quantum limit. *Science.* **306** 1330–1336.

Gómez-Ruiz, F. J., Acevedo, O. L., Rogriguez, F. J., Quiroga, L., and Johnson, N. F. 2018. Pulsed generation of quantum coherences and non-classicality in light-matter systems. *Frontiers Phys.* **6** 92.

Grover, L. K. 1996. A fast quantum mechanical algorithm for database search, in *Proc. Ann ACM Symp. Theory Comp.* (New York: ACM) pp. 212–219.

Handel, P. H. and Chung, A. L. 1993. *Noise in Physical Systems and '1/f' Fluctuations.* (New York: AIP Publishing).

Hollenberg, L. C. L., Dzurak, A. S., Wellard, C., Hamilton, A. R., Reilly, D. J. Milburn, G. J., and Clark, R. G. 2004. Charge-based quantum computing using single donors in semiconductors. *Phys. Rev. B.* **69** 113301.

Horodecki, R., Horodecki, P., Horodecki, M., and Horodecki, K. 2009. Quantum entanglement. *Rev. Mod. Phys.* **81** 865–942.

Iqbal, M., Tantivasadakarn, N., Gatterman, T. M., Gerber, J. A., Gilmore, K., Gresh, D., Hankin, A., Hewitt, N., Horst, C. V., Matheny, M., Mengle, T., Neyenhuis, B., Vishwanath, A., Foss-Feig, M., Verresen, R., and Dreyer, H. 2024a. Topological order from measurements and feed-forward on a trapped ion quantum computer. *Commun. Phys.* **7** 205.

Iqbal, M., Tantivasadakarn, N., VerresenCampbell, S. L., Dreiling, J. M., Figgatt, C., Gaebler, J. P., Johansen, J., Mills, M., Moses, S. A., Pino, J. M., Ransford, A. Rowe, M., Siegfried, P., Stutz, R. P., Foss-Feig, M., Vishwanath, A., and Dreyer, H. 2024b. Non-Abelian Topological order and anyons on a trapped-ion processor. *Nature.* **626** 505–511.

Jaynes, E. T. and Cummings, F. W. 1963. Comparison of quantum and semiclassical radiation theories with applications to the beam maser. *Proc. IEEE.* **51** 89–109.

Jelezko, F. and Wrachtrup, D. 2006. Single defect centers in diamond: A review. *Phys. Stat. Sol. (A).* **203** 3207–3225.

Jozsa, R. 1991. Characterizing classes of functions computable by quantum parallelism. *Proc. Roy. Soc. London A* **435** 563–5

Johnson, J. 1928. Thermal agitation of electricity in conductors. *Phys. Rev.* **32** 97–109.

Kane, B. E. 1998. A silicon-based nuclear spin quantum computer. *Nature.* **393** 133–137.

Kim, I. and Iafrate, G. J. 2004. Entanglement in the interaction between two quantum oscillator systems. *Found. Phys. Lett.* **17** 507–534.

Kish, L. B. and Ferry, D. K. 2018. Information entropy and thermal entropy: Apples and oranges. *J. Comp. Electron.* **17** 43–50.

Koch, J., Yu, T. M., Gambetta, J., Houck, A. A., Schuster, D. I., Majer, J., Blais, A., Devoret, M. H., Girvin, S. M., and Schoelkopf, R. J. 2007. Charge-insensitive qubit design derived from the Cooper pair box. *Phys. Rev. A.* **76** 042319.

Krøjer, S., Dahl, A. E., Christensen, K. S., Kjaergaard, M., and Flensberg, K. 2024. Fast control of a flux qubit via exponentially tunable wave-function overlap. *Phys. Rev. Res.* **6** 023064.

Landauer, R. 1961. Irreversibility and Heat Generation in the Computing Process. *IBM J. Res. Develop.* 5 183–191

Licata, I. and Fiscaletti, D. 2014. A Fisher-Bohm geometry for quantum information. *Electron. J. Theor. Phys.* **31** 71–88.

Lloyd, S. 1993. A potentially realizable quantum computer. *Science.* **261** 1569–1571.

Loss, D. and DiVincenzo, D. A. 1998. Quantum computation with quantum dots. *Phys. Rev. A.* **57** 120–126.

Lyding, J. W., Albein, G. C., Shen, T.-C., Wang, C., and Tucker, J. R. 1994. Nanoscale patterning and oxidation of silicon surfaces with an ultrahigh vacuum scanning tunneling microscope. *J. Vac. Sci. Technol. B.* **12** 3735–3740.

Majorana, E. 1937. Teoria Simmetrica dell'Elettrone e del Postrone. *Il Nuovo Cim.* **14** 171–184.

Marinescu, D. C. and Marinescu, G. M. 2004. *Approaching Quantum Computing.* (London: Pearson).

Matsukawa, T., Fukai, T., Suzuki, S., Hara, K., Koh, M., and OHdomari, I. 1997. Development of single-ion implantation–controllability of implanted ion number. *Appl. Surf. Sci.* **117**/118 677–683.

Mead, C. A. and Conway, L. 1979. *Introduction to VLSI Systems.* (Reading, MA: Addison-Wesley).

Monroe, C., Meekhof, D. M., King, B. E., Itano, W. M., and Wineland, D. J. 1995. Demonstration of a fundamental quantum logic gate. *Phys. Rev. Lett.* **75** 4714–4717.

Mooij, J. E., Orlando, T. P., Levitov, L., Tian, L., van der Wal, C. H., and Lloyd, S. 1999. Josephson persistent-current qubit. *Science.* **285** 1036–1039.

Moyal, J. E. 1949. Quantum mechanics as a statistical theory. *Proc. Cambridge Phil. Soc.* **45** 99–124.

Nakamura, Y., Pashkin, Y. A., and Tsai, J. S. 1999. Coherent control of macroscopic quantum states in a single-Cooper-pair box. *Nature.* **398** 786–788.

Neuhauser, W., Hohenstatt, M., Toschek, P. E., and Dehmelt, H. 1980. Localized visible Ba^+ mono-ion oscillator. *Phys. Rev. A.* **22** 1137–1140.

Nielsen, M. A. and Chuang, I. L. 2000. *Quantum Computing and Quantum Information.* (New York: Cambridge University Press).

Norozhny, N. B., Sanchez-Mondragon, I. I., and Eberly, J. H. 1981. Coherence versus incoherence: Collapse and revival in a simple quantum model. *Phys. Rev. A.* **23** 236–247.

Nyquist, H. 1928. Thermal agitation of electric charge in conductors. *Phys. Rev.* **32** 110–113.

O'Brien, J. L., Schofield, S. R., Simmons, M. Y., Clark, R. G., Dzurak, A. S., Curson, N. J., Kane, B. E., McAlpine, N. S., Hawley, M. E., and Brown, G. W. 2001. Towards the fabrication of phosphorus qubits for a silicon quantum computer. *Phys. Rev. B.* **64** 161401.

Planck, M. 1900. Zur Theorie des gesetzes der Energievertheilung im Normalspektrum. *Verhand. Deut. Phy. Gesell.* **2** 237–245.

Schrödinger, E. 1935. Die gegenwärtige situation in der quantenmechanik. *Naturwiss.* **23** 807–812, 823–828, 844–849.

Shannon, C. S. 1948. A mathematical theory of communications. *Bell Sys. Tech. J.* **27** 379–423.

Sharifi, J. 2023. Superconductor qubits hamiltonian approximations effect on quantum state evolution and control. *Sci. Rep.* **11** 12791.

Shiffer, M. 1991. Shannon's information is not entropy. *Phys. Lett. A.* **154** 361–365.

Shor, P. W. 1994. Algorithms for quantum computation: Discrete logarithms and factoring, in *Proc. 35th Ann. Symp. Fndtns. Comp. Sci.* (New York: IEEE Press) pp. 124–134.

Turing, A. 1937. On computable numbers, with an application to the Entscheidungsproblem. *London Math. Soc., Ser II.* **42** 230–265.

Vernon, Z. and Sipe, J. E. 2015. Spontaneous four-wave mixing in lossy micro-ring resonators. *Phys. Rev. A.* **91** 053802.

Von Klitzing, K., Dorda, G., and Pepper, M. 1980. New method for high-accuratacy determination of the fine structure based on quantized hall resistance. *Phys. Rev. Lett.* **45** 494–496.

von Neumann, J. 1935. *Mathematische Grundlagen der Quantenmechanik* (Berlin: Springer), Tr. by R. T. Beyer. 1955. *Mathematical Foundations of Quantum Mechanics* (Princeton: Princeton University Press).

von Neumann, J. 1945. *First Draft of a Report on the EDVAC.* (Philadelphia, PA: Moore School of Engineering).

You, J. Q., Tsai, J. S., and Nori, F. 2002. Scalable quantum computing with josephson charge qubits. *Phys. Rev. Lett.* **89** 197902.

Yu, J., Retamal, J. C., Sanz, M., Solano, E., and Albarran-Arriagada, F. 2022. Superconducting circuit architecture for digital-analog quantum computing. *EPJ Quan. Techn.* **9** 9.

Index

Note: Page numbers followed by f and t indicate figures and tables, respectively.

For Product Safety Concerns and Information please contact our EU representative GPSR@taylorandfrancis.com
Taylor & Francis Verlag GmbH, Kaufingerstraße 24, 80331 München, Germany

www.ingramcontent.com/pod-product-compliance
Ingram Content Group UK Ltd.
Pitfield, Milton Keynes, MK11 3LW, UK
UKHW050058200926
475702UK00019B/2896
* 9 7 8 1 0 4 1 1 9 4 6 6 8 *